The Politics
of Paradigms

SUNY series in American Philosophy and Cultural Thought
―――――――
Randall E. Auxier and John R. Shook, editors

THE POLITICS OF PARADIGMS

*Thomas S. Kuhn, James B. Conant,
and the Cold War "Struggle for Men's Minds"*

GEORGE A. REISCH

Cover photograph of James B. Conant, HUP Conant, James B. (83), courtesy of Harvard University Archives. Cover photograph of Thomas S. Kuhn by Philippe Halsman © Halsman Archive.

Published by State University of New York Press, Albany

© 2019 George A. Reisch

All rights reserved

No part of this book may be used or reproduced in any manner whatsoever without written permission. No part of this book may be stored in a retrieval system or transmitted in any form or by any means including electronic, electrostatic, magnetic tape, mechanical, photocopying, recording, or otherwise without the prior permission in writing of the publisher.

For information, contact State University of New York Press, Albany, NY
www.sunypress.edu

Library of Congress Cataloging-in-Publication Data

Names: Reisch, George A., author.
Title: The politics of paradigms : Thomas S. Kuhn, James B. Conant, and the Cold War "struggle for men's minds" / George A. Reisch (State University of New York).
Description: Albany : State University of New York Press, [2019] | Series: SUNY series in American philosophy and cultural thought | Includes bibliographical references and index.
Identifiers: LCCN 2018021849 | ISBN 9781438473673 (hardcover) | ISBN 9781438473666 (pbk.) | ISBN 9781438473680 (ebook) Subjects:
LCSH: Kuhn, Thomas S. | Kuhn, Thomas S. Structure of scientific revolutions. | Science—Philosophy. | Science—Social aspects. | Conant, James Bryant, 1893–1978. | Cold war—Social aspects.
Classification: LCC Q143.K83 R45 2019 | DDC 501/.092—dc23
LC record available at https://lccn.loc.gov/2018021849

10 9 8 7 6 5 4 3 2 1

To M. O., Who Tolerated It.

Contents

LIST OF ILLUSTRATIONS	xi
PREFACE	xiii
BOMBS AND BOOKS: AN INTRODUCTION	xv
TIMELINE OF EVENTS AND DOCUMENTS	xli
CAST OF ADDITIONAL CHARACTERS	xliii

Part I. War and Crisis

CHAPTER 1 Progress and Revolution in the Suburbs of New York	3
CHAPTER 2 War and General Education at Harvard	27
CHAPTER 3 History of Science in a Divided World	45

Part II. "The Struggle for Men's Minds"

CHAPTER 4 The Cold War Conversions of Thomas S. Kuhn and James Bryant Conant	65
CHAPTER 5 Sidney Hook and the Anticommunist Inquisition	81

CHAPTER 6
Brainwashing, or the Structure of Philosophical Revolutions 111

CHAPTER 7
The Necessary Dangers of Consensus and Unity 133

Part III. The Cold War Origins of *The Structure of Scientific Revolutions*

CHAPTER 8
The Language, Psychology, and Psychoanalysis of Scientific "Reorientations" 153

CHAPTER 9
"Attention Senator McCarthy": The Perils of Methodology in Totalitarian Times 171

CHAPTER 10
Ideology and Revolution in the *International Encyclopedia of Unified Science* 197

CHAPTER 11
Progress, Ideology, and "Writing History Backwards" 215

CHAPTER 12
From "Ideology" and "Consensus" to Paradigmania 235

Part IV. The New World of Paradigms

CHAPTER 13
"If Mr. Kuhn Is Right . . .": Paradigms and Dogmas in Cold War Science Education 255

CHAPTER 14
The Magic of Paradigms 271

CHAPTER 15
Spies, Prisons, Mobs, Bandwagons, and Beasts 293

CHAPTER 16
The Thomas Kuhn Experience 317

CHAPTER 17
A Revolution and a New Ideology   337

EPILOGUE
Writing and Rewriting History   355

ACKNOWLEDGMENTS   367

ABBREVIATIONS   369

NOTES   371

BIBLIOGRAPHY   419

INDEX   441

Illustrations

Figure I.1	Schine, Cohn, and McCarthy circa 1953.	xviii
Figure I.2	McCarthy confronts Conant at a Senate Appropriations Committee hearing, June 15, 1953.	xxi
Figure I.3	Thomas Kuhn, Harvard University graduation photograph, 1943.	xxiii
Figure I.4	"The Struggle for Men's Minds" and variations in popular and scholarly media, 1947–1954.	xxviii
Figure 1.1	Elizabeth Dilling's *The Red Network* dust jacket cover image.	13
Figure 1.2	Images from *School: A Film About Progressive Education* (1939) featuring the Hessian Hills School about two years after Thomas Kuhn graduated.	17
Figure 1.3	William Remington confers with his accuser, Elizabeth Bentley, July 1948.	20
Figure 2.1	Military officers at Harvard, Oct. 9, 1943.	34
Figure 2.2	Harvard Commencement, 1943.	43
Figure 5.1	Harlow Shapley.	88
Figure 5.2	Sidney Hook lecturing at the counterconference he organized to protest the Waldorf conference.	100
Figure 6.1	"Too Slow for Me," by Charles Henry Sykes, reproduced in *Literary Digest*, July 5, 1919.	112
Figure 9.1	Lowell Institute full-page flyer advertising Kuhn's upcoming series of lectures, ca. February 1951.	172

Figure 9.2	*The American Legion Magazine*, November 1951.	175
Figure 16.1	HUAC protestors dragged down stairs by police in San Francisco City Hall.	325
Figure 17.1	Nixon's address to the nation on the bombing of Cambodia, April 30, 1970.	337
Figure 17.2	Elizabeth Moos at Women Strike for Peace protest in Washington, March 22, 1973.	339

Preface

I have long been fascinated by the cold war and its effects on American thinking about science and knowledge. But until several years ago, I assumed that Thomas Kuhn's *The Structure of Scientific Revolutions* stood apart from those effects. In my book, *How the Cold War Transformed Philosophy of Science*, Kuhn appeared as a bystander because his enormous stature and influence seemed to belong to a later, different time. After its publication in 1962, that is, *Structure* offered an alternative to the perils that American philosophers of science faced in the late 1940s and '50s. Concepts such as normal science, puzzle solving, anomalies, crises, and paradigm shifts were not only new and exciting. At a time when topics in philosophy of science included Marxism, the social functions of scientific research, and the quest to unify the sciences across disciplinary and suspiciously "international" boundaries, they were safer—especially on campuses investigated by anti-communist politicians or J. Edgar Hoover's FBI.

As I saw it then, the dividing line between these two eras was announced by another classic from the early 1960s, the sociologist Daniel Bell's collection of essays titled *The End of Ideology*. Here Bell described a wave of professionalism and scholarly detachment washing over the academic landscape. Wholehearted, active, and existential commitments to Democracy, Capitalism, Marxism, Socialism and other capital-letter ideologies were now as dead as Joseph Stalin and Joseph McCarthy. In their place, eclecticism, skepticism, and scientific rigor had taken over, Bell noted, and the academy's future lay in "hypotheses, parameters, variables, and paradigms."[1]

Bell was right about paradigms. That concept was the heart of Kuhn's new theory of science that scholars and laypeople soon embraced. But *Structure* itself did not really end our fascination with ideology. For Kuhn developed his signature theories in the coldest depths of the postwar standoff with the Soviet Union, and he borrowed from this political landscape ideas and assumptions that were crucial for his theory to work. They concern the nature of the human mind, its susceptibilities to external control and propaganda, and the power of ideology to control thought and perception. Instead of heeding Bell's declaration, Kuhn acknowledged these powers, renamed and reconfigured them, and applied them brilliantly and persuasively to our understanding of science and its history.

Embedded as they are within Kuhn's terminology and his focus on science education and scientific communities, these ideological features of Kuhn's theories are admittedly hard to see. Having first read *Structure* in 1982, and dozens of times since, I began to recognize its affinities to the cold war only as I learned about Kuhn's life, the highly politicized and politically connected world he was born to, and how *Structure* itself took shape beginning in the late 1940s. Before this research, the politics within Kuhn's paradigms seemed to me natural and uncontroversial, so I took them for granted. Others I simply never noticed. As I have come to understand better through writing this book, these are some of the ways that ideology works.

Instead of presenting and defending this view of *Structure* in the form of claims backed by evidence and explicit arguments, however, I chose to write this book mainly as a narrative, a story of Kuhn's collaboration with his mentor, Harvard president James Bryant Conant, and the famous book that collaboration produced. Based in large parts on Kuhn's archival papers, I believe this format shows two things better and more vividly than any systematic exposition can—the young Thomas Kuhn embedded in the anxieties and suppositions of American politics as much as his elders, and his famous book, a classic of academic scholarship, firmly rooted in the turbulent political world he and Conant knew inside and outside the ivory tower.

Bombs and Books

An Introduction

Boarding his plane in Germany in the spring of 1953, James Bryant Conant knew there might be trouble when he arrived in Washington. An aide handed him a note. He opened it to see the words ". . . Senator McCarthy. . . ."

Ever since his speech at the Women's Republican Club of Wheeling, West Virginia, in February 1950, the junior senator from Wisconsin was on a rampage. Other politicians had discovered that a strong, militant stance against communism pleased their constituents and supported their careers. By the time McCarthy made headlines, Hollywood screenwriters had been scrutinized by Congress for years. But McCarthy outdid his anticommunist colleagues by announcing that communism had infiltrated Washington itself. From then on, he held hearings, conducted investigations, intimidated enemies and colleagues alike, and ruined careers by calling citizens to Washington to ask, often under the heat of klieg lights, "Are you now, or have you ever been, a member of the Communist Party?" Being called before McCarthy's committee, even mentioned by him or his staffers in the wrong way, could mean public and professional disgrace. Nobody liked suspected communists, and McCarthy made a career of exposing them.

Conant himself had nothing to fear. His ideological credentials were sterling. He was a famous educator, the president of Harvard University, and a chemist who helped oversee the Manhattan Project that introduced the atomic bomb and the modern "atomic age" of scientific progress. The problem that lay ahead of Conant was not *his* politics, but those of his colleagues, his friends, and those who worked for him. For years his chief physicist in the Manhattan project, J. Robert Oppenheimer, fended off rumors and suspicions that would not go away despite firm support from Conant and others who testified to Oppenheimer's patriotism. As president of Harvard, Conant knew how much trouble anticommunists could make by demanding the scalps of "suspicious" or "pink" members of his faculty.

McCarthy had been on Conant's trail for years. In 1949, when three leftist professors at the University of Washington lost their jobs, universities became fair game for anticommunist politicians. In his Wheeling speech, McCarthy had put the nation's elite, East Coast universities on notice. He named the Harvard astronomer and proud leftist Harlow Shapley (albeit as "Howard Shipley"). McCarthy's ally and admirer William F. Buckley, then an undergraduate editor of the *Yale Daily News*, had already taken the Harvard-Yale rivalry into anticommunist politics by waging war against the *Harvard Crimson*, whose editors had exposed FBI agents and their friendly, talkative informants on Yale's campus. In his book *God and Man at Yale*, Buckley explained that nearly all the Ivies were careening toward atheism and socialism.[1]

Just months before, in early 1953, Conant's Harvard—the "Kremlin on the Charles," as its conservative detractors put it—faced investigation by McCarthy and other anticommunists in Washington. As faculty, trustees, and students buzzed about the impending inquisition and how Conant would handle it, he stunned everyone by resigning from Harvard. Dwight Eisenhower, Conant's counterpart at Columbia University, had been elected President of the United States and asked Conant to become the new High Commissioner of Germany.[2]

By accepting Eisenhower's offer, Conant escaped McCarthy, who would soon denounce Harvard as a "smelly mess" of subversion and communist indoctrination.[3] That problem now belonged to Nathan Pusey, Conant's successor. But Conant remained in McCarthy's sights and was now even more vulnerable. No longer ensconced in a private university, he was responsible to the State Department and the American public that McCarthy represented—a public that was especially nervous and fearful. For despite the nation's victories in war, Americans felt increasingly surrounded by international communism—in the U.S.S.R., China, North Korea and, of course, East Germany.

As the new high commissioner, Conant would be in charge of keeping the invader at bay. Still, McCarthy smelled blood. Yes, Conant was a national hero, and he was on record opposing communist faculty in colleges and universities. But Conant belonged to the elite, East Coast establishment that McCarthy believed had already betrayed the nation to the ideological enemy. Its members included Alger Hiss, whose trials convinced many Americans—including Conant—that the anticommunist paranoia in Washington might be justified after all. And it included Hiss's friend and colleague Dean Acheson, the secretary of state who had publicly defended Hiss. McCarthy loathed Acheson and called him a "pompous diplomat in striped pants, with a phony British accent."[4] He had attended Groton, Yale, and the Law School at Harvard. *Conant's* Harvard.

When Congress interviewed Conant to approve his new diplomatic post, McCarthy suppressed his misgivings and kept quiet. That was a personal favor to Eisenhower who asked McCarthy to hold his fire. But now that Conant was on the job in Germany, McCarthy considered him fair game, and the note in Conant's hands was proof. Arriving through CIA communications, and not through the State Department channels that McCarthy and his aides monitored, it tipped him off that two McCarthy staffers, Roy Cohn and G. David Schine, were heading to Europe to investigate a new and disturbing report: United States Information Service libraries, the State Department's outposts of information and culture across the war-torn continent, were circulating pro-communist materials.[5] Having already ushered communist ideology into Washington, McCarthy believed, the State Department was now allowing it to spread throughout Europe in the form of broadcasts, films, and books.

Of Treason and Trousers

When his plane took off, Conant again escaped McCarthy's grasp. He would not deal with Cohn and Schine because he was on his way to Washington to meet German chancellor Conrad Adenauer (who would arrive by sea) and escort him on a short tour of the nation. But depending on what Cohn and Schine discovered and who they talked to at the commission, this was almost surely going to be a headache for Conant. As high commissioner, he was the top man in charge of the Information Service that Cohn and Schine set out to inspect.

Before the pain set in, however, there was some amusement. Reporters throughout Europe were fascinated by Cohn and Schine. Cohn was a twenty-six-year-old attorney from New York City best known (and, for many, reviled) for helping to prosecute Julius and Ethel Rosenberg for passing information about the atomic bomb to Soviet agents. Schine, the handsome scion of a California hotel fortune, was merely twenty-five (see

Figure I.1). Neither had expertise in libraries, publishing, broadcasting, or European culture and history. And they acted like children. When the duo came to the headquarters of Conant's commission in Bad Godesberg (just outside Bonn), the Frankfurt *Abendpost* described their antics. They "came to Europe in order to study 'waste and mismanagement in the American Information Program,'" the paper said, but they enjoyed an expensive, two-hour dinner at the Hotel Adler. The next day, after interviewing Glenn Wolfe, whom Conant had left in charge, Schine "announced that he put on the wrong trousers" and demanded that a driver retrieve a different pair from his room at the Adler. Once properly trousered, Schine suddenly discovered to greater alarm that he had lost or misplaced his notebook. He and Cohn frantically returned again to the hotel where

> it was observed that Mr. Schine batted Mr. Cohn over the head with a rolled up magazine. Then both of them disappeared into Mr. Schine's room for five minutes. Later, the chambermaid found ash trays and their contents strewn throughout the room. The furniture was completely overturned.

Figure I.1. Schine (left), Cohn (center), and McCarthy (right) circa 1953. (Image courtesy of Wisconsin Historical Society, image number WHi-8003)

Things did not go much better in Berlin, Munich, Vienna, Belgrade, Paris, or England. They became testy and combative with reporters who could not ignore the irony at hand: two brash, young Americans waving the flag of freedom from tyranny had come to Europe, blacklist in hand, to snoop through card catalogs of American libraries. They forced librarians to remove books from circulation and even to burn them, rumor would soon have it. "McCarthy's 'Young Snoopers' Hurt U.S.," the *Frankfurter Rundschau* explained. The paper implored McCarthy to "spare both his fellow citizens and us such practices" and to think twice about intellectual and cultural freedom. "We Germans know what freedom of opinion means from bitter experience—but also through those people whom McCarthy today attacks."[6]

If Conant was amused, he was nonetheless under orders to take Cohn and Schine's mission seriously. Like everyone else in Washington, his new boss, Secretary of State John Foster Dulles, was unwilling to anger or cross McCarthy. He instructed Conant and the other commissioners in Europe to cooperate with their traveling investigation, to keep them away from personnel files, and by all means to handle the situation carefully—"This is not something which can be delegated to a junior member of the staff."[7]

When he returned to Germany, Conant learned from Wolfe the trouble Cohn and Schine had created. One problem was Theodore Kaghan, the commission's director of information services in Bonn, who insulted Cohn and Schine by calling them "Junketeering Gumshoes." Cohn then telephoned McCarthy and announced to reporters that Kaghan would soon be returning to Washington to testify before the Senate. With one phone call, Kaghan's career began to unwind, even though he had "loyalty and security clearances and an anti-communist record reaching from Vienna to Berlin," as he later told the *New York Times*. Soon there would be trouble for Kaghan's counterpart in Munich, Lowell Clucas.[8]

Into the Ring

In June, Conant returned to Washington to present his annual budget to Congress. The task was difficult enough because funding for postwar operations in Europe had been reduced, but Conant also worried that McCarthy, a member of the Senate Appropriations Committee, might use the occasion to launch an attack. Fortunately, Conant was told, this was unlikely. McCarthy was only a member of the committee, not in charge.

And the meetings would be closed to the press, whose cameras and bright lights McCarthy craved.

"Imagine, then, my consternation," Conant later wrote. Walking to the Senate Committee room, he saw klieg lights, reporters, and photographers milling about. The reason was in the morning papers, which Conant had not yet seen, having just arrived after traveling overnight from Boston. Eisenhower himself had weighed in on Cohn and Schine's traveling investigation and new reports, confirmed by Dulles himself, that some librarians had not only pulled books from the shelves. They burned them.[9]

The symbolism was too rich for Eisenhower to ignore. The very idea that Americans burned offensive books played into criticism that McCarthy's America was becoming just as dogmatic, illiberal, and intolerant of dissent as Nazi Germany or Soviet Russia. Eisenhower did not want to look like Joseph Stalin. But he could not deny that critics of the United States had some basis for the comparison. In just a few days, Julius and Ethel Rosenberg would be electrocuted in upstate New York. Despite sustained protests and appeals from leading citizens and intellectuals around the world, Eisenhower refused to grant them clemency.[10] So when the reports of book burning came in, Eisenhower saw an opportunity to claim some moral high ground over McCarthy as well as Cohn, the Rosenberg prosecutor. Speaking at Dartmouth's convocation, he turned to these reports and urged his audience, "Don't join the book burners." No American, he insisted, should be afraid of any book in American libraries.[11]

McCarthy deflected the insult. "He couldn't very well have been referring to me," he told a reporter. "I have burned no books." Privately, however, McCarthy fumed. This was a public insult from a president who in fact owed him at least one favor for not opposing his nominee for high commissioner of Germany. So McCarthy—the former amateur boxer and boxing coach—took off his gloves and confronted Conant at his budget presentation.[12]

About a half hour into the session, McCarthy strolled in, took a seat across the table, and dominated the meeting (see Figure I.2). He asked Conant to specify how much of his budget would support books and other informational materials. Before Conant could begin to answer, McCarthy resumed talking. And talking. This was his trademark technique—to produce a waterfall of vague accusations, ominously large numbers, and loaded questions to fluster, confuse, and exhaust his targets. "May I ask you this," he said to Conant:

Figure I.2. McCarthy (far left) confronts Conant (far right) at a Senate Appropriations Committee hearing, June 15, 1953. Glenn Wolfe is seated to Conant's right. (Image courtesy of Getty Images)

> Our Committee has recently exposed the fact that there are some 30,000 publications by communist authors on information shelves. Many of them in Germany . . . by Communist authors on our shelves with our stamp of approval—some 30,000. May I ask what your attitude toward that is? Do you favor taking those books off the shelves? Would you favor leaving them on the shelves? Would you favor discontinuing the purchase of those books or the continuation of that purchase?[13]

Conant tried to avoid McCarthy's paranoid generalizations. Were he to choose which belonged on the shelves, he explained, he'd examine each book and author individually and carefully—"to see who our Communist author was, what his point of view was, and whether the reading of that book by the Germans would do us more good than harm."

McCarthy knew, however, that Conant could not defend this tolerant, open-minded stance for long. When he was president of Harvard, Conant was on record that Communists were dangerous and had no business teaching in American universities. He had to agree that books and the ideas they contain, just as much as communist teachers, were a part of the ongoing communist conspiracy to control the globe. So McCarthy steered Conant into that part of the ring:

> McCarthy: Let's see what the point of view of the author is. The Communist is under Communist Party discipline, and the point of view is furthering the Communist conspiracy. There is no doubt about that, is there?
>
> Conant: With such a man, I would not want his books on the shelves.
>
> McCarthy: Such a man, I think—and *every* Communist—we can agree has the task of furthering the Communist cause; otherwise, he is not a Communist; is that not correct?
>
> Conant: Quite so.
>
> McCarthy: And one of your tasks over there is to fight communism, so . . . would you favor using part of [this budget's funds] to buy the works of Communist authors and put them on your bookshelves?

"No," Conant replied firmly. He was no supporter of the communist cause.

What about the communist volumes on the shelves *now?* McCarthy rehearsed again Cohn and Schine's ominous report and pressed Conant to answer: "Would you favor removing from the bookshelves the works of Communist authors?" Having agreed not to add books by Communists, Conant logically had to disapprove of those already there. Yes, he conceded, "I would be in favor of taking them off."

Now McCarthy reached for his prize. "You would not call that '*book burning*' if you took them off, would you?" Unaware of the national significance the phrase had acquired overnight, and that McCarthy pummeled him to get back at Eisenhower, Conant granted the point. "I suppose you wouldn't," he replied, "but I wouldn't suppose that you would burn them."

McCarthy scored more points as the day wore on. To Conant's objection that the books in question should have been removed discreetly, "without too much publicity," McCarthy grilled him about why anyone could object to the public exposure of Communist authors and what specific objections

Conant had. None at all, Conant conceded: "I certainly don't object to anything that congressional committees do, sir."

At Conant's attempt to defend Lowell Clucas, the officer in Munich, as "a good man for the job," McCarthy angrily denounced Clucas, Kaghan, and Conant himself as unreliable in the fight against communism:

> Then I say—and this is definitely on the record—I feel and think that if you feel you should have men like Kaghan and Clucas spending money over there on the information program, I do not think this senate should give you one penny. I think you have done infinite damage if you continue to keep men like that running the program, and they will continue to do damage.

He threatened to block Conant's budget and keep a spotlight on Conant's operation until he performed a thorough "house cleaning over there." To Conant's great alarm, McCarthy hinted that Glenn Wolfe, sitting at Conant's side, might be next.

Until he resigned from the presidency, one of Conant's colleagues at Harvard was the young historian of science, Thomas S. Kuhn (see Figure I.3). Kuhn

Figure I.3. Thomas S. Kuhn, Harvard University graduation photograph, 1943. (Image courtesy of Harvard University Archives)

was familiar with the bare-knuckle politics of anticommunism. He had greatly admired the founder of his elementary school who was forced to resign her position for her political activities. Her son-in-law, William Remington, belonged to the first wave of suspected Communists in Washington prosecuted after the war. McCarthy probably had Remington in mind when he boasted at Wheeling of having "in my hand fifty-seven cases of individuals who would appear to be either card carrying members or certainly loyal to the Communist Party, but who nevertheless are still helping to shape our foreign policy."[14] When McCarthy called Harvard a "smelly mess," he was in the midst of interrogating the Harvard physicist Wendell Furry who ten years before had been one of Kuhn's teachers.

There is no record that Kuhn paid close attention to Conant's travails this particular day. But he probably did, because the encounter made front pages around the country and Kuhn admired Conant intensely. He learned eagerly from him, and seemed always proud to be his associate. As an undergraduate at Harvard, he praised Conant's wisdom and leadership from the editorial pulpit of the student newspaper, *The Crimson*. After the war, when he returned to Harvard for graduate study in physics, he favorably reviewed Conant's latest venture in the alumni magazine—a new undergraduate curriculum dedicated to general education. Kuhn would soon join Conant's general education program as a member of the faculty and, under his tutelage, leave behind his training as a physicist to become an expert in the history of science.

After Conant left Harvard to become high commissioner and then ambassador to West Germany, he and Kuhn stayed in touch. Conant wrote a laudatory foreword to Kuhn's first book, *The Copernican Revolution* and, in 1962, Kuhn dedicated his second book to Conant, the man who taught him much and transformed his life and career. This book is called *The Structure of Scientific Revolutions*. Its dedication reads,

> To James B. Conant, who started it.

Conant started two things, in fact—Kuhn's career as a historian of science as well as the postwar revolution in the scholarly understanding of science for which Kuhn and his book became famous. Giving credit where it is due, Kuhn's dedication nods to Conant's efforts to teach Americans how science really works. Conant deplored the schoolhouse view of science as collections of observations and facts and emphasized instead that science is driven by ideas. Its past and its future, he explained in his book *On Under-*

standing Science, is a progression of "conceptual schemes" through which scientists understand nature, build instruments, conduct experiments, and move knowledge forward.

Today, Conant is remembered for many things—his expertise in chemistry, the growth of Harvard over his twenty-year presidency, the development of the atom bomb, his diplomacy in Germany, his support of the Educational Testing Service and its Scholastic Aptitude Test, and at the end of his career his crusade to improve public education. But these "conceptual schemes," the central pillar of his view of science, are no longer so visible. They were eclipsed by the terms Kuhn introduced in *The Structure of Scientific Revolutions*—"paradigms" and "paradigm shifts."

Paradigms, Kuhn explained, are the lifeblood of science. A paradigm exists at the heart of every scientific community. To become a scientist means joining a community and internalizing its paradigm through years of education and professional practice. To have a scientific career means curating and contributing to the paradigm—expanding its range of application, refining its parameters, and solving the conceptual and experimental puzzles it presents. Though most outlive the scientists whose careers unfold within them, paradigms have a life cycle of their own. Eventually, every paradigm will break down and fail. Some of its puzzles will refuse solution, and those scientists trained within it will be at a loss to understand why. Doubts about its truth will emerge and some forward-thinking scientists will explore new options. Debate will split the community and throw it into crisis. At that point, the stage is set for the emergence of a new paradigm to replace the old and a new kind of "normal science" to form around it. Paradigms shift and science undergoes a revolution.

Kuhn knew that *The Structure of Scientific Revolutions* would make a big splash. Its first sentence promised "a decisive transformation in the image of science by which we are now possessed." Other historians and philosophers had studied scientific revolutions, but Kuhn believed that most scholars did not fully understand their effects and their philosophical implications—how radically scientific knowledge changes through revolutions, how paradigm shifts lead to entirely new worlds of scientific understanding and perception. He put it this way:

> Led by a new paradigm, scientists adopt new instruments and look in new places. Even more important, during revolutions scientists see new and different things when looking with familiar instruments in places they have looked before. It is as

if the professional community had been suddenly transported to another planet where familiar objects are seen in a different light and are joined by unfamiliar ones as well.[15]

Even Conant did not understand this, for he insisted that knowledge grows and accumulates over time; that science learns more and more about the world as one conceptual scheme is replaced by another. But that simply isn't true, Kuhn argued in *Structure*. When a new paradigm transports them to a different planet, scientists leave behind their old paradigm and the knowledge it once made possible. In a real sense, he wrote, "when paradigms change, the world itself changes with them."[16]

Upon their publication in 1962, Kuhn's insights spread far and wide. *Structure* soon became required reading for philosophers, historians, and sociologists of science. But it also captivated scholars in fields across the academy and then moved into the worlds of business, politics, economics, and even the self-help industry. In Stephen Covey's *The Seven Habits of Highly Effective People*, the most effective people bravely scrutinize their own, personal paradigms to discover their blind spots and better navigate the worlds around them. *Structure* became a touchstone in the so-called culture wars of the 1980s and '90s and it continues to inspire activists and would-be revolutionaries. "Paradigm shift: the great machine of capitalism starts to heave," the magazine *Adbusters* announced in 2013.[17]

Revolution, Ideology, and "The Struggle for Men's Minds"

This book explores the origins of *The Structure of Scientific Revolutions* and how Kuhn's ideas were shaped by his powerful relationship to Conant and the cold war politics that surrounded their collaboration. To contemporary eyes, politics—especially the ranting and bullying of McCarthy—may seem irrelevant to a scholarly book such as Kuhn's. As many historians and philosophers have pointed out, Kuhn's main interests lay in scientific theory, science education, epistemology, sociology, linguistics, and history—areas of study that today seem unrelated to national and international politics. But that was not always so. In different ways, and especially during the years of their collaboration, Kuhn and Conant paid close attention to politics. For Conant, as his cat-and-mouse relationship with McCarthy shows, science, education, diplomacy, and politics were never far apart. For the students in his general education program destined to become doctors, lawyers, politicians, or business professionals, Conant believed there were no better lessons in the virtues of liberalism and intellectual freedom—and conversely the perversions of Nazi and Soviet totalitarianism—than case studies in the history of astronomy, chemistry, or physics. With Kuhn, the connection became stronger and he turned the lesson plan around: the best way to understand momentous changes in science's history, he argued, is to see them not only as products of reason, logic, and the careful design and analysis of experiments. We should see them additionally through political lenses, he argued in *Structure*, as something like political revolutions that occur within science. They follow the same temporal schema, give rise to factions who may have difficulty understanding each other, and their outcomes seem to transform the world and our knowledge and perceptions of it.

In July 1953, weeks after Conant's battle with McCarthy, Kuhn wrote a letter to his editor at the *International Encyclopedia of Unified Science*. He had earlier promised to write an essay on the history of science and was now following up to describe what he planned to say. The main idea, he explained, was that a scientific theory functions as an *ideology* in scientific communities and in the minds of the scientists who belong to them. It directs attention to certain kinds of scientific problems. It sets experimental and logical standards that professional scientists must meet. In some ways, it even tells scientists what to think—"it dictates preferred techniques of interpretation," Kuhn wrote—and discourages creativity and imagination that might lead scientists to think outside their ideological box.[18]

Kuhn's wide range of intellectual interests and his working relationship with Conant put him a position to see this intriguing and provocative connection between politics and science, that scientific and political communities are subject alike to the world-shaping powers of ideology. These powers were on display during the momentous political events of Kuhn's youth and they surrounded his postwar collaboration with Conant. Though the bombs and artillery of World War II in Europe and Asia had ceased, when Kuhn joined Conant's general education project the nation was fast descending into ominous, uncharted geopolitical territory. Conant called it an "armed truce" between the United States and the Soviets. But it would soon be known as the "cold war" of ideology pitting liberalism and democracy in the West against totalitarianism in the East. In the gendered language of its time, this winner-take-all contest of political philosophy was often called "the war for men's minds" or, usually, the "struggle for men's minds." The summer session at the University of Wyoming in 1951 was dedicated to this struggle, as were speeches, editorials, magazine articles and films from the late 1940s well into the 1950s (see Figure I.4).[19] The drama and high stakes were perhaps most vividly symbolized by the brainwashing sensation that swept over the nation in the early 1950s as Americans learned about GIs in Korea who had been forcibly transformed into obedient, true-believing communists. Backed by Chinese and Russian psychologists, the story went, their North Korean captors almost literally "washed their brains," replaced one ideology with another, and turned them into fundamentally different kinds of human beings.

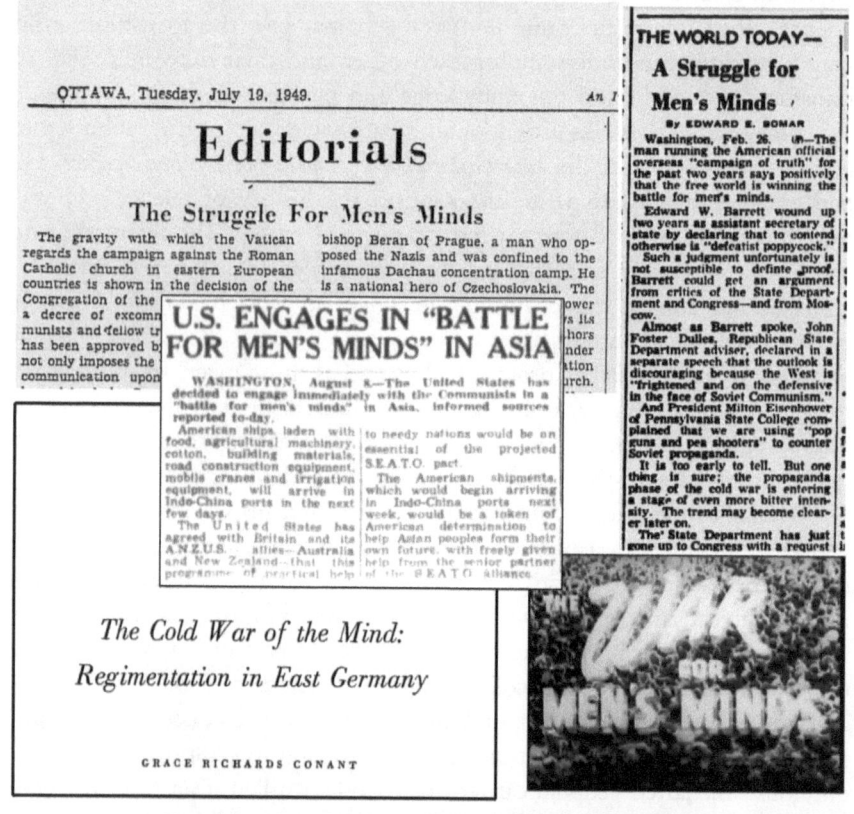

One of the popular clichés of our time is that the struggle between the free world and the Communist bloc is a "battle for men's minds," an ideological contest for the emotional and intellectual endorsement of the so-called uncommitted peoples of the world. The cliché is partly true, but like most catchy phrases is

Figure I.4. "The Struggle for Men's Minds" and variations in popular and scholarly media, 1947–1954.

The struggle for men's minds helped to fuel McCarthy's crusade, Cohn and Schine's sensational tour of Europe, and McCarthy's attack on Conant that day in Washington. For in this contest of ideologies vying to control the human mind, books and the ideas they contain became heavy artillery. The same logic propelled the philosopher Sidney Hook's crusade against communism in colleges and universities. Hook warned the nation that communist faculty were under orders to inject their ideology into lectures and classes. Just as McCarthy urged Conant to "have a house cleaning over there" in Germany, Hook urged Conant and other university presidents to clean the halls of academia at home and remain vigilant against so-called fellow-traveling professors who seemed secretly captivated by Soviet ideology.

McCarthy and Hook became leading public figures during the struggle for men's minds. But the majority of intellectuals, scientists, and university administrators who never made headlines acknowledged the enormous power of ideas to control the mind and the enormous stakes of the ongoing struggle against communism. In this respect, neither James Conant nor Thomas Kuhn was unique in being keenly interested in the political implications of how science is understood. Yet Kuhn's view was original and provocative. While most intellectuals believed that politics and political ideology threatened to *interfere* with science and its progress, Kuhn saw a different, more constructive connection. If the struggle for men's minds would determine the future of humanity and the nature of civilization, then perhaps similar, winner-take-all contests between ideologies in science—struggles for *scientists'* minds—had all along determined the historical course of science and would continue to do so into the future.

The Invisibility of Politics

In the myriad expositions, analyses, and criticisms of Kuhn's ideas that have been published in the wake of *Structure*, the role of politics is rarely addressed. One explanation for this circumstance begins with Kuhn's provocative letter to his editor. Thinking out loud about what his future essay would be called, he said the word *ideology* should naturally appear in the title. But he knew that the word had controversial associations. So he suggested instead "The Structure of Scientific Revolutions." This and other changes in Kuhn's terms and concepts, especially the eventual replacement of "ideology" by "paradigm," put distance between *Structure* and the ideologically charged politics that surrounded its birth.

Another reason is that the American academy did not fare well during these early years of the cold war. Having come of age during the Great Depression of the 1930s, many intellectuals and professors of the 1950s and '60s had once been sympathetic to socialism, if not the Communist Party itself. Even Sidney Hook, the era's most outspoken and determined foe of all things Stalinist, was once a proud and vocal Marxist who rubbed elbows with party Communists. So when anticommunism became the order of the day in the late 1940s, the stage was set for professional and personal carnage. Those who did not convert to anticommunism often faced an unhappy choice between two kinds of public disgrace—that of the soft-minded sympathizer who remained captivated or "duped" by Stalinist ideology, or the stool pigeon or snitch who saved themselves by "naming names" of former comrades in front of FBI agents or congressional committees. Some, such as Kuhn's teacher Wendell Furry, refused to perform this ritual and invoked Fifth Amendment protections against self-incrimination when called to testify. But they often fared no better and were commonly denounced as "Fifth Amendment Communists."

As they were for Hollywood screenwriters, the late 1940s and early '50s were something of a collective nightmare that many teachers and professors simply wished to forget in subsequent decades. In addition, the progressive and tolerant academic culture that formed later in the 1960s around ideals of free speech, civil liberties, and social justice rendered bitter memories of these years somewhat antique and irrelevant to new generations of intellectuals. Not until the 1980s would historians such as Ellen Schrecker discover how violent and transformative this era had been for American colleges and universities, as topics such as Marxism lost academic respectability and professors disengaged from politics and public dissent in favor of scholarly objectivity.[20] Still, the McCarthy era's effects on intellectual life have usually been seen as repressive and destructive—making it seem all the more unlikely that themes and debates from this difficult and readily ignored era creatively sparked and encouraged Kuhn's revolutionary theories about science.

A third explanation for the near invisibility of politics is the enormous and lasting influence of *The Structure of Scientific Revolutions* itself. It remains a classic textbook for aspiring historians of science, philosophers, sociologists, anthropologists, and scholars in many other fields. For graduate students in the humanities and social sciences, fluency in "paradigm shifts" and "paradigm maintenance" is a familiar scholarly credential. Crucially, however, as Kuhn himself emphasized in *Structure*, textbooks are forward-looking. They focus students' attention on their paradigm's future and how it can be further applied and refined. At the same time, they discourage critical inquiry into the past by supplying readers—usually in the introductory chapter, Kuhn noted—with a ready-made history that is easy to understand but invariably false (at this point in *Structure* Kuhn invoked George Orwell's *1984*).[21] Instead of winning a revolutionary struggle between formidable adversaries, each of whom possessed credible arguments on its side, textbooks lead students to believe that the reigning paradigm simply corrected the oversights and blunders of the past. Though it is physicists, chemists, biologists, and other kinds of normal scientists for whom revolutions become "invisible" in this way, Kuhn began *Structure* with a comparable story about the blindness of historians and philosophers to the historical record of "research activity itself"; a story about the dynamics of paradigms liberating readers from the false "image of science by which we are now possessed."[22] As its fame and influence grew, this revolutionary view of *Structure* and its achievement took hold. With a single, relatively short book, it is often said, Kuhn swept away decades of mistakes attributable to philosophical bias (most of it foisted on scholars by the movement known as logical empiricism) and persistent neglect of historical and sociological forces within science itself. Having cleared away so much debris, *Structure* invited its readers to ignore this error-laden past and orient themselves to future research and teaching enlightened by an understanding of paradigms and all they explain.

In recent decades, historians of philosophy have questioned just how original and revolutionary Kuhn's theory of science really was. Historical interest in logical empiricism revealed that Kuhn's views were not so very different from some held by Rudolf Carnap, a leading logical empiricist who co-edited the *International Encyclopedia of Unified Science* with Charles Morris.[23] Instead of rejecting his intellectual forbearers, Kuhn is now seen by most scholars as belonging squarely within certain philosophical, sociological, and psychological traditions.[24] Recent books by Alexander Bird and K. Brad Wray, for example, are concerned less with the origins of Kuhn's ideas than with the place of Kuhn's ideas in the intellectual firmament and their yet-untapped potential for understanding science and the growth of knowledge. Bird does offer a historical account of Kuhn's ideas, but it is exclusively intellectual and scholarly. "There are seeds of Kuhn's own revolution in such historians and sociologists as Ludwik Fleck, Karl Mannheim and Robert Merton, as well as philosophers such as Toulmin and Hanson," he writes. But Kuhn's personal interest in politics and ideology plays no role in Bird's account because it includes, as he put it, only "the sparsest mention of Kuhn's biography."[25]

Philosophers who notice political and cultural themes in *Structure* tend to push them aside as incidental, if not misleading. In his systematic analysis of *Structure*, for example, the philosopher Paul Hoyningen-Huene noted that German-speaking scholars were puzzled by Kuhn's talk of a paradigm shift as a "conversion" or *Bekehrung*. This made it seem as though scientific progress involved not only reason, observation, and careful experimentation but something like religious faith or personal transformation. For careful scholars such as Hoyningen-Huene, this was a distraction because "this proximity to the cliché of religious conversion was never really part of Kuhn's theory."[26] The philosopher John Earman once put it more colorfully by noting that *Structure* abounds with "purple passages"—not only about the "conversion experiences" Kuhn placed at the heart of momentous paradigm shifts, but about the intensity of paradigmatic crises and revolutionary debates between factions who live almost literally in incompatible worlds of thought and experience. Like Hoyningen-Huene, Earman saw these parts of *Structure* as distractions that interfere with the philosopher's task of rendering Kuhn's theories in clear, objective, black and white prose.[27]

This book sees these passages differently—not as distractions or matters of style that, like thin ties of the 1960s or bell bottom pants of the '70s, can be discarded without loss. Along with *Structure*'s sustained interest in scientific revolutions—a word that rings with widely recognized political implications—its purple passages point to the struggle for men's minds, itself a festival of dramatic, purple prose about the overwhelming powers of ideology and the reality-changing stakes of cold war politics. As Kuhn's interests in "conversions" might suggest, these roots sometimes extend through cold war fears about brainwashing into more distant and original features of American culture, such as the nation's Puritan devotion to religious conversion and purification. Hoyningen-Huene is right that religious conversions are not themselves a part of Kuhn's theory of science, but they were a fixture within Kuhn's intellectual heritage. He and Conant greatly admired William James, for example, the intellectual giant who strode Harvard yard decades before and who celebrated the life-changing power of religious transformations in his classic book *The Varieties of Religious Experience*. "A fine and truly beautiful book," Kuhn noted after he first read it in the early 1940s. As argued in these chapters, the event that evidently sparked Kuhn's interest in the inner, cognitive experience of scientific revolutions, an event he called his "Aristotle Experience," echoes both James's account of sudden, life-changing experiences as well as contemporary interest in political conversions and brainwashing.

A few intellectual historians have nonetheless recognized the politics behind Kuhn's purple prose. David Hollinger noted that, unlike most other writers about science in the 1950s and '60s, Kuhn "focused on the political dynamics within scientific communities" and portrayed his normal scientists behaving "more in a totalitarian than in a democratic manner." Peter Novick took it for granted that Kuhn's gloss on revolutions as "a choice between incompatible modes of community life" was a "metaphor drawn from politics."[28] But few historians seeking to understand *Structure*'s origins have followed these threads into political realities. In his book *Working Knowledge*, for example, Joel Isaac argues that *Structure* is best understood reflecting the interdisciplinary intellectual landscape at Harvard at mid-century. Features of this landscape include the case-study method of teaching that Conant and Kuhn utilized in the general education program, Kuhn's eclectic uses of psychology, sociology, and science education to explain how normal science and paradigm shifts come to be, and the "conceptual schemes" that Conant and others working at Harvard took for granted as they sought to understand precisely how science works and how knowledge grows. This broad interest in conceptual schemes was sparked and sustained by the biochemist Lawrence J. Henderson, who distilled conceptual schemes and their importance from the sociological writings of the Italian Vilfredo Pareto. Henderson taught courses on Pareto, founded the "Pareto Circle" (in which Henderson and his students discussed Pareto's ideas and formulations in the 1930s), and laid a foundation on which Kuhn would later build his new image of science. That foundation included Conant, who knew Henderson not only as an influential Harvard professor, but a member of his family (Henderson was his wife's uncle). And it included the Harvard Society of Fellows, which Henderson helped to create in the 1930s. Kuhn later joined the society and bounced his ideas off several talented intellects who were students of Henderson, Pareto, or both.[29]

As Isaac sees it, the scope of this influence was narrow, intellectual, and disciplinary. Pareto offered to Henderson and others "the technical understanding of epistemology," which they believed added rigor and prestige to the ways they theorized human groups. Yet Depression-era politics saturated Henderson's enterprise. These were years when many intellectuals, especially New Yorkers like Sidney Hook, took it for granted that American capitalism was broken and bound to be replaced by either socialism or fascism. This was a prognosis that most wealthy Bostonians could not ignore and it explains something of their attraction to Pareto, his searching critiques of Marx, and his very different view of human history. In place of Marx's vision of history driven by class conflict to socialism, for example, Pareto believed that social and economic systems were formed of mutually dependent parts whose functional relationships disposed them to seek equilibrium and stability in the wake of external or internal shocks such as the ongoing Depression. "As a republican Bostonian who had not rejected his comparatively wealthy family," wrote George Homans, who studied under Henderson and knew Kuhn within the Society of Fellows, "I felt during the thirties that I was under personal attack, above all from the Marxists. I was ready to believe Pareto because he provided me with a defense."[30]

A Role for Politics

What is to be gained by examining *Structure*'s origins against this political backdrop? At the very least, we stand to gain a more complete and realistic picture of Kuhn, his famous book, and where it came from. To be sure, Kuhn was a scholar and *Structure* reflects his education, his intellectual curiosity, his extensive reading in philosophy and history, and his desire to enlighten historians and philosophers whom he saw caught within a faulty, inaccurate picture of science's history. But American politics, as much as theories of language, epistemology, psychology, history, and other scholarly and scientific interests, was among *Structure*'s formative ingredients. Isaac sees Kuhn "formed almost exclusively by the Harvard complex" and "in many ways the ultimate product"[31] of it, but Kuhn was no blank slate when he arrived Harvard in the fall of 1940, or when he returned for his graduate training after the war. During his elementary and secondary education, he and his family were keenly interested in politics and the rise of Nazism—especially menacing for Jewish families such as the Kuhns. In prep school, if not before, Kuhn was fascinated by the powers of ideology and propaganda and he brought these interests to Harvard. Alongside his reading, teaching, and research, they helped him to develop and refine the ideas that would come together in *Structure*'s theories of normal science and paradigms.

Taking politics into account also sheds light on Kuhn's originality, unorthodoxy, and intellectual daring. As its dedication declares, *Structure*'s origins cannot be understood apart from James Bryant Conant. Yet Isaac and other scholars tend to reduce *Structure* to an expression or product of Kuhn's relationship to Conant, to his university, or both. The historian Jamie Cohen-Cole, for example, has examined how some of the Harvard sociologists and psychologists surrounding Kuhn paid close attention to cold war politics and understood it as Conant did—as a contest of ideologies that threatened freedom of thought around the world if communism proved to be victorious. As creativity and open-mindedness became important subjects within their research, Cohen-Cole sees Kuhn also taking for granted this ideal of the creative, open mind.[32] Perhaps the most striking example of this tendency is the book *Thomas Kuhn: A Philosophical History for Our Time*, in which Steve Fuller portrays *Structure* as "an exemplary document of the Cold War era." But the politics that count in Fuller's treatment are mainly Conant's, and Kuhn himself is portrayed as translating those politics into a popular philosophy of science.[33]

By the time Kuhn completed his first draft and unveiled his new theory of paradigms, however, he was in open revolt against Conant's political and educational agendas. For example, given the formidable powers of ideology, he came to doubt the premise of Conant's general education program that nonscientists—outsiders to these all-important ideological worlds—could really come to understand how scientists think and work. As for scientific progress, not only was Conant incorrect to believe that knowledge accumulated across paradigm shifts; Kuhn had decided that this nearly universal ideal of open-mindedness was exaggerated and potentially misleading for our understanding of science. During times of crisis and revolution, to be sure, creativity plays a crucial role. But the "normal" scientists Kuhn described were dogmatic, uncreative, and conservative and *they had to be* if a future revolution were to usher in a new paradigm. Both conclusions rubbed directly against Conant's highly politicized view of science and its place in the struggle for men's minds. As chapter 14 shows, this helps to explain why Conant, whom one might have expected to applaud *Structure* when he first read it, was frankly annoyed and frustrated by this word *paradigm* that appeared on nearly every page of Kuhn's manuscript.[34]

Finally, the politics of paradigms bears on *Structure*'s reception and how it became an intellectual and popular juggernaut. Among scholars, Conant's puzzled and critical reaction to paradigms proved to be typical. For despite the many problems and flaws he first detected in its pages, Conant warmed up to *Structure* and soon praised it as brilliant, important, and not to be ignored. A similar drama—perhaps even a collective "conversion experience"—played out across the academy. As one observer trenchantly put it, Kuhn produced "an irritating, naive, confused, and provocative work" that combined history, psychology, sociology, and philosophy "all together in one glorious explosion" that scholars across the academy found irresistible. Kuhn himself pointed out that *Structure* was not unique in pursuing a humanized and historical understanding of science. Writings by the American philosopher Norwood Russell Hanson, the Hungarian chemist-turned-philosopher Michael Polanyi, the French historian Alexander Koyré, and Conant himself had said often similar things about how science works and helped to fuel and ignite this glorious explosion.[35]

But it was *Structure*, and none of these others, that became an enduring, influential classic. One reason for that, this book contends, is the cultural and intellectual power of the politics embedded within Kuhn's discussions of crises, conversions, and paradigm shifts. As he explained in the chapter "The Nature and Necessity of Scientific Revolutions," scientific and political revolutions unfold according to a similar logic; and in both cases they necessarily involve battles—whether in the laboratory or in the streets—"between incompatible modes of community life."[36] For those who read *Structure* in the 1960s and '70s, especially in the United States, these and similar passages could not have failed to echo the sustained public warnings from public anticommunists such as Sidney Hook, J. Edgar Hoover, McCarthy, and, at times, Conant himself that communist ideology threatened to fundamentally transform the nation; that in confronting communism citizens faced an existential choice between incompatible "ways of life." If all that was true—and many scholars and the vast majority of Americans believed it was—then Kuhn's insight seemed all the more fundamental. Scientists, after all, are human beings. So why shouldn't the history of science involve dramatic, ideological conversions and revolutionary transformations like those at stake within the struggle for men's minds?

But it wasn't true, at least for long. By the mid-1950s, communism in America was moribund. Within a few years politicians like John Kennedy would call for peace and cooperation between the superpowers and American college students would begin to rebel against the paranoid, duck-and-cover world their elders had created. When writing his autobiography in the 1960s, Conant looked back with fresh eyes on his brutal encounter with McCarthy. The politician he had struggled to accommodate that day in 1953 now seemed no less a demagogue than Hitler himself, whose "big lie" about subversive Jews in Germany anticipated McCarthy's big lie about Communists in Washington and the State Department. Conant all but confessed that he bought into this big lie more than he should have. But then again, he observed, almost everyone did during those difficult and unhappy "days of inquisition."[37]

This about-face reflects the nation's transition from McCarthy, universally feared but tolerated, to McCarthyism, universally despised and denounced. And it suggests another telling difference between Kuhn and Conant. For Kuhn either never realized or never admitted how much his great book, its purple prose, and its eager reception had to do with the dramatic politics of these years. Unlike Conant, the administrator and public intellectual who balanced scholarship and intellectual progress against the passions of public life, Kuhn became a professional scholar who avoided public spotlights and wrote his books and papers mainly for other scholars. When he looked back on *Structure*, he described what most philosophers and historians see today—an influential but flawed classic that just happens to speak of crises, reality-changing revolutions, conversion experiences, and rewritten history à la Orwell's *1984*. Kuhn knew that those who live through events are not reliable historians of them; that decades if not centuries of intellectual, cultural, and emotional distance are required for historians to understand the past.[38] So he may have been too close to the struggle for men's minds and *Structure* to see a connection, or to see his purple prose as anything more than mere coloration or the exuberance of a young scholar determined to take the academic world by storm.

On the other hand, the Thomas Kuhn that emerges in these chapters from his letters, notes, and years of struggle to craft *The Structure of Scientific Revolutions* seems too smart and observant to revisit passages like this—"the transfer of allegiance from paradigm to paradigm is a conversion experience that cannot be forced"—and miss, or fail to remember, the dramatic and urgent politics of loyalty and allegiance to American ideals that once swirled around his childhood school; wrenching debates over American intervention in World War II; the brainwashing sensation; anticommunism at Harvard; and Conant's bruising encounter with McCarthy over the power of books and ideas to determine the very texture of modern civilization.

A Story in Four Parts

Though it explores selected episodes in their lives, separately and together, this book is not a biography of Kuhn nor of Conant. It is the story of *The Structure of Scientific Revolutions* and how its signature ideas took shape within the complex relationship between them. That relationship was rooted in shared interests, professional collaboration, and mutual respect, if not quite friendship. But it also included debate and clashing beliefs about science, how it works and what scientific progress means, and—of vital import for the ongoing struggle for men's minds—the nature of scientific communities and the psychological, ultimately political, makeup of scientists themselves.

The chapters in Part I, "War and Crisis," document the radicalism and pacifism of Kuhn's early education, his interests in politics, his personal struggle over the nation's intervention in World War II, and his abundant admiration for Conant, his future mentor and collaborator. They also examine the political agendas behind Conant's writings about the history of science, his liberal vision of scientists as free, unfettered minds (which Kuhn would later challenge), and the urgent need recognized by Conant and other prominent cold warriors for all Americans to understand something about the nature of modern science in the new, postwar world of atomic weapons, scientific research, and ideological competition with the Soviets.

Part II, "The Struggle for Men's Minds," explores the contentious landscape of the early cold war that surrounded Kuhn and Conant and helped define the cultural climate into which *The Structure of Scientific Revolutions* would be later introduced. It begins with the theme of conversion and follows it from Kuhn's momentous "Aristotle experience," through Conant's shifting responses to the crises he experienced shortly after the war concluded, and his decision in 1949 to take a public stand against communist faculty. It continues with Sidney Hook's crusade to convince the nation and its educational leaders—including Conant—of the powers of communist ideology and its powerful, dangerous hold on some Americans' minds. Hook's crusade anticipated the brainwashing sensation, explored in the next chapter along with those ways that future readers of *Structure* could detect Hook's and others' claims about mind control within the logic of Kuhn's theories of normal science and conversion experiences. "The Necessary Dangers of Consensus and Unity" then argues that Conant's style of leadership, especially his respect for consensus and unity of purpose during times of war, made an impression that set the stage for Kuhn's enthusiasm—as well as Conant's initial dismay—for the theory of paradigms that Kuhn would later unveil.

Against the background of the struggle for men's minds and Kuhn's Aristotle experience, the chapters in Part III, "The Cold War Origins of *The Structure of Scientific Revolutions*," examine how Kuhn thought about *Structure* and how he wrote it. They begin with his first theoretical explorations into the nature of science in the late 1940s and conclude with *Structure*'s publication in 1962. They show how Kuhn's developing theory of science drew on concepts of mind and language, as well as formative but lesser-known influences, such as developmental psychology and psychoanalysis, that linked in various ways to the ongoing ideological struggle. Chapter 9 examines Kuhn's nervous debut as a public intellectual onstage at Boston's Lowell Institute, where he lectured about the nature of science against the backgrounds of McCarthy's ongoing crusade to ferret out subversive intellectuals as well as contentious public debate over the nature of scientific research and the creation of the National Science Foundation. The next chapter focuses on the academic study of ideology and its place in the *International Encyclopedia of Unified Science*, the forum for which *Structure* was first commissioned. It would not finally be published, however, before Kuhn finally reconciled his commitment to revolutionary breaks in scientific thought with some credible notion of scientific progress. This struggle is examined through Kuhn's encounters with the historian Eugen Rosenstock-Huessy as well as George Orwell, who evidently helped Kuhn grasp the immense cognitive powers of historical understanding and the roles they could play in his theory of science. As the last chapter in this section shows, Kuhn did not solve all of the problems he saw in his monograph until he incorporated his new theory of scientific paradigms which compounded the array of philosophical, psychological, sociological, and political themes that had so fascinated him up to that point.

The last part, "The New World of Paradigms," explores how *Structure* was received by various audiences in the 1960s. Some early readers, such as the philosopher Paul Feyerabend and the geneticist H. Bentley Glass, condemned Kuhn's ideas explicitly for the unorthodox politics they recognized within them. "The Magic of Paradigms" examines Conant's almost furious initial reaction to Kuhn's paradigms, his gradual acceptance of them, and—the chapter argues—the ways he strategically cloaked his misgivings to effectively oppose Kuhn's new image of science in his late writings. "Spies, Prisons, Mobs, Bandwagons, and Beasts" examines the politics of paradigms from an international vantage by revisiting well-known critiques of *Structure* by British philosophers in the 1960s. Against the backdrop of the ongoing cold war, it suggests that *Structure* can be seen through their eyes as a distinctly American theory of science whose influence was not fully independent from America's cultural and economic power. In an age that embraced the feminist slogan "the personal is political," however, some of the most remarkable readings of *Structure* belonged to individuals who sought a personal connection with Kuhn and confessed that *Structure* had changed them, sometimes profoundly, for the better. These audiences as well as the student movement of the 1960s and its shared enthusiasm for revolutions are sampled in "The Thomas Kuhn Experience."

The last chapter, "A Revolution and a New Ideology," examines Kuhn's return in 1970 to the antiwar politics of his youth. At Princeton University, he joined with students and faculty to investigate faculty research and its possible relationship to the Vietnam War. The tumultuous events of 1970 also marked a new relationship between Kuhn and his now-famous book as he renounced paradigms and reconfigured his working theory of science. If Conant had belatedly won their debate over the wisdom of introducing paradigms, as the chapter suggests, Conant's evolving view of the cold war struggle and how it was being mismanaged in Vietnam and Cambodia suggests that Kuhn won their debate in this different context: as Conant saw it, Nixon's generals seemed almost brainwashed, paradigm-bound, and unable to perceive the facts on the ground that contradicted their rosy predictions about the war's progress.

These four sets of chapters move chronologically from Kuhn's childhood and education through his collaboration with Conant and his composition of *Structure*, and conclude with his break with his theory of paradigms in 1970. However, this book's dual focus on the origins of *Structure* as well as its later reception among selected audiences requires each chapter to occasionally move forward or backward in time—forward, for example, to the forms that Kuhn's ideas would take later in *Structure*'s pages, or backward to retrace newly introduced themes, such as conversion, indoctrination, ideology, propaganda, or social and political consensus, that situate Kuhn's theorizing within the struggle for men's minds. Additionally, some chapters pause to examine remarkable individuals such as Elizabeth Moos, George S. Counts, Sidney Hook, Edward S. Greenbaum, Roy Cohn, and others introduced in the Cast of Additional Characters. Most of these figures make no appearance in *Structure* or Kuhn's other writings. But they played roles in Kuhn's life, Conant's life, and in some cases both of their lives. They are strands woven into the politicized culture in which Kuhn and Conant lived and worked, and in which Kuhn's purple prose seemed a credible, natural way to describe how science works. If fears of brainwashing, of communist subversion through libraries and classrooms, and of the power of ideology to control the human mind seem exaggerated or irrelevant to us today, this interconnected world of scholars, educators, and public figures reminds us that several decades ago there was no shortage of intelligent, capable people—including Thomas Kuhn and James Conant—who took these concerns for granted.

Timeline of Events and Documents

World War II and Cold War		Thomas Kuhn	James Bryant Conant
Hitler named Chancellor of Germany	1933	Hessian Hills School	Elected President of Harvard
	1935	lectures on peace	
Moscow show trials			
Nazi-Soviet Pact announced	1939	"The Crisis in Democracy"	
Trotsky assassinated	1940	Enters Harvard University	Rallies nation for interventionism
Pearl Harbor		"The War and My Crisis"	Militarizes Harvard campus
	1943	Military radar specialist	Forms General Education Committee
U.S. at War			
	1944	Witnesses DeGaulle in Paris	
Hiroshima and Nagasaki bombed	1945	Reviews *General Education in a Free Society*	
Churchill's "Iron Curtain" speech	1946	Psychoanalysis	
Kennan's Long Telegram	1947	Joins General Education program	Russian Research Center established
		Aristotle experience	
William Remington indicted for espionage	1948		Education in a Divided World
Waldorf peace conference	1949	Reads Ludwig Fleck	Conant/NEA prohibit Communist faculty
University of Washington faculty fired			
Korean conflict; "brainwashing";	1950		Alger Hiss case
Klaus Fuchs case			Committee on the Present Danger
McCarthy's speech at Wheeling, W. Va.			
	1951	Hessian Hills School closes	
		Lowell Lectures	Writes to future president of Harvard
Heresy Yes—Conspiracy, No! by Sidney Hook	1952	Invitation to write for *International Encyclopedia of Unified Science*	
Ethel and Julius Rosenberg executed	1953	Unveils "theory as ideology"	Resigns presidency
McCarthy: Harvard "dirty, smelly mess"		Furry, Gorham Davis testify	Debates McCarthy in Washington
William Remington murdered	1954		
	1955		Ambassador to West Germany

continued on next page

World War II and Cold War		Thomas Kuhn	James Bryant Conant
	1956	Arrives U.C. Berkeley	*The Citadel of Learning*
Sputnik satellites	1957	"Sputnik & American Public Mind"; *The Copernican Revolution*	Arrives New York City; Begins studies of American education
	1958	Stanford Center	
The Manchurian Candidate (novel)	1960	Theory of paradigms; draft of *Structure*	
SDS Port Huron Statement	1961	"The Function of Dogma in Scientific Research"	Criticizes *Structure of Scientific Revolutions*; returns to Germany
The Manchurian Candidate (movie)	1962	*The Structure of Scientific Revolutions*	
	1963	Arrives Princeton University	
			Two Modes of Thought
	1965	Popper, Lakatos conference	
Tet offensive in Vietnam	1968		Questions military leadership and decision making
	1969	"Disciplinary Matrices"	
U.S. bombing of Cambodia announced	1970	Princeton committee on military research	

Cast of Additional Characters

James Burnham (1905–1987). Philosophy professor at NYU alongside Sidney Hook in the 1930s. Later worked for the CIA and William F. Buckley's *National Review*. (Image courtesy of New York University)

George S. Counts (1899–1974). Progressive education theorist at Teacher's College, Columbia University who inspired Elizabeth Moos, the head of Hessian Hills School, where Thomas Kuhn studied. Later converted to anticommunism. (Image courtesy of Special Collections Research Center, Morris Library, Southern Illinois University Carbondale)

Robert Gorham Davis (1908–1998). Thomas Kuhn's composition instructor at Harvard. Abandoned the Communist Party in 1939 and later testified cooperatively to HUAC. (Image courtesy of Smith College Archives)

Elizabeth Dilling (1894–1996). Author of *The Red Network* (1935), in which progressive educators, including George S. Counts and Sidney Hook, are vilified as communist and unpatriotic. Shown testifying before the Senate on Jan. 11, 1939 to oppose the nomination of Felix Frankfurter to the Supreme Court. (Image courtesy of Library of Congress)

Paul Feyerabend (1924–1994). Austrian philosopher of physics, student of Karl Popper, and colleague of Thomas Kuhn's at the University of California, Berkeley. He objected to the political implications he found upon reading an early draft of *Structure*. (Image courtesy of Grazia Borrini-Feyerabend)

xliv | Cast of Additional Characters

Philipp Frank (1884–1966). Original member of the Vienna Circle of philosophers. After 1938, he lectured on physics and philosophy at Harvard and headed the Institute for the Unity of Science that sponsored *The International Encyclopedia of Unified Science* in which *The Structure of Scientific Revolutions* was published. Investigated by the FBI in the early 1950s. (Image courtesy of Harvard University Archives)

Wendell Furry (1907–1984) Harvard professor of physics who taught Kuhn undergraduate physics and was later investigated by the FBI and congressional committees for his political affiliations. (Image courtesy of Harvard University Archives)

Hiram Bentley Glass (1906–2005). Geneticist, educator, columnist, and biology education reformer in the wake of the Sputnik crisis. Criticized the presentation of paradigms in Kuhn's essay "The Function of Dogma in Scientific Research" in 1960. (Image courtesy of Sheridan Libraries, Johns Hopkins University)

Edward S. Greenbaum (1890–1970). Prominent New York City attorney, author, court reformer and Brigadier General. Friend of the Kuhn family and member with James Conant of The Committee on the Present Danger. (Photo in public domain)

Tom Hayden (1939–2016). Social and political activist who helped lead Students for Democratic Society and drafted the Port Huron Statement outlining "participatory democracy." (Image courtesy of Alamy)

Edward Hunter (1902–1978). Journalist whose writings and congressional testimony helped create and sustain public and academic interest brainwashing in the 1950s. (Image from the book jacket of his *Brain-Washing in Red China: The Calculated Destruction of Men's Minds* [New York: Vanguard Press, 1953])

George F. Kennan (1904–2005). Diplomat, historian, and author of the "long-telegram" from Moscow to the State Department that outlined the cold war policy of "containment" of Soviet Communism. (Photo courtesy of Library of Congress)

Cast of Additional Characters | xlv

Lawrence Kubie (1896–1973). Psychoanalyst, professor, author. Family friend of the Kuhn family and the Conant family. (Photo courtesy of Smith College Special Collections)

Elizabeth Moos (d.1985). Cofounder of the progressive Hessian Hills School, in Croton-on-Hudson, New York which Kuhn attended from 1933 to 1937. (Image courtesy of The Croton Historical Society)

Charles Morris (1901–1979). Philosopher and editor of *The International Encyclopedia of Unified Science* which commissioned Kuhn to write *The Structure of Scientific Revolutions*. (Image courtesy of University of Florida Archives, George A. Smathers Libraries)

Karl Popper (1902–1994). Austrian philosopher of science, famous for his theory of "falsifiability" as a criterion for science, who later taught at the London School of Economics and debated Kuhn in 1965 over his theory of "normal science." (Image courtesy of University of Klagenfurt/Karl Popper Library)

William Remington (1917–1954). Economist and son-in-law of Elizabeth Moos. Convicted of perjury after being investigated as a Soviet spy. (Federal Bureau of Prisons photo)

Julius and Ethel Rosenberg (1918, 1915–1953). American couple convicted of passing documents and information to the Soviet Union and electrocuted at Sing Sing Prison, June 19, 1953. (Image courtesy of the Library of Congress)

Eugen Rosenstock-Huessy (1888–1973) German historian and social theorist who fled Nazi Germany to teach at Harvard and then Dartmouth. His theory of national political revolutions influenced Kuhn's theory of scientific revolutions. (Image courtesy of Mariot Huessy, Eugen Rosenstock-Huessy Fund)

Part I

War and Crisis

1

Progress and Revolution in the Suburbs of New York

On Saturday November 16, 1935, in the small town of Croton-on-Hudson, New York, students at the Hessian Hills School performed a day-long assembly for their parents and their community. The theme was peace, and it was topical. Only months before, Congress had passed the first Neutrality Act to help Americans avoid future wars. Hessian Hills had its finger on the pulse of the nation, and other schools and organizations joined in. Students from area schools attended, as did representatives of the Women's International League for Peace and Freedom and the Federation of Children's Organizations—known previously as the Communist Party–sponsored Young Pioneers of America.

The most exciting development of the day concerned a new organization called the American Student Union. Inspired by students at Oxford University who vowed earlier that spring that "this House will in no circumstances fight for its King and Country," the students at Hessian Hills, the oldest roughly equivalent to the eight grade, decided that they too would stand against war and take the Oxford Pledge.

Leading off the festivities was thirteen-year-old Thomas Samuel Kuhn. He gave a remarkable speech that local papers quoted in their editions the following Monday.

> CROTON, Nov. 18—A boy about fourteen years old stood confidently on the platform: "Who profits by our national possessions?" he asked. "Not you, Not I, nobody but the capitalists. If, to protect the interest of those people, we must spend the bulk of our national income on armaments, then I say, 'damn our possessions!'"

During afternoon debates—in which adults were not allowed to participate—the reporter was struck by the intelligence and maturity these children. They

> talked of jingoism, propaganda and trade unionism with a familiarity few adult audiences could surpass. One girl who didn't seem to be over 10 asked to have explained how the German Socialists justified their participation in the World War—she said it has never been completely clear to her.

A student from New York's public Patrick Henry School was impressed by this private, progressive school and quickly sensed that his education fell short. He scolded his teachers for failing to discuss matters of war and peace and warned that unless they stepped up, he and other public school students would be educated by "soap-box orators on corners" spouting propaganda. At this point, another boy responded to remind this young man that "the word 'propaganda' must be used carefully since propaganda is used for constructive purposes as well as destructive."[1]

Six months later Kuhn was on stage again. Hessian Hills now had its own chapter of the American Student Union and it assembled to celebrate the union's international student strike. The festivities included reminiscences from a Hessian Hills teacher who had lived in Belgium in the aftermath of the Great War. Students sang the gospel standard "Down by the Riverside" (with its decisive refrain, "Ain't Gonna Study War No More") and they presented a play they had written. "Let Dread War Cease" was about a young girl struggling to persuade her older brother to refuse the call-up. Despite her arguments, he went to war and was killed in combat. In the final scene, his ghost tells the audience that his sister was right after all: "War is hell." At the end of the assembly, those students who had just formed the new chapter took the stage holding aloft a banner. It read, "Unite! Fight Against War. Organization Will Lead to Peace."[2]

Many years later, Kuhn recalled that, despite his ardent pacifism, he did not join his fellow students in taking the Oxford Pledge. Were the nation invaded, he reasoned, then surely he would fight to defend it.[3] But his handwringing about the pledge did not mean that he bowed out of this day's events. At eleven o'clock, when students around the world left their classrooms, Kuhn read his report to the audience:

> Today is the day elected by the American Student Union for their strike against war. All over the country between 11 and 1 o'clock, High school and college students are striking against

war. The reasons why [include that] these students are beginning to realize that they are the ones who will have to fight in a future war.

They know that "war clouds are gathering in Europe" and "naval and army appropriations are growing larger."

War was horrible and futile, Kuhn argued, and it was extremely expensive. For the cost of one battleship, he had calculated, "you would be able to get thirteen million books" for needy school children, or you could "feed a needy student three meals a day . . . for fifty thousand years" or, more practically, "fifty thousand men for one year."

At thirteen, Kuhn seemed to know that pacifism was not only a thesis to be defended with economic arguments, but also something like the *constructive* propaganda that came up at the assembly months before. The strike showed not only that these students objected to war, but that their opposition was rooted in a larger set of cultural values and expressions. Yes, he admitted, there were probably some who just carried signs without knowing why. But most, he pointed out, "really are in ernest [*sic*]." Most

> are assembling either in their school or some other place and having speakers just as we are having speakers. Songs, just as we are going to have songs. Skits, just as we are going to have a skit. These things are to build up a feeling, and a unity against war.

He also knew that this growing sense of unity was not popular everywhere: "But let me tell you that in some public high schools the principal had the school surrounded by police, and had the teachers hold the doors, and that if a student still got out he was expelled from the school!"

Kuhn agreed with the local reporter. These were important adult matters that only the older children at the school could begin to engage with. But soon enough, he pointed out in closing, the younger students would come to understand what was at stake—nothing less than "the future of the world and of the Human race."

Thomas Kuhn never shied away from bold, sweeping claims. His youthful confidence that the world could be made better and more peaceful had at least two sources: his parents and his school. His mother, Minette Stroock, was a New Yorker of Dutch descent. Her husband, Samuel L. Kuhn, belonged to a family whose roots in merchandising and banking stretched from Germany to Cincinnati to New York City. The Manhattan banking firm Kuhn & Loeb (later, Lehman Brothers Kuhn & Loeb) established by Abraham Kuhn and Solomon Loeb became one of the nation's prominent investment banks in part by helping to finance the expansion of the nation's railroads.[4] While Abraham retired to Germany after founding the firm in New York, his brother Samuel remained in Cincinnati to found his own banking house, S. Kuhn & Sons. His son Simon remained in banking as he and his wife, Setty Swartz Kuhn, became prominent figures in Cincinnati business and civic affairs. Setty was once voted "outstanding woman of the half century" by the Better Housing League in Cincinnati, one of the civic and philanthropic causes she founded and supported before her death in 1920.[5]

Their son Samuel served in World War I and studied at Harvard and MIT to become a hydraulic engineer. After Thomas was born in 1922, Samuel and Minette moved to Manhattan, where they lived as prominent, liberal Jews who supported an array of educational, cultural, and economic causes. These ranged from Junior Achievement and the Jewish Board of Guardians, to the public woodworking shop that Samuel created decades later on East Seventy-Fifth Street. Samuel was so devoted to woodworking, the *New York Times* reported at the time, that he opened a storefront business called Your Workshop where amateurs could use his shop's tools and machinery. The business was essentially philanthropic, the reporter suggested. The hourly rates one would pay to build a chair, repair a picture frame, or get expert advice from Samuel covered only some of his expenses.[6]

Kuhn's parents were generous and progressive. For Thomas and his younger brother Roger, they believed in the new, American style of "progressive education" taking root in the nation's larger cities. Before Hessian Hills, Kuhn attended the experimental Lincoln School in Manhattan then recently founded by the Rockefeller-funded educational philanthropy, the General Education Board. The board supported the reform and establishment of high schools throughout the nation, often enlisting local universities to run the schools along progressivist lines. The Lincoln School was run by the Teachers College at Columbia University, home of the philosopher John Dewey, the nation's foremost authority on and supporter of progressive education. Dewey, his star student Sidney Hook, and his colleague George S. Counts helped make New York the center for the theory and practice of progressive education.

Chicago, however, was a close rival. Colonel Francis W. Parker had founded a progressive school there and Dewey had earlier created the Laboratory School next to the University of Chicago campus. Parker and Dewey insisted that classrooms must not be sites of indoctrination and memorization, where rows of atomized student-soldiers await educational orders in reading, writing, and arithmetic from their superiors. As Dewey argued in his books *Education and Democracy* and *School and Society*, schools should be microcosms of egalitarian and democratic society. Students learn and grow best, Dewey wrote, when school offers a kind of "embryonic community life, active with types of occupations which reflect the life of the larger society, and permeated throughout with the spirit of art, history, and science."[7] Human learning is not only a cognitive or intellectual process, Dewey insisted. It is social and proactive—it comes when curiosity and intelligence have room to discover the world, the reality of other points of view, and the dynamics of debate and criticism. Progressive educators therefore organized classrooms for active collaborations. Where the youngest learned to stack blocks, the oldest could build musical instruments or plan an imaginary city.

Dewey himself never shied away from bold, ambitious claims on behalf of his innovations. Progressive education reflected the epic story of human learning, through which pioneers in science or literature drew lessons from experience without the benefit of textbooks or teachers. Progressive educators did not pretend to know exactly how these processes worked—Parker was known to say that in education "the road to success is through constant blundering." At the Lincoln School, its director Otis Caldwell once remarked that there is "a kind of frankness about what we don't know about education."[8] None believed that each and every child could be shaped to become an Isaac Newton or a William Shakespeare; but they knew what the startled reporter observed that November day in 1935 at Thomas Kuhn's school: most children have a potential they are unlikely to realize in schools that control behavior tightly and train the mind to memorize poems, names, dates, and multiplication tables.

George S. Counts and the Revolutionary Classroom

With the stock market crash of 1929, progressive education and its ideals became more popular. Articles in the *New York Times* enthused over the reforms at Lincoln and other schools, the virtues of learning-by-doing, and the replacement of the Three R's by new, progressive methods. One article described how students founded their own, imaginary bank.[9] But progressive educators were never themselves unified by shared goals, methods, and values. The more radical New Yorkers insisted that the progressive, child-centered school of the 1930s was not adequate to the immense social and political challenges created by the Depression. Columbia educator George S. Counts, speaking at educational meetings in 1932, provoked and challenged his audiences with lectures titled "Dare Progressive Education Be Progressive?," "Education Through Indoctrination," and "Freedom, Culture, Social Planning and Leadership." He chided his fellow progressives for not being progressive enough and failing to understand and grasp the revolutionary opportunities at hand. He called on them to reject

> the viewpoint of the members of the liberal-minded upper middle class who send their children to the Progressive Schools—persons who are fairly well-off, who have abandoned the faiths of their fathers, who assume an agnostic attitude towards all important questions, who pride themselves on their open-mindedness and tolerance, who favor in a mild sort of way fairly liberal programs of social reconstruction, who are full of good will and humane sentiment, who have vague aspirations for world peace and human brotherhood . . .

His sermons pointed at families like the Kuhns who believed in progressive education, social and economic reform—but not in revolutionary change. Yet that was no longer good enough, Counts explained. For "in spite of all their good qualities," progressives such as these could not be counted on given what lay ahead: "We live in troublous times; we live in an age of profound change; we live in an age of revolution."[10]

Like Dewey, Hook, and other intellectuals of his generation, Counts had visited Russia and found compelling potential solutions to American problems. The year before he delivered these challenging lectures, he translated a Russian schoolbook about Stalin's first five-year plan for accelerated social and cultural reform. He also wrote the book *The Soviet Challenge to America* to show Americans that education was a powerful force for social and political change.[11] There was nothing wrong with "indoctrination" in Counts's eyes. As he told an audience at the National Council of Education meeting in Washington, children are inevitably indoctrinated and shaped—the question was whether this would be done by the interests and institutions that "actually rule society" and create and sustain inequality and injustice, or by progressive educators who, if they seized the opportunity, "might become a social force of some magnitude."[12]

In Baltimore, Counts read "Dare Education Be Progressive?" at a meeting of the Progressive Education Association and left his audience stunned. "There was a silence when he finished. A silence that speaks far more eloquently than applause," one witness recalled. Counts won over his more cautious peers and colleagues, the itinerary for the rest of the conference was scrubbed, and the association's board of directors called special meetings. They formed a new committee on economics and sociology to take up Counts's challenge and formulate a revolutionary plan of action for the association.[13]

The Hessian Hills School

Elizabeth Moos, the forty-year-old Director of the Hessian Hills School, was among those in Baltimore electrified by Counts's speech. She had grown up in Chicago, attended the Francis Parker School, and later came to New York to intern at progressive schools and summer camps. After settling in Croton, about forty miles north of Manhattan near the banks of the Hudson, she and her colleague Margaret Hatfield taught their own children in the garage of Moos's home. As word spread and they accepted other children, Moos's garage became a school.

Moos and her husband, the violinist Robert Imandt, built a schoolhouse to replace the overflowing garage. But in January 1931 it burned to the ground. Amid the calamity, however, the school's profile and reputation rose higher. Progressive learning continued as teachers and students improvised for classroom space and supplies, while local parents raised funds to buy land for a bigger and better school. Among the local intellectuals who stepped up to support the rebuilding were influential New Yorkers and Greenwich Village bohemians who either lived or summered in Croton. These included Floyd Dell and Max Eastman, editors of the socialist magazine *The Masses*, the economist Stuart Chase (future author of *The Tyranny of Words*), the pro-Soviet journalist Lincoln Steffens, the pro-Soviet chemist and philosopher William Malisoff, who founded the Philosophy of Science Association in 1934, and the lawyer Jerome Frank. Besides these intellectuals, Croton was also home to many artists, such as the muralist George Biddle and Moos herself, who specialized in teaching dance, and who upheld the arts as antidotes to the alienation of modern life and tools to explore the natural (but ordinarily suppressed) creativity of the Freudian, unconscious mind.[14]

Though they were experts in different areas, these supporters and benefactors came together around Hessian Hills in the early 1930s because of their shared hopes for radical political change. The year of the stock market crash, one teacher—who routinely wore a political button in support of Socialist Party presidential candidate Norman Thomas—staged a student debate over the pacifist proposition that "gold makes war." Another soon installed a bulletin board on which students were encouraged to post clippings about social issues. Many students welcomed social engagement and the local notoriety it sometimes brought. The editor of the student newspaper, *The Crier*, made news in early 1932 by single-handedly recruiting Norman Thomas and the former governor of New York, Alfred Smith, to contribute essays about international disarmament to the paper.[15]

Moos herself joined the ranks of Counts's revolutionary educators when she was invited to reply to Counts's provocative speech. Printed along with "Dare Progressive Education Be Progressive?" in the journal *Progressive Education*, her essay, "Steps Toward the American Dream," candidly endorsed Counts's radicalism. The speech captured what she, her faculty, and her students had been thinking in the past year, she said. Her only reservation was Counts's remark that educators must never—not "for a single instant"—ignore pressing social realities in the world. Moos reassured her readers that, at least at Hessian Hills, it was possible to attend both to the child's individual educational needs and at the same time "help the child function as an integrated member of society."[16] At Smith College, her alma mater, she argued in June 1932 that "[g]overnment, capitalism, family, religion are either in flux or already in collapse, depending on our point of view." The task of the modern educator, she explained, was to downplay the social and economic individualism of the Hoover years that were widely believed to have caused this collapse, so that "My country right or wrong" becomes "My country is myself—I am responsible for her when she goes wrong." Above all, she added, future citizens must be taught to recognize bias and propaganda in their textbooks, to read and reason critically, and never to be "stampeded into mass murder" as they had been by participating in World War I.[17]

Moos's enthusiasm and growing notoriety probably helped the effort to rebuild her school. John Dewey himself became honorary chairman of the building fund committee, while *The New York Times* admiringly reported on fundraising dinners, exhibits of student artwork (one held at Edward Steichen's gallery, An American Place), and the exciting, modernist design commissioned for the new building. After it was built in 1932, Hessian Hills joined the ranks of the nation's most prestigious and well-supported schools. Moos and her board of influential parents and professors could attract figures such as the acclaimed pilot Amelia Earhart, champion boxer Gene Tunney, and the Marxist muralist Diego Rivera to speak at fundraisers in Manhattan.[18]

With its new building, Hessian Hills grew and prospered. But not all observers supported the school's political radicalism. The conservative *Croton-on-Hudson News*, reflecting the often reactionary sensibilities that reigned outside New York City, was sometimes skeptical and suspicious of the unorthodox school. To escape the professional whirlwind of her school and her notoriety, Moos and her family lived for the school year 1932–33 in Bali, where she studied indigenous dance. But her sabbatical year was no vacation from American politics. Life in Indonesia made her increasingly aware of racism and critical of American foreign policy for its control of the Philippines. Moos's convictions were aligning ever more with Counts's. Were it not for her distance, she might have compared notes directly with him when he arrived in Croton that year to speak to local parents.[19]

Progressive Education or Communist Subversion?

The Kuhns moved from Manhattan to Croton-on-Hudson at the peak of the school's growth and prestige. Thomas attended Hessian Hills from 1933 until 1937, when he was graduated from the highest "group" of students (roughly the eighth or ninth grade). He then finished his secondary education by attending the progressive Solebury School in New Hope, Pennsylvania, for one year, and then The Taft School in Watertown, Connecticut. Yet Taft and Solebury paled in Kuhn's recollections next to his glowing memories of Hessian Hills. His brother Roger had no doubt many years later that the school was also the crucible of his own life-long progressive and pacifist values.[20]

By all accounts, the Kuhn family engaged with the school and Croton's progressive politics. Minette wrote for the Hessian Hills School Association Bulletin, for example, and Roger remembers a pivotal event in the history of the American Left refracted through his father's friendship with Max Eastman. In his growing disgust with Stalin, Eastman predicted that the Soviet dictator would eventually make common cause with Hitler. Samuel found this simply unthinkable and made a friendly bet with Eastman (the loser would hold a dinner party for the winner). After the Nazi-Soviet non-aggression pact became known in 1939, Samuel's invitations to the party featured a sketch of himself eating humble pie.[21]

This pact was the beginning of the end for the era in which progressive education and politics thrived in communities such as Croton. Even though it proved temporary, the alliance between Hitler and Stalin fed suspicions that progressivism vitally threated American, democratic values. One example of this suspicion was Elizabeth Dilling's *The Red Network*, a best-seller of 1934 that helped to strengthen the isolationist (and, with Dilling's support, anti-Semitic) political Right in the years before Pearl Harbor (see Figure 1.1). Shortly after the PEA's new committee on economics and sociology issued its report in 1933, "A Call to the Teachers of the Nation," Dilling denounced the committee that produced it in the pages of her book.[22]

Dilling was a one-time University of Chicago student who lived in a spacious, five bedroom home in the leafy suburb of Kenilworth, north of Chicago.[23] She and her husband Albert, a sanitary engineer with the city of Chicago, traveled often and widely, and it was a trip to Russia that awakened Dilling to the Soviet menace. At Moscow's Museum of the Revolution, she recounted, she was shocked to see photographs of bloody, beaten revolutionaries and maps of a future world order encompassing the (former) United States. All this "strikes terror to the heart," Dilling exclaimed.

The Red Network spoke directly to Dilling's fellow "super-patriots" who were eager to protect Americanism and Christianity from the hairy, Russian hands depicted on the book's dust jacket. Years before McCarthy and other anticommunist politicians sounded the alarm, she denounced the conspiratorial goals of international communism and insisted that radicalism in its myriad forms—including atheism and free love—had already broken through the gates via the Communist Party newspaper, *Daily Worker*, and even *The New York Times*.[24]

Besides the radical educators in the PEA, Dilling singled out Albert Einstein as one of the masterminds within the red network. Dilling obsessed over his influence and prominence and returned to Einstein throughout the book (his name appears on about 10 percent of *The Red Network*'s pages). Behind the mind-bending theories of space and time for which Einstein became famous, she knew there lurked a godless, dangerous social agenda. He cleverly promoted it from his "self-erected scientific throne" and easily raised funds from wealthy donors sympathetic to his crusade and his corrupt, "relativistic" values.[25]

In the coming years, one of those donors would be Thomas Kuhn's grandmother Setty. As founder and head of the Cincinnati branch of the Foreign Policy Association, she was among the prominent citizens whom

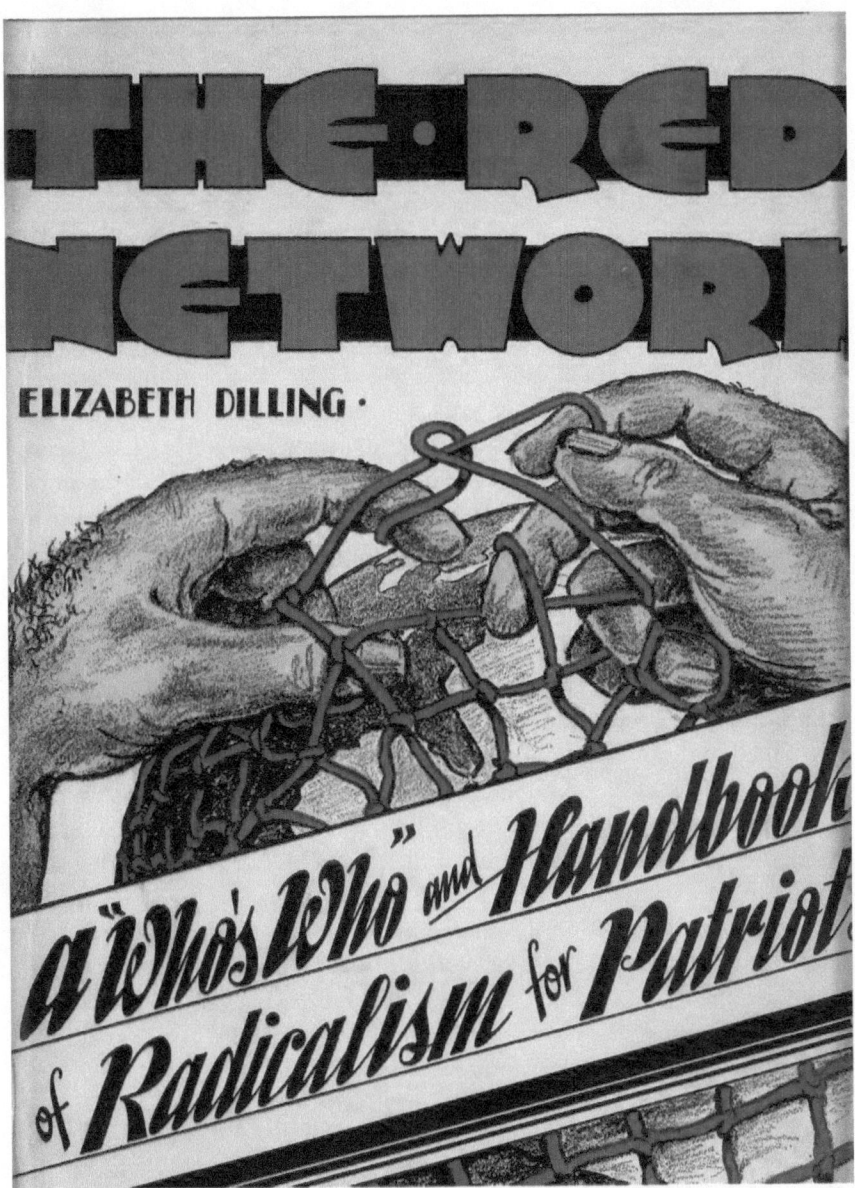

Figure 1.1. Elizabeth Dilling's *The Red Network* dust jacket cover image.

Einstein called on in 1946 to help educate the nation about the perils of nuclear weapons. Einstein and the Federation of American Scientists needed two hundred thousand dollars to "let the people know that a new type of thinking is essential if mankind is to survive and move toward higher levels."[26] Setty immediately telegraphed her reply and her pledge to help. Five days later she wrote again to tell Einstein that she was "having one thousand copies of your telegram mimeographed and mailed as soon as possible to citizens in responsible positions." She enclosed a check of her own for one hundred dollars and promised more support in the future. Later that summer Einstein sent her a copy of John Hersey's article "Hiroshima" (later expanded into Hersey's famous book) and invited her to attend a luncheon in Princeton.[27]

"This is how the red network operates," Dilling might have said. So-called luminaries such as Einstein and persuasive orators like George Counts seduce their wealthy followers only to prepare them and the rest of the nation for the Soviet takeover. The nation had to keep a close eye on these subversives, Dilling insisted, which is why the last two-thirds of *The Red Network* is an alphabetized directory of organizations and people—from the Abraham Lincoln Center settlement house in Chicago to one Max Zuckerman ("born Russia, 1868; came to Am. 1891; Workmen's Circle; org. Pioneer Youth Am; N.Y."). For each entry, Dilling listed its suspicious connections to pacifist, labor, or communist organizations.

The Progressive Education Association received a full page of copy. Dilling denounced it as "a radical left-wing teachers group" and quoted from the new report by Counts's committee.[28] The report was arguably more radical than Counts's speech in Baltimore. It called for teachers to "foster in boys and girls a profound devotion to the welfare of the masses," to support economic planning to redistribute wealth and achieve "material security for all." The PEA must also equip itself for revolutionary battle to meet these goals. That would require funding, lawyers, and strategic expertise "to wage successful warfare in the press, the courts, and the legislative chambers of the nation."[29] "Note the opposition to patriotic societies," Dilling interjected. And note who had dared to print this subversive treatise—the publisher John Day and Company, she noted, had also published Einstein's "The Fight Against War."

Counts, Hook, and other members of the PEA's new committee whose names Dilling listed were no strangers to this bare-knuckle political combat. Since the late 1920s, they had contended with the Daughters of the American Revolution, the American Legion, the Hearst press, and similar organizations that disliked and feared progressive educators. Many called for loyalty oaths for teachers as conditions of employment in the nation's school. Legislation promoted by the D.A.R in 1929, for example, intended to reassure parents that "instructors in your communities are of the right calibre and are teaching sound Americanism instead of instilling pernicious doctrines into the minds of their pupils."[30]

At the 1936 meeting of the National Education Association, Counts offered a muscular defense. To a cheering audience of one thousand, the *New York Times* reported, he denounced the D.A.R., William Randolph Hearst, The American Legion, the radio commentator Father Charles Coughlin, and others. These reactionaries offered nothing but "all pretend patriotism," Counts insisted, either because they did not understand the Constitution and Bill of Rights or because their strident protests about communism were merely pretexts to gain more social and economic control.[31]

The Defection from the Left

Despite Counts's vigorous defense, anyone who was a member of the PEA or a close observer of schools like Hessian Hills could sense that reactionaries such as Dilling were not only holding their own in these culture wars; they were gradually winning. Dilling's *Red Network*, one historian wrote, "branded the stigma of radicalism on the PEA . . . a stigma destined to exert growing influence as the decade progressed."[32] After the summer of 1939, when the Soviet-Nazi pact was announced, progressives such as Moos could not easily defend themselves and their work as standing firm against Nazism. Until Hitler turned on Stalin and invaded Russia in June 1941, leftists who promoted socialism were easily tagged as Nazi sympathizers, if not collaborationists.

Max Eastman was not the only one who saw this debacle coming. Counts, Hook, and other New York intellectuals watched their hopes for worldwide socialism fade behind the rise of Stalin, his dictatorial methods, and his relentless destruction of those he considered enemies. In the late 1930s, many cast their lot with the exiled Bolshevik Leon Trotsky, whom they hoped might yet lead the world toward a healthy democratic socialism. But after the pact and Trotsky's assassination in 1940, many leftists moved to the right and transformed their fight for education and social and economic justice into a fight against Stalin and his influences, wherever, whenever, and in whomever they seemed to detect them.

For Hook and Counts, the transformation was total. With his books *Towards the Understanding of Karl Marx* and *From Hegel to Marx*, Hook had once aspired to be the nation's leading authority and interpreter of Marxist theory—something like America's own Vladimir Lenin. In his book about

Stalin's five-year plan, Counts swooned over "the great social vision . . . of that strange new society which is rising on the ruins of imperial Russia" and urged American textbook writers to follow the Russian example.[33] As late as 1947, Counts translated another book for American educators titled *I Want to Be Like Stalin* in which he urged peace and international understanding.[34] But Counts soon joined Hook and the majority of once-radical intellectuals who now agreed with Elizabeth Dilling: Soviet communism posed an urgent, existential threat to the nation.

After the news of the Hitler-Stalin pact, and especially after Hitler's invasion of Poland in 1939, pacifists such as Elizabeth Moos and Thomas Kuhn found this climate difficult and distressing. As Kuhn explained to his sophomore composition professor at Harvard, it was extremely difficult for him to maintain his pacifist values and beliefs as the interventionist consensus grew and spread across the nation.[35] He passed through his crisis and eventually adopted an interventionist position, but Moos remained not only a pacifist but a political radical. And for this she paid a steep price.

Moos's school board and faculty had often argued over just how much political activism rightly belonged in Hessian Hills's classrooms. Speaking in Langhorne, Pennsylvania, in the spring of 1937, she insisted nonetheless that "building up the anti-war spirit is one of our jobs." Students must be made to understand the economic causes and interests that cause wars and to "build up resistance" to pro-war propaganda in the media. According to some observers, her ever-growing political confidence was in part a kind of refuge from personal turmoil as her second marriage fell apart.[36]

For a while, Moos's sensibilities reigned. For a school-wide pageant in early 1939, students created a Diego Rivera–style mural depicting American laborers next to a celebratory poem: "Workers first of all—all kinds/Workers on the belt line—workers in the mine/ . . . /Don't Forget the Unemployed." A voiceover described the history of labor in America while additional scenes covered the legacy of slavery in the South, child labor in Northern factories, and the threat of domestic fascism in the United States.[37] Even *The Croton-on-Hudson News* was impressed. This was "patriotism in the truest and deepest sense," it reported. But after the Nazi-Soviet pact, the paper editorialized against Croton's "Reds and Parlor Pinks" and refused to publish announcements for events or organizations it deemed pro-communist, such as the local chapter of the Women's International League for Peace and Freedom, which Moos joined. Those who wrote to the editor to protest this blackout found only their names printed in its pages, not their letters or the arguments they presented. The *Croton-on-Hudson News* became more like *The Red Network*.[38]

Global and local politics conspired to end Moos's career. Some who were not bothered by her communist reputation and supposed eagerness to indoctrinate her students grew impatient with her domineering style. Though she had been its founder, many parents and faculty increasingly resented "her proprietary attitude" toward school affairs. After Moos took a group of older students to New York City for the May Day parade of 1940, however, her professional fate was sealed. By this time, the once-popular event was now widely believed to be controlled exclusively by communists whose leader in Moscow had made a friendly deal with Hitler. Believing that Moos had gone too far, some parents sent their children to other schools. By the end of 1941, she was asked to resign.³⁹

The Remingtons

Visitors from around the world could learn about the Hessian Hills School at the 1939 World's Fair, held that year at Flushing Meadows in New York City. Filmmaker Lee Dick debuted her film *School: A Film About Progressive Education* that was commissioned by the PEA (see Figure 1.2). "This

Figure 1.2. Images from *School: A Film About Progressive Education* (1939) featuring the Hessian Hills School about two years after Thomas Kuhn graduated. The math teacher Leon Scaiky whom Kuhn later described appears in the lower right, middle.

is a film about a school—the Hessian Hills School at Croton-on-Hudson, New York," the opening screen reads. "The principal actors are the children of the ten-year-old group, but they are not actors at all, for this is just one day in their everyday life which you will see." In narrationless, cinema vérité style, the twenty-three minute film shows a dozen students and occasionally some teachers (including the math teacher Leon Scaiky, whom Kuhn fondly remembered) going about the business of progressive education. Students operate their school store, make shelves for books in the woodworking shop, rehearse their lines for an upcoming pageant about democracy, and landscape the school grounds. "Out of real situations," the text reads, "learning develops naturally—school work becomes a part of everyday life" and helps "teach these boys and girls to become responsible members of a democratic community."[40]

Beginning in 1948, however, Americans would pay attention to Hessian Hills for a different reason as Moos's daughter Ann and her son-in-law William Remington became front-page news. Like her mother, Ann was a devout pacifist who had carried her convictions from Hessian Hills to Bennington College, where she joined the American Student Union and Bennington's own Peace Committee. Attending an intercollegiate peace conference at Dartmouth in 1938, she met Remington, a picture-perfect big man on campus widely known for his athletic and scholarly prowess. The couple eloped but kept their marriage a secret until they were officially wed at Moos's home in Croton the next summer.

"The guests were a strange mix," Remington's biographer wrote. There were "Croton communists, artists, philosophers; writers for *New Masses* and the *Daily Worker*; even a veteran of the Abraham Lincoln brigade." That veteran was Alvin Warren, who had become Elizabeth Moos's lover after the breakup of her second marriage and who, along with *New Masses* editor Joe North and Ann Remington herself, helped to persuade Moos to join the Communist Party officially in the late 1930s. On the groom's side were Remington's parents from the affluent, Protestant suburb of Ridgewood, New Jersey, as well as his classmates from Dartmouth who were puzzled by William's choice of Ann for a bride. Two of them deliberately made trouble by praising Herbert Hoover as they made small talk with other guests—until, that is, Elizabeth Moos asked them to leave.[41]

Remington joined the Communist Party before his marriage and graduation. But he delighted in concealing his politics, sometimes even fashioning himself as a conservative, depending on the moment and how it suited his academic and professional aspirations. With his degree and good looks, Remington was a man in control of his career. He won a scholarship to study economics at Columbia University in New York City, where Ann took secretarial courses and worked for the main office of the American Youth Congress, an umbrella organization for Left-leaning and communist youth organizations. Both were full-fledged Communists, but they considered themselves independent-minded. Ann kept current with Party literature, for example, but unlike the stereotype of the obedient Party member, the Remingtons did not fall in line behind every new pamphlet or brochure. They chose their beliefs and associations as they wished.[42]

In May 1940 they moved to Washington. William worked as an economist in the War Productions Board and Ann continued to work for peace organizations. In nearby Virginia they built a new home, but they missed the debates, discussions, and political engagement they enjoyed in New York and in Croton. Remington knew he had to be careful, for he was in the midst of obtaining government security clearances, but he and Ann desired a closer connection to the Party. Joe North put them in touch with a couple, "John" and "Helen," whom they first met in New York City. Helen later turned up in Washington to meet the Remingtons again and eventually to meet several times with William.

He would later testify that he knew her as a journalist asking about the government's perceptions of the ongoing war in Europe. But "Helen" was Elizabeth Bentley, the sensational "Red Spy Queen" who made headlines in 1948 by naming names of government officials she had known as comrades when she was a spy for the Soviet Union.[43] "John" was the Russian spymaster Jacob Golos. He and Bentley were lovers until Golos died of a heart attack in 1943. Bentley filled his shoes and took over his contacts but was soon asked by Party officers to turn them over to others in the network. When this happened, she later recounted in her autobiography, "a wave of revulsion and nausea swept over me." All this time, she realized, the Russians has just been using her. Communism was "just a dirty racket."[44]

With Bentley's change of heart over communism, she contacted the FBI and convinced them to hear her tale. Three of the Communists she had worked with, they learned, remained active in government—Remington, Treasury Department economist Harry Dexter White, and Roosevelt's one-time advisor Lauchlin Currie. For the next five years, Remington struggled to survive investigations, loyalty board hearings, the impending breakup of his marriage, and repeated attempts by the FBI to find a communist smoking gun in his background. His case helped set the stage for McCarthy's memorable debut in Wheeling, where he claimed to have a long list of such "card-carrying members of the Communist Party" in Washington; and it helped establish Bentley's future career as a well-paid celebrity lecturer who repeated her story to right-wing groups around the nation.

In the long run, though, Bentley may have regretted her celebrity. While the FBI assured her that her reputation would be protected, many hours of depositions, public interviews, and cross-examinations portrayed her as an unsteady alcoholic prone to automobile accidents and violent boyfriends. In his defense, Remington sued her for libel and won a settlement because her lawyers discouraged her from testifying.[45] The resulting layers of intrigue were irresistible for the national press: the spy queen and her lurid past, the handsome government intellectual who may have lied about his politics, Ann Remington's psychotherapy, her damning public testimony against her soon-to-be former husband, and their mother-in-law Elizabeth Moos—frequently described as the former director of the Hessian Hills School in Croton.

Moos was central to Remington's defense. Sure, he admitted, he had flirted with leftism as a student—who hadn't? But by his third year at Dartmouth, he said, he experienced "the dawn of reason" and never again believed in or promoted communism. It was his wife and her mother, he said—they were Communists and made him look bad. It was through "Mrs. Elizabeth Moos of Croton-on-Hudson, N.Y.," the Associated Press reported, that he first met Elizabeth Bentley.[46]

For a while, Remington's defense served him well. He publicly met his accuser (see Figure 1.3) and an initial review by a government loyalty review board cleared him of espionage charges. But Hoover's FBI agents

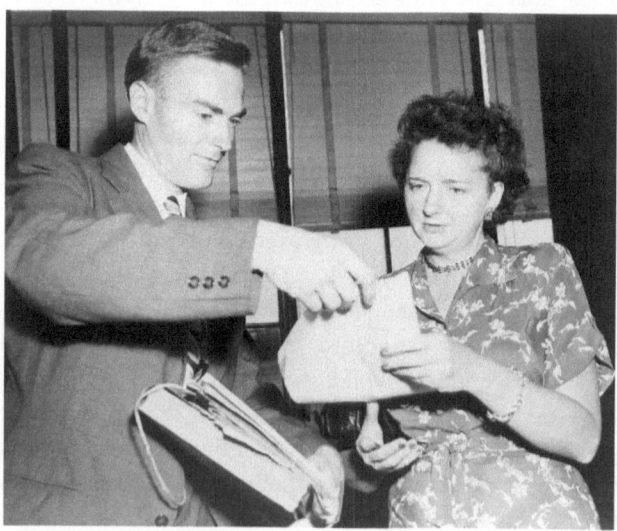

Figure 1.3. William Remington confers with his accuser, Elizabeth Bentley, July 1948. (Image courtesy of Alamy)

and federal prosecutors, one of whom was Roy Cohn, were sure there was more evidence to be found. If Remington had not actually passed classified information to Bentley, then they were sure that he must have lied under oath about something during his many interviews, depositions, and testimonies. Remington was eventually found guilty on two counts of perjury. In April 1953, he surrendered himself to serve a five-year term at the federal penitentiary in Lewisburg, Pennsylvania.[47]

Elizabeth Moos and the Perils of Pacifism

Later in her life, Elizabeth Moos continued to promote pacifism and other progressive causes. In the 1950s and '60s, she wrote books about Soviet education and published them through the National Council of American-Soviet Friendship. In the 1970s, she twice traveled to China on behalf of the U.S. China Peoples Friendship Association and returned to praise the progress Chinese women had made since the 1949 revolution. Though her writings and actions made it clear that Moos remained a defiant leftist and pacifist, the Remington case helped to cement her national reputation as a dangerous subversive.[48]

In early 1951 she was indicted by a Federal Grand Jury along with W. E. B. Du Bois and other officers of the Peace Information Center—an organization Moos helped create to support the World Peace Council's "Stockholm Appeal" for worldwide nuclear disarmament. By this time, McCarthyism and the Korean conflict were in full swing and support for peace was widely perceived as a public cover for aiding and abetting the spread of international communism. First among those who gave pacifism this bad name were Moos's former colleagues in progressive education, George Counts and Sidney Hook. Two years before in 1949, Counts helped Hook stage a successful and widely publicized campaign against the famous "Waldorf Conference" for international peace. Though Harvard's Harlow Shapley, the famous astronomer, had organized the conference, Hook and Counts persuaded many Americans that Shapley and others were either secret communists or "Communist dupes" who had been fooled into hosting Stalinism in the heart of New York City.

Even *The New York Times*, which fifteen years before had adored Moos and her progressive, experimental school, covered her arrest with disdain. The paper described her Peace Information Center as a suspicious redoubt of communist activity: based in two rooms at the Chelsea Hotel, it worked with "Communist dominated unions" to circulate pacifist propaganda adapted from Moscow's party line for public consumption in the United States. The plea for world peace it circulated, the paper wrote, was widely "attacked as a Soviet Trick." The Center itself, it implied, was illegal because of the indictment filed against Moos, Du Bois, and other officers: they had not registered as agents of a foreign power under the Foreign Agents Registration Act.

Moos was in Europe when the indictment was filed. Upon returning to New York she was arrested at Idlewild Airport (now John F. Kennedy International Airport). "Ex Kin of Remington Seized on Foreign Agent Charge," the *Brooklyn Eagle* blared on its first page. Moos appeared in a photograph smiling and taking her arrest in stride. She held up her wrists and shouted to the photographers present, "Be sure to get the handcuffs in." Seven months later, a federal judge acquitted Moos and Du Bois because he saw no substantive link between the activities of her Peace Information Center and any foreign government or agent.[49]

A year later, Moos appeared in Herbert Philbrick's best-selling memoir *I Led 3 Lives: Citizen, "Communist," Counterspy*. Philbrick described meeting Moos in her "lavish apartment" in Boston when he was a rising organizer in communist politics working for Henry Wallace's Progressive Party and his presidential campaign in 1948.[50] Moos and her son-in-law Remington were mentioned only once, but Philbrick's book and its eye-catching cover were displayed in each episode of the eponymous television series that ran from 1953 to 1956. It reminded Americans that even an ordinary advertising executive such as Philbrick, or a schoolteacher like Elizabeth Moos, could be a stealthy conspirator working to spark a communist revolution—at least until Hoover's agents, taking crucial tips from the counterspying Philbrick, nabbed them at the end of each episode. During the television run of *I Led 3 Lives*, Moos's association with Remington was reinforced one last time: after serving a year and a half of his sentence as a model prisoner, Remington was beaten and killed by three inmates who may have shared McCarthy's animus against Ivy League communists.[51]

In the early 1960s, Moos was called to testify when the House Un-American Activities Committee investigated alleged communist infiltration of pacifist groups such as the Women's International League for Peace and Freedom and Women Strike for Peace. Before questioning her, HUAC Counsel Alfred M. Nittle had gone carefully over her record—including Remington's accusations and her publications on Soviet education—and took it for granted that Moos supported pacifist causes because Moscow directed her to do so. After a difficult series of questions, Moos managed to have the last word before she took refuge behind the fifth amendment:

> MR. NITTLE. Is your present participation in the peace movement a response to Communist directives to engage in such?
>
> MISS MOOS. Before answering that, I think I must make a statement that I think this committee is doing a terrible disservice to America and to everyone in the world—
>
> MR. NITTLE. Now, will you please answer that question?
>
> MISS MOOS. I will be heard—when they try to attribute every act, every conscious act that is done for peace to the Communists. Are they the only ones? Do you think they want peace more than we do? Having made this statement, I decline to answer your question on the grounds of the fifth amendment.
>
> MR. NITTLE. The staff has no further questions.[52]

While Moos remained an unreconstructed leftist, her former colleagues and her former school tried but failed to effectively reinvent themselves. As for the Progressive Education Association, "members deserted in droves" during World War II. After trying to rebrand itself "The American Education Fellowship," it finally dissolved in 1955.[53] The Hessian Hills School hired a succession of directors in Moos's absence, but none lasted for more than a year or two and none stemmed steadily decreasing enrollments.[54]

In 1948, hope for progressivism flickered briefly during Wallace's Progressive Party campaign, and Moos herself was invited back to the school to speak that October. But in politics and education, this was progressivism's last gasp. Wallace endured hostile, violent crowds on his tours through the South and won only a sliver of the popular vote as Harry Truman was reelected president. The next summer in nearby Cortland Manor, New York, an anticommunist mob attacked audiences who had come to hear singer and peace activist Paul Robeson. Moos was present at the now-infamous "Peekskill Riot" and became widely associated with Robeson and his defenders. The current director Winifred Dahlberg did her best to distance Hessian Hills from this unsavory reputation by inviting prominent anticommunists to speak. The first was writer, critic, and former Trotskyite Dwight MacDonald, who spoke in 1950 on "The Dream World of Soviet Totalitarianism." In early 1951, however, the school board decided to close the school and offer its once-celebrated modern building for sale to other schools. Finding no takers, it was eventually sold to a nearby synagogue.⁵⁵

∽

Decades later, Thomas Kuhn confessed readily that he and his classmates at Hessian Hills were "quite radical." "I was more radical than my parents," he recalled, "but they did not look down on it." He also remembered well his moment of celebrity in November 1935, noting that the national possessions he denounced from the school's stage were not material but imperialist. "Who profits by our national possessions? Not you, not I, nobody but the capitalists," he recalled, before adding a phrase that did not appear in the local published report: "Let the Philippines Go!" About the Remington affair, Kuhn mentioned only that Moos was William Remington's mother-in-law, and that "he was somebody who was ultimately put away for having been a Communist courier."⁵⁶

Did Hessian Hills have much to do with Kuhn's scholarship and his theory of science? It did, Kuhn would say, because the independent politics of progressive education and life in Croton contributed to his intellectual independence. Progressive beliefs and ideas, however, had no direct bearing. "It's a little absurd to say that if I hadn't had those years at Hessian Hills I wouldn't have done what I've done, and it's probably false," Kuhn said in 1982. But he completed his thought with an intriguing caveat: "But I wouldn't bet on its being just false." Whatever influences Hessian Hills had, Kuhn emphasized that neither he nor other students were themselves indoctrinated or brainwashed. "There were various radical left teachers all over," he said, "except that we were all encouraged to be pacifists. There was no Marxist training or anything of the sort."⁵⁷

Yet influence takes many forms besides indoctrination or political training. As Kuhn's theory of science developed and evolved in later years, it is not difficult to hear multiple echoes of his Hessian Hills experience, of the school's commitment to progressive, collective, and collaborative learning, and the reputation for indoctrination that ultimately destroyed the school. Kuhn would argue in *The Structure of Scientific Revolutions* that modern science is necessarily social and that all of the important and essential ingredients for scientific change and progress exist within scientific communities, just as Dewey had long insisted that human learning always takes place through socially shaped experiences. Kuhn's independent, unguided exploration of sociology and psychology in the late 1940s suggests another feature of progressive education, as did his eventual discovery of Jean Piaget, the Swiss child psychologist whose theories of learning beginning in the 1930s would update Dewey's.

For some early readers of *Structure*, Kuhn's claims would spark another round of the debate over education and "indoctrination" that consumed Elizabeth Moos decades before. In order to function properly, Kuhn would claim, scientific communities must educate their new members to be dogmatic and closed-minded. But it would be Kuhn's picture of scientific revolutions that most vividly recalls these highly charged disputes of the 1930s. For whether on the political Right, with characters such as Elizabeth Dilling and Joseph McCarthy, on the radical Left with George Counts and Elizabeth Moos, or in the Hessian Hills' auditorium where Kuhn and other students addressed the fates of democracy or the human race, it was universally agreed that one system of political ideas, one dominant ideology, would determine the very texture and character of the future. Either socialism, fascism, or laissez-faire capitalism would win the day. After Hitler's defeat, a similar assumption held firm during the cold war: *either* liberalism or Stalin's international communism would prevail in the end, but never parts or aspects of both.

This is the image of political revolution in *The Structure of Scientific Revolutions*. Each of the scientific communities and traditions Kuhn described there recapitulates—in its own, professional and scientific way—this grand and highly charged narrative. If humanity moves forward into its future through winner-take-all contests of political ideologies, so too, Kuhn would explain in his first drafts and outlines, the path of modern science is determined by winner-take-all contests between the "ideologies" that operate within scientific communities. Before he would refine and finally unveil his theory, however, Kuhn spent years refining and filling in this narrative by reading history, philosophy of science, sociology, psychology, and by collaborating with James Bryant Conant.

2

War and General Education at Harvard

When James Bryant Conant became assistant professor of chemistry at Harvard, he wanted to accomplish much, much more. Shortly after he became engaged to his wife, Grace ("Patty") Thayer Richards, whom he met at Harvard in 1916 while finishing his doctorate, Conant confided his ambitions: "The first was to become the leading organic chemist in the United States; after that I would like to be president of Harvard; and after that, a Cabinet member, perhaps Secretary of the Interior."[1]

Conant called his autobiography *My Several Lives* for good reason. Even before he embarked on this ambitious career plan, he had been a military scientist during World War I. He developed chemical weapons, specifically, poison gases at government laboratories in Washington state. Characteristically, Conant aimed to create a superior kind of mustard gas—lethal, inexpensive, and capable of being efficiently mass produced—and he succeeded. By the end of the war, Army plants were producing tons of gases that Conant had helped devise, including the highly toxic Lewisite. He was promoted to Major and, by war's end in 1918, officially commended by the Chemical Warfare Service.[2]

The end of the war promised lucrative opportunities for chemists like Conant who were sought by corporations eager to do business in Germany and Europe. But Conant, a mere twenty-five years old, stuck to his plans and returned to Cambridge. By the end of the 1920s he was a full professor and a national leader in chemistry with contracts to advise foundations and corporations. Four years later, in 1933, he was elected president by the Harvard Corporation. Many of Conant's initiatives and reforms raised eyebrows (perhaps none more than eliminating the requirement to study Latin). But Conant wowed most of his doubters and skeptics, some of whom wondered how a mere chemist could fill the learned shoes of his predecessors.[3] Conant was no "narrow gauge chemist," as his biographer James Hershberg put it. He was keenly interested in almost everything in science as well as the humanities. He could not resist stepping inside any nearby bookstore and made no secret that as president he aimed to elevate the university's intellectual status, to make its central mission the advancement of learning and the growth and spread of human knowledge.[4]

At the same time, Conant strengthened the university's connections to the nation. He instituted fellowships and scholarships for students who fell outside its traditional base of New England wealth and privilege. He supported standardized testing to measure the effectiveness of institutions and to help recognize young intellectual talent in whatever state or region it might be found.[5] These ambitions rested on an insight that Conant had gained in Germany, where he and Patty traveled for eight months shortly after they were married. Patty had been there before, but they both fell in love with Germany. Conant filled his notebooks with "names, addresses, sketches and diagrams of experimental setups, and formulas studded with exclamation marks." He adored German culture, its history, and especially its scientific achievements. Why was it, he wondered, that modern chemistry flourished and grew at such a pace in Germany and not in the United States or other nations? His trip and his growing knowledge of German culture offered an answer. "If you really want to know how organic chemistry is really discussed by those who do it and hot from the stove," he told a colleague, "attend a meeting of an organic section of the German Chemical Society." "You will blush for your country and its organic chemists from then on!" he exclaimed.[6]

On this and subsequent trips to Germany, Conant confirmed his devotion to Germany's cosmopolitan, pluralistic, and highly competitive intellectual culture. It inspired many of his presidential reforms (including his "up or out" policy by which faculty are either rapidly promoted or let go). When the time came, it would also inform his criticism of Kuhn's theory of "normal science" and its picture of scientists devoted to a single, shared paradigm. Not agreement but competition and debate moves both science and society forward, he believed. It made Germany a powerhouse in chemistry and it would make Harvard, free to draw intellectual talent from the vast and diverse population of the United States, a powerhouse in all kinds of knowledge and learning. For the Germanophile Conant, *Wissenschaft*, humanity's forward march to greater understanding, would be alive and well after World War I.

Planning Science and Building Bombs

Conant's liberalism did not impress everyone in the 1930s. To his right, traditionalists and conservatives resented his reformist zeal and confidence; for the socialists and New Dealers to his left, only government planning and organization, not the free and unpredictable interplay of competition and rivalry, offered a path out of the Depression. Even though Conant had voted for and would ultimately side with Roosevelt about the need to intervene in Europe, his liberal insights about the growth of knowledge put him at odds with a growing movement that sought to plan science. Britain's John Desmond Bernal, for example, rejected Conant's free-market approach to knowledge as bourgeois, naively capitalist, and just wrong. A scientist and crystallographer by training, Bernal and his followers believed that Marxism illuminated science's dependence on economics and social realities. Instead of leaving the progress of science up to scientists who might fail to understand, or simply ignore, these realities, modern science should be guided and planned to maximize its social functions and benefits.[7]

Bernal's book *The Social Function of Science* was published in 1939, an important year in Conant's career. On Friday, September 1, headlines announced "German Army Attacks Poland." Two days later Conant addressed an international congress of American and European philosophers who also cared deeply about the future of science and the nature of scientific research. Calling themselves "The Unity of Science Movement," they included Harvard faculty as well as Europeans who had fled Europe for new careers at American universities. Many of those speaking had enlisted to write for the new *International Encyclopedia of Unified Science*, unveiled two years before by the University of Chicago Press. The movement's leader, Otto Neurath, shared Bernal's interests in Marxism, science, and society; while others focused more narrowly on the logical structure of scientific theories, the characteristics of scientific language, or particular topics in psychology, physics, or biology.[8] Others, such as the University of Chicago's Charles Morris and the Viennese philosopher and physicist Philipp Frank dedicated themselves to reconciling these interests within the movement and promoting it to the public.

Conant welcomed this group to Harvard not only because he was its president. He was keenly interested in philosophy of science and debates about scientific method, the nature of science, and science's relationship to its history (one session at the meeting was held jointly with the History of

Science Society, and the highlight of the next Wednesday afternoon was an exhibit of Harvard's collection of historical scientific instruments). The most volatile debate surrounded the unity of science movement's vision of an intellectually connected world now threatened by war and the toxic ideology of Nazism. Could a better understanding of science and its inherent unity have somehow maintained international peace? Or, as the philosopher Horace Kallen suggested in his presentation, was this unity of science movement itself suspiciously totalitarian in its scientific and philosophical ways? Neurath, who had fled his native Austria to Holland and would later find refuge in England, had his hands full rebutting Kallen's insinuations as well as trying to convince skeptics that planning and organizing scientific research were *more* scientific, intelligent, and responsible than allowing unorganized chaos to reign. Conant regarded himself as unqualified to spar with these experts, but he enjoyed these debates and agreed with Neurath, Frank, and Morris that their work was important—that intellectuals as well as the American public must better understand modern science and how it works. Months before the event, Conant went to bat for Frank who was on a lecture tour of American colleges when Hitler annexed Czechoslovakia. Unable to return to his life in Prague, Frank faced unemployment and possibly deportation before Conant helped arrange a part-time position that Frank would occupy for the rest of his career.[9]

In more ways than one, that tumultuous week of September led Conant to think about whether and how scientific research could be planned, how scientific growth could be deliberately guided or controlled. There were Neurath, Frank, and this congress; and there was Hitler whose rampage would return Conant to military science. At top levels of Roosevelt's National Defense Research Committee and the Office of Scientific Research and Development, Conant would oversee research to test what Einstein and Leo Szilard had told President Roosevelt that August in their famous letter—that Einstein's equivalence of mass and energy, $e = mc^2$, heralded the development of a new kind of weapon of unprecedented power.

Was it true? How should scientists be organized to find out? If the answer were yes, and the supremely talented Germans were aware of it too, Conant wondered, how could American scientists and engineers be sure to win the race to build the bomb? And if the quest were successful, how would the new atomic weapons and technologies be managed after the war? Wrestling with these questions alongside MIT's Vannevar Bush, the physicist Robert Oppenheimer, and military officers such as General Leslie Groves, Conant became something of a scientific planner. For guidance he could draw partly on his earlier war experiences developing nerve gases, but there was a crucial difference. Then, the chemistry behind these inventions was well known. But the physics of splitting the atom rested on a recent scientific revolution, Einstein's revolution. Unlike true-believing Marxists who presumed that dialectical materialism was a reliable theoretical guide to research, Conant was at sea. He had no theory to tell him whether the nation's scientists could achieve these goals or, if so, how exactly to proceed. To make things more complicated, the questions involved basic theory, technological know-how, as well as "a whole army of specialists engaged in advancing science in a spectacular fashion," as he later described it. This was not just a scientific puzzle but an experiment in the sociology of science. And Conant was at the center, charged with making it all work.[10]

Competition drove Conant forward—his own sense of competing with the Germans in a race for the bomb, and competition among the scientists he supervised. Given uncertainties over how to most efficiently and quickly gather enough fissionable material to build a bomb, for example, he created separate teams to utilize different techniques. Besides Enrico Fermi's efforts to breed plutonium through a controlled nuclear reaction at the University of Chicago, Conant assigned other teams to work on gaseous diffusion, electromagnetic, and centrifugal separation of Uranium 235. Despite the costs of multiple teams, Conant kept them all working as hard and as fast as possible, for no one knew which method would work best or prove fastest. In 1941, he summed up the general situation in a private review of the situation. "For eighteen months this highly secret war effort has moved at a giddy pace. New results, new ideas, new decisions and new organization have kept all concerned in a healthy state of turmoil."[11] Even when he tried to guide it and move it along, this kind of turmoil, the clashing of competing ideas and proposals, was the heart of science as Conant understood it.

The Run-Up to War

Years before, as Europe became increasingly unstable, Conant neither wanted nor expected to return to military science. He and other high-profile Americans (such as the transatlantic aviator Charles Lindbergh) believed that the nation must stay out of Europe's conflict. But by the summer of 1940, he supported President Roosevelt's efforts to aid threatened nations of Europe. Conant went on something of a public relations blitz, making the case to Congress and veterans' organizations that the nation must immediately begin building up its armed forces. "We as a people have awakened to the immanence of the threat," he told the Jewish War Veterans at New York's Waldorf Astoria Hotel, "that human liberty on this continent is now in danger." "Our entire way of life," was at stake, he told Congress as he urged mobilizing for national defense.[12]

The carnage of World War I made many, perhaps most, Americans skeptical of interventionists like Conant. Conant's campus was divided as well. A month after Hitler invaded Poland, the student newspaper, *The Harvard Crimson*, reported that 250 students joined the American Independence League to oppose intervention. A spokesman for the league said, "[T]he greatest single obstacle which this league must combat is an irrational and fatalistic way of thinking that is startlingly prevalent in America now." A month later on Armistice Day, the League and the Harvard Chapter of the American Student Union held rallies, issued appeals, took polls, and organized events with civic- and labor-minded groups in Boston—all in support of the nation's broadly anti-interventionist consensus. The night before, the Student Union organized a free screening of Jean Renoir's antiwar epic *The Grand Illusion*.[13]

For almost two years, students, faculty, and administration debated the issues. *The Crimson* covered the events and arguments eagerly, its editorial voice usually tending to isolationism. Like their counterparts at rival Yale, where students created the isolationist group America First,[14] many Harvard students sided with noninterventionist faculty. *The Crimson* featured Donald C. McKay, assistant professor of history, who believed that intervention was unnecessary (Russia, he predicted, would keep Germany in check) and that prosecuting another war "might definitely weaken American democracy." McKay was skeptical about interventionist arguments and *The Crimson* followed his lead. The paper worried when Conant and some four hundred members of the faculty allied themselves with one of the nation's leading interventionists, Kansas newspaper editor William Allen White. White's Committee to Defend America by Aiding the Allies, an editor opined, "is probably the most powerful propaganda agency America has ever seen." Despite its name the committee called for full-blown military intervention, so when the time came, he predicted, "the United States will be blasted into war by a barrage of skillful propaganda."[15]

On other campuses, the American Student Union's international strikes continued to grow in size and impact. In the spring of 1940 more than one hundred thousand students planned to strike on 110 campuses. But Harvard's chapter would not participate, *The Crimson* reported, because it was experiencing "internal dissension over policy." Hitler's advances were making pacifism less and less defensible and the student body, perhaps following the lead from Conant, inched toward intervention.[16]

In May, after Norway, the Netherlands, and France fell, Conant delivered a national radio address sponsored by White's committee. The situation in Europe "actually threatens our way of life," he said. Roosevelt pressed the same argument masterfully in his State of the Union speech to Congress the next January. Promoting what would become the Lend-Lease Act, he urged the nation to support Britain by contributing munitions, ships, and airplanes. As Roosevelt framed the choices facing the nation, there was really no choice to make. Never before among the many wars the nation had fought was the enemy "aiming at domination of the whole world." Hitler's crusade was destroying democratic life throughout Europe and his army was still on the march. Contrary to the noninterventionists, the nation and its interests were already "overwhelmingly involved" in the conflict and citizens must prepare to cooperate and pay additional taxes to support the war efforts.[17]

Most distressing to the editors of *The Crimson* was a remark by Roosevelt about those who might disagree with him. "The best way of dealing with the few slackers or trouble-makers in our midst," he said, "is first to shame them by patriotic example, and if that fails, to use the sovereignty of government to save government."[18] This vague but unmistakable threat directed at dissenters must have struck anyone who had heard Conant's very different Baccalaureate Address to the class of 1937. Conant denounced the rise of totalitarianism in Germany, Russia, and Spain and urged graduating students to exercise individuality, independence of mind, and to resist the powerful pressures of conformity that Roosevelt had just appealed to. Shortly after Hitler invaded Poland, Conant predicted publicly that the coming turmoil may test the nation's faith in "modern civilization" and challenge "man's belief in a life of reason."[19] Indeed, reason seemed to have little to do with it, as Roosevelt branded dissenters as "slackers" and "trouble makers" and took it for granted that each citizen must adopt the same views.

The Crimson's editorial staff expressed a queasy, sickening realization that propaganda and national passions were making war inevitable. "If it must be a war," the editorial page replied, "let it be entered grimly, not as a matter of cheering and singing but as something to be gotten over with." And let us not pretend, as Roosevelt's rosy images of a world without fear and tyranny suggested, that war does not risk creating new problems for the nation. It may mean "the destruction of 'our way of life' as well."[20]

The National Conversion

The debate ended abruptly on December 7, 1941. White's Committee shortened its name to the Committee to Defend America and the vast majority of Americans, in shock and outrage over Japan's surprise attack in Hawaii, embraced the new consensus. Within hours Congress granted Roosevelt his request for a declaration of war against Japan. Within a month Congress declared war against Germany and Italy.

Conant immediately enlisted Harvard in the effort. The day after Pearl Harbor, he announced to a full house at Sanders Theatre that "the United States is now at war" and "we are here tonight to testify that each one of us stands ready to do his part in insuring that a speedy and complete victory is ours. To this end I pledge all the resources of Harvard University." To help keep the war effort democratic, Conant urged the armed forces to recruit future officers from the nation's high schools so that Harvard, in turn, could educate them to be effective military officers. By the next summer, the campus was home to thousands of naval personnel who had come for training prior to deployment (see Figure 2.1).[21]

Figure 2.1. Military officers at Harvard, Oct. 9, 1943. (Image courtesy of Harvard University Archives)

At times like this, Conant thrived. As Hershberg put it, he relished national consensus, "the unalloyed 'common cause,' the sense of a nation firmly united, the exhilaration of flexing new sinews of American power." He saw little conflict between the university's intellectual mission and the new, militarized campus he had cultivated, complete with students in uniform, daily reviews, accelerated plans of study, and the replacement of liberal arts courses with a military curriculum. As he saw it, the competitive engines of human knowledge, the creative frictions of clashing ideas, were still at work, only now on a larger, global scale. Yes, the pluralism of reasoned debate over intervention had given way to conformity and regimentation, but the goal and purpose remained to advance *Wissenschaft* and learning by defeating the enemies of political and intellectual freedom.[22]

Some students and faculty refused to accept the new consensus. As many as six hundred rallied on campus against intervention in the spring of 1941. But as inspections and parade-like reviews became common, the ranks of dissenters dwindled.[23] The Harvard chapter of the American Student Union dissolved and holdout pacifists, Quakers, and conscientious objectors became oddballs on campus. *The Crimson* covered them sympathetically— "Pacifist Group Holds Fast Amid Quickening War Fever" it headlined in August, 1942[24]—but they received no encouragement from Conant or his administration.

On October 6, 1942, Conant spoke to the incoming class of 1946 and warned that within a matter of months, most likely, the draft age would be lowered and their educational experience would be transformed. The narrow, military education in store for them as soldiers would be quite unlike the free-ranging liberal arts education most had signed up for. As Roosevelt's remark about "slackers" and "trouble makers" had forewarned, however, every able student was expected to join the war effort. "I feel sure that . . . none of you would wish to be left behind as your contemporaries march off to war," Conant said. "No able-bodied young man will wish to be placed in the position of having consciously avoided the risks of battle." Provost Paul Buck, who would handle day-to-day business while Conant was otherwise engaged, seconded Conant and told the incoming freshmen that their "first responsibility" was the war effort.[25]

∽

Thomas S. Kuhn was in the audience at Sanders Theatre. He had arrived as a freshman two years before and was soon elected to *The Crimson*'s editorial board. The board prided itself on its critical analyses of the intervention debates and its willingness to "raise its voice in a protest against the sentiments and ideas voiced by the University's chief."[26] Yet to the occasional consternation of his fellow editors, Kuhn supported Conant eagerly and consistently. Earlier that April, he praised Conant as a "far-sighted" leader with a "complete . . . grasp of the University's relationship to the war effort." Kuhn had once been an ardent pacifist, but now he scolded those who had staged an "unfortunate masquerade portraying President Conant as marching the student body off to the wars at the point of a bayonet." As the nation prepared for war, Kuhn wrote, too many students failed to acknowledge Conant's wise leadership. Too many continued to cling to the illusion that college sheltered and isolated them from the national effort. Too many failed to show up to hear Conant speak.[27]

When Conant delivered the news that the university would pledge itself to the war effort, Kuhn was the paper's editorial chairman. In an editorial that appeared the next day, he admitted that Conant seemed to be turning his back on Harvard's intellectual mission. Conant had even sugar-coated or understated the magnitude of what was about to happen. "The President's plan is not designed to preserve Harvard's liberal tradition," Kuhn wrote. Nor would it preserve the liberal tradition in American high schools if, as Conant suggested, half of the nation's graduates receive only shortened or condensed courses in science and the humanities. This was less an educational compromise than an educational capitulation, for Conant proposed "that the colleges as we have known them close their doors until the end of the war, and that their classrooms, dormitories, and staffs be devoted to 'indoctrination' at a pre-commission level."

By using the word *indoctrination,* Kuhn made Conant's plans sound positively totalitarian. But, Kuhn went on to explain, Conant was positively correct. The perilous world situation demanded nothing less than an overthrow of the old, "peacetime ideal" of liberal education. "This is a conversion as complete and necessary as that of American industry," he wrote. Just as the nation's automobile and appliance factories had been transformed to produce airplanes and guns, Harvard's buildings, grounds, and facilities must be converted to meet "the demands of total war."[28]

"The War and My Crisis"

Years before, as a precocious, talented, and emphatically pacifist young student at Hessian Hills, Kuhn would not so easily have accepted this interventionist consensus. By his second year at Harvard, however, Kuhn took this confident, public stand in favor of Harvard's war effort. His conversion had begun in the late 1930s as he finished high school, and it reached a difficult crescendo during his freshman year. As a sophomore, Kuhn wrote about it for his English professor, Robert Gorham Davis. In an essay titled "The War and My Crisis," he offered some scholarly analysis, some psychotherapeutic introspection, and some pure, sophomore drama.

His story began with his childhood in a liberal, rational, and progressive household. The atmosphere often

> led to discussions of the entire family over important decisions, discussions in which my brother and I were able to feel a sharing of responsibility. And in the realm of political society, a realm into which I was led early and which expanded for me as I grew, it stood for peace and security, for the right to work and to earn. It held for the release of the Scottsborough boys, freedom for Communists, support of New Deal Policy . . . and for student political organizations.

Conant would have approved what Kuhn wrote next: this liberalism prized change and growth. This applied to Kuhn's intellectual and political outlook as well as his maturing literary tastes ("I moved from the pulps . . . to Turgenev and, later, Proust," he added). In the Kuhn household, "Development was taken for granted. All this was liberalism for me."[29]

The foundation of Kuhn's growth was reason. He believed without doubt "that there was no means but reason to reach a valid conclusion and, conversely, that reason was adequate to the proper solution of all questions." But reason, he discovered, was not adequate to the debate over intervention. For Kuhn had been a committed pacifist who saw war as evil and unproductive. Though he reserved the right to defend his nation, he could not imagine good reasons to wage war. Yet soon before coming to Harvard he started to glimpse reasons in favor of intervention, such as the deep, historical and cultural connections between Europe and America.

When he arrived on campus, he wrote, "the problem struck me with its full force." Reason had led him in two, incompatible directions into "an antinomy of pure reason." His interventionism was supported by the growing national consensus. But his pacifism remained strong, too. It rested on his history, his sense of himself, and his pride in his intellectual and moral development. "For a person whose entire rational life revolved around these concepts," he explained, this was a "major crisis." One of these incompatible ideas "would have to be completely destroyed if my faith in reason was to be retained."

Kuhn did not mention the Hessian Hills School by name. He called it simply the "progressive school" where his ardor for pacifism meshed with his growing intellect and respect for reason. Nor did he name Elizabeth Moos, the school's guiding, pacifist light. But she remained on Kuhn's mind and a presence in his essay. He described meeting with her in the summer of 1940—shortly before he came to Harvard and shortly after Moos had cemented her reputation as a radical by taking students to the May Day Parade in New York City. Hoping that her lifelong pacifism and wisdom as a teacher held some insight or a solution to his crisis, he put to her one half of his antinomy, his new conviction that America was inseparably bound to Europe. "She sympathized with my view and even followed it to a certain point," Kuhn reported. But, she insisted "that American boys should not, under any conditions, be sent over-seas again."

Kuhn had always been wary of dogmatism and absolutism. Even as a thirteen-year-old pacifist, he once explained, he stopped short of taking the Oxford Oath because it seemed too absolute. Kuhn had grown intellectually not by ignoring but by wrestling with complications, as his personal crisis over the war illustrates. But his former schoolmaster seemed to have boxed herself in. Her pacifism blinded her to a logical conclusion that followed from premises she seemed to accept. It "stopped the logical chain" of reasoning in her mind.

The reunion with Moos offered no clear solution. But it taught Kuhn something about the faculty of reason: it is not the only source from which beliefs and actions emerge. As campus politics and editorials in *The Crimson* demonstrated when he first arrived at Harvard, beliefs were shaped as well by propaganda, fear, and social pressures demanding conformity. He did not use the word *ideology* in this essay, but he would rely on it in his future theorizing about science—perhaps because of what Kuhn learned as he struggled. As he put it to Robert Gorham Davis,

> In the last year I have found it possible to consolidate my position somewhat. . . . I still see reason as the only source of judgement, but I realize that all decision is composed of more than judgement.

It is composed as well of "irrational" and "subjective" features of events and situations. Kuhn's own decisions, he had realized, would seem valid, objective, or reasonable to others only when those other's lives involved similar events and situations; only when "I resemble those others in this irrational subjective element." Lacking these commonalities, Kuhn could not persuade his former teacher to accept his position; and he could not accept hers. In terms he would later use in *The Structure of Scientific Revolutions*, they were "talking past each other" without much hope of finding agreement about the wisdom of American intervention.[30]

Whether or not Kuhn was aware of it, his composition professor had a large stake in these issues. Unlike Moos, who remained steadfast in her radicalism, Robert Gorham Davis was a Communist who moved rapidly to the right at the end of the 1930s. He would later testify to the House Un-American Activities Committee that he no longer respected the Communist Party or the irrational dogmas to which he once subscribed.[31] In his comments on Kuhn's essay, however, Gorham Davis ignored the politics and personal turmoil Kuhn described. He commented only on formalities like sentence construction and grammar. He graded the essay a "C" and warned Kuhn about "too-involved sentences" and relative clauses beginning with "which."

Liberal Arts and General Education in a Time of War

Kuhn's sophomore crisis illustrates that it is not always possible to reconcile the demands of reason and logic with the passions of politics and, especially, war. Conant had seen this before during the angry debates surrounding American intervention in World War I. He knew that the rules of civic and intellectual life changed dramatically during times of war, that reason and cool, intellectual analyses were sometimes forced to accommodate the passions and irrational quirks of circumstance. As an educator, his job was to be flexible—to adjust to the realities of wartime when necessary, even if that meant turning Harvard into a militarized campus, but never to let emergency conditions become permanent. This bothered some educators who feared that those universities that lent their campuses to the war effort might never be able to turn things around at war's end. Wendell Willkie, the popular 1940 Republican candidate for president, warned against "wartime obscuration of the study of the liberal arts in the colleges." Instead of suppressing the liberal arts, Willkie and others believed that during war universities should take special measures, such as admitting women or creating select groups of soldiers to study Shakespeare and Chaucer, to ensure that the "conversion" for war (as Kuhn had described it in his editorial) would be neither total nor irreversible.[32]

In the Sunday *New York Times*, beneath a photograph of uniformed students standing for inspection, Conant attacked Willkie's argument. "Personally I have not the slightest doubt that the study of the liberal arts will not only survive this war but prosper in the days of peace," he explained. With future victory, he predicted, students would tire of science and technology and yearn for different, more literary or artistic challenges. This was not only a matter of educational fashion and the swing of culture's pendulum, he explained. For while the liberal arts must now take a back seat to the rigors of military education, they remained vital to the cultural patterns and the freedoms the nation had to preserve. "Those problems of human nature and human destiny which man has assembled under the headings of literature, history, and philosophy" remain a "deep-moving force" in our Western, liberal tradition. Any student who is never exposed to these problems and never at least glimpses their importance will not really understand what the war is all about, Conant insisted.[33]

Even worse, Conant explained, students who overlook the essential complexity of humanity—"the profound truth of Pascal's saying that human nature is both the glory and the scandal of the universe," he wrote—will be more susceptible to the simplistic ideologies that oppose liberalism. Devotees of Marxism, Conant knew, exalt it as a comprehensive science of all things that promises a future utopia of peace and contentment. That is what Hitler and his big lie promised to the German people, as well. "A student reared on deficient spiritual and intellectual diet is an easy prey for proponents of the totalitarian view," he explained.

As Conant saw it, this debate over the liberal arts in the colleges was a dangerous distraction. Don't worry about "young gentlemen born to the purple" at Harvard and similar institutions, he urged. The real educational challenge facing the nation concerned "the other 90 percent" of students around the nation in high schools and colleges. Given the frightening possibility that the nation as a whole might succumb to an ideology like the one that had taken over Germany—the last nation on earth in which totalitarianism would flourish, Conant might once have predicted—this majority and the quality of its education "will be the controlling factor in our ability to resist an attack of totalitarian ideas."[34]

Two weeks before his *New York Times* article appeared, Conant set in motion the "vigorous exploration" of American education he called for in the article. It would be undertaken by a new faculty committee, the University Committee on "The Objectives of a General Education in a Free Society," with Buck, his right-hand man, its chairman. After two years of meetings and research, the committee published its final report shortly after the allied victory in 1945. The report, known as "the redbook," explained that the main threat to modern education was the nation's "explosive growth."[35] Ever-larger numbers of Americans experienced life locally and regionally, without contact with their fellow citizens hundreds or thousands of miles away. As a result, school systems in different states formulated curricula and educational goals in strikingly different ways—as different as rural Mississippi and urban, industrialized New York City. Even within individual states, geography, history, religion, and opposing ideals of government created cultural divides that threatened national unity. After an allied victory, the redbook predicted, this growth would surely accelerate and place even more pressure on the nation's fragile unity and, in turn, its political and economic stability.

The children who would inherit this national cultural chaos, the report observed, were themselves divided into groups by magazines, movies, and radio. "Like the high-school curriculum," the committee wrote, "the movies and radio, not to speak of magazines and newspapers have adapted themselves to the enormous range of taste and intelligence which exists in the general public." Of course, this pluralism and variety belongs in a free, creative, and unregimented society. But if this fragmentation continued unchecked, the committee insisted, the nation could become susceptible once again to civil war, to fascist ideology, or class warfare and, possibly, revolution because in different areas of the nation "tastes will have been so differently formed."[36]

Unity and difference—the one and the many—were out of balance in the United States of America. The redbook proposed an educational, curricular solution. But the committee was quick to emphasize that curricular reform alone was not enough. "Even a good grounding in mathematics and the physical and biological sciences, combined with an ability to read and write several foreign languages," Conant wrote in its introduction, "does not provide a sufficient educational background for citizens of a free nation." For that, he wrote (echoing the words he used in his *New York Times* piece) students must see the larger "cultural pattern" and the wisdom of the nation's liberal tradition. If the rise of Nazism had demonstrated the fragility of freedom and humanity in Germany, general education in the United States would help preserve them and the "cultural pattern" in which they thrive.[37]

The Other Unity of Science

Neither Conant nor his committee minimized the difficulties ahead. The nation's enormity and variety had always cultivated contradictions and paradoxes. "The United States" itself was a kind of oxymoron. Rolling out general education cross the country would excite old, unresolvable debates over the proper size and scope of government, between Jeffersonian versus Jacksonian ideals of democracy, between rural and urban values and beliefs. But Conant had a unifying trick up his sleeve. For as difficult as these problems were, science education—and in particular study of science's history—promised a path by which the nation as a whole could grasp this "broad basis of understanding" on which it could thrive as a unified *and* liberal, pluralistic whole.

First, the report explained, science education must include history of science. Momentous episodes in the past—the discovery of oxygen, the Copernican revolution, the invention of the air pump—could "illuminate and vitalize" science for students. Second, and more importantly, it would illuminate this cultural pattern of liberalism on which modern science and modern America depend. Science and its history, in other words, are parts of "the total intellectual and historical process" that citizens must grasp as an essential part of their liberal tradition.[38] They illuminate the fragile but essential compromises that allowed America to survive and flourish this far—the delicate balance between freedom of thought and firm, principled conviction; between Jacksonian ideals of openness and tolerance, and Jeffersonian confidence in hard-won truths. As the committee put it,

> The ideal of free inquiry is a precious heritage of Western culture; yet a measure of firm belief is surely part of the good life. A free society means toleration, which in turn comes from openness of mind. But freedom also presupposes conviction; a free choice unless it be wholly arbitrary (and then it would not be free) comes from belief and ultimately from principle. A free society, then, cherishes both toleration and conviction.

Toleration and conviction may seem incompatible, the report admitted: "If I am convinced of the truth of my views, on what grounds should I tolerate your views, which I believe to be false?" But this sense of incompatibility can and must be tempered by education, the report explained. Students will learn that to be human means embracing an "intellectual humility" that is grounded in history; in the understanding that "wise men have made endless mistakes in the past."[39]

So understood, the historical case studies that Conant and his collaborators proposed as models for their new science curriculum were studies in intellectual achievement as well as humility. They were stories of "wise men" humbled at the same time that they moved human knowledge forward. As Conant would soon explain in his popular book *On Understanding Science*, science is no repository of permanent truths or facts; it is instead a restless progression of theories and experiments proposed and executed by skeptical scientists seeking to refine, if not refute, each others' views. This competitive turmoil often leads to *new* theories and *new* experiments that could not have been imagined beforehand. "Almost by definition," Conant put it, "science moves ahead."[40] Like modern, liberal society, science is an ongoing social and intellectual process, a complex march of clashing ideas, methods, experiments, and "endless mistakes."

Progressive Education Meets General Education

A few months after Conant and his committee began their work in 1943, Kuhn graduated summa cum laude with a degree in physics (see Figure 2.2). He then began military work on radar installations—first on campus and

Figure 2.2. Harvard Commencement, 1943. (Image courtesy of Harvard University Archives)

then in England and France. On returning to pursue his PhD in physics, Kuhn again joined public debate on campus by writing a review of the just-published report for the *Harvard Alumni Bulletin*. Impressed by his editorials for the *Crimson,* Paul Buck commissioned Kuhn to evaluate the report specifically from the undergraduate's point of view.[41]

For the most part, Kuhn liked the report. By no means, he began, were the recommendations "a death blow" to the current system of electives that undergrads were used to, for they largely ran parallel to the existing system. Some of the changes would be welcome, he opined, while others, such as phasing out minor concentrations, might not. Kuhn also wondered aloud if Harvard had enough "great teachers" to make the system work and achieve its goals. But he had little doubt that general education was needed given the increasing abundance of knowledge to be learned. Overall, Conant's committee had tackled this task admirably, he wrote. "No better reforms than the Report's have yet been proposed."[42]

Kuhn's review said nothing about the committee's recommendations for the study of history of science. Nor did he examine the committee's concerns about the possibility of class warfare or other kinds of division and disunity in "a free society." Then pursuing his doctorate in physics, he viewed the redbook through the narrow lenses of electives, teachers, and reading assignments that populate the undergraduate experience, as Buck had asked. Soon, however, Kuhn would become an insider to Conant's quest for a national program of general education and a national understanding of modern science and its history.

3

History of Science in a Divided World

When the objectives report was published, Thomas Kuhn and his younger brother Roger played important roles. Tom reviewed the report in the *Alumni Bulletin* while Roger chaired a student council committee to poll students taking the new courses. Professor of government Benjamin Wright, the head of the new General Education Committee, praised Roger's work in evaluating the new courses, and he turned to Tom to help teach them.[1]

In February 1947, Wright sent a short note asking Kuhn to "drop by my office."[2] When he arrived, Kuhn learned that Conant needed teachers for one of the new courses in the general education program. Kuhn would have read about it in the objectives report—a course about "the basic physical principles and concepts and the methods by which these have been developed."[3] Would Kuhn be willing to accept the job?

The timing was perfect. For Kuhn was in the midst of another crisis. He was losing interest in a future career in physics and was increasingly drawn to philosophical problems about knowledge and the growth of science. As an undergraduate, his curiosity ranged widely. Besides his courses in physics and electrical engineering and his editorials for *The Crimson*, he reviewed novels, movies, and books for the paper. He was president of the exclusive undergraduate Signet Society, a position usually held by scholars drawn to the arts and humanities. As a graduate student specializing in physics, however, he was frustrated and somewhat bored. Even while working on radar during the war, he read philosophy and explored the logical and philosophical foundations of scientific knowledge in books by Bertrand Russell, Percy Bridgman, and Philipp Frank. When Kuhn returned to Harvard, he even registered to take some philosophy courses as part of his work toward his PhD. But those were not working out. Kuhn needed more undergraduate training in philosophy under his belt; and he recoiled at seeing himself, a veteran, taking courses alongside teenagers.[4]

Wright's proposal offered a path forward, a chance to develop and teach courses that were part science, part philosophy, and part history. Adopting the case-study methods pioneered in the law and business schools, students would find themselves immersed in particular episodes in science's history that would illuminate the nature of science better than learning fact-filled chronicles of scientific achievement. They would explore how the sciences were connected to the historical worlds surrounding them, the economics, the available technologies, and the prevailing beliefs and culture. With Conant himself involved, Kuhn could hardly imagine saying no.

"I have never forgotten the first time I met him," Kuhn later said about Conant.[5] When they first discussed their collaboration in the new program, they must have been struck by coincidences. Conant, too, was elected to the editorial board of the *Crimson* as an undergrad, a position he dearly wanted (and for which he twice endured the initiation ritual at the hands of senior editors known as "comping"). He had also been a member of the Signet Society, and they may have discovered that each had inherited pacifist convictions as children that they later overcame as American intervention in foreign wars became likely—World War I, for Conant, and World War II, for Kuhn.[6] They were on opposing sides of a long-standing tribal divide between chemists and physicists, but that would feed a friendly rivalry that enlivened their collaboration. And they shared interests in literature, art, history, and philosophy of science, where Conant stayed abreast of trends by reading his faculty members Bridgman, Frank, and Quine (Kuhn was reading them, too). As for history of science, they shared a disappointment with how it was pursued by Harvard's George Sarton. In hoping to revitalize the field, Conant had just started a new degree program in history of science.[7]

A handful of other young scientists at the university were recruited as well, including the astronomer Fletcher Watson and chemist Leonard Nash. In various combinations, Conant, Kuhn, Watson, and Nash taught *Natural Sciences 4*, titled "The Growth of Experimental Sciences," and eventually other courses in the program. Conant and Watson taught the course that fall, while Kuhn completed his work in physics. But Kuhn's education in history of science—Conant-style—began immediately as Conant handed him page proofs for his new book, *On Understanding Science*. It would serve as a preliminary textbook for the course while the team developed additional course materials.[8] Conant also instructed Kuhn to put his knowledge of physics to use and study classic texts by Plato, Aristotle, and their contemporaries. This would help to expand the range of case studies that Conant had already sketched in *On Understanding Science*.

For the next three years, Conant and his assistants met regularly, often for lunch, to discuss the courses they taught and the texts they were developing.[9] Conant left them in 1950 to attend to his administrative and national duties, but he occasionally returned to the classroom (in 1952, for instance, to teach a seminar in philosophy of science) and continued a friendly, scholarly relationship with Kuhn. Late in 1950, for example, Kuhn wrote to Conant with some new thoughts about Galileo's and Torricelli's theories on the atmosphere surrounding the earth. If he found them useful, Kuhn said, Conant was free to incorporate them into his own writings. Conant was grateful for Kuhn's "keeping in touch with me about any new developments arising out of our joint efforts" and for all those "who had lunch with me so often in the last three years" to discuss their collaboration. Those discussions had played a part in Conant's then-forthcoming book, *Science and Common Sense*—"my latest excursion into popularizing the methods of science"—one that Kuhn would surely find interesting, Conant said.[10]

On Understanding James B. Conant

Kuhn dove into the collaboration, reading eagerly in the history of physics and theorizing himself about his "joint efforts" with Conant. In May 1947, shortly after his initial meeting, he drew up notes titled "Objectives of a General Education Course in the Physical Science."

"The following is not intended as a course outline," he typed. It was just a listing of topics that Kuhn expected a course like *Natural Sciences 4* to discuss. But it was not really what Conant had in mind. Philosophically, Kuhn pictured science proceeding from empirical particulars—the colors, sounds, and feelings of "immediate sense data"—to the "establishment of a system," such as Newton's physics. There were stages of physical science, Kuhn outlined, including "pure classification" and "perceptual objects," then "successive abstraction" as "perceptual objects" turn into fully fledged "scientific objects." Finally, the scientific mind arranges these objects into a system, a "conceptual structure," that makes successful predictions about nature.[11]

This is a classic picture of science. It distills the British empiricist philosophers David Hume and John Locke, mixes them with the Viennese physicist Ernst Mach, and arrives at Philipp Frank's office at the Institute for the Unity of Science on Massachusetts Avenue. As Frank understood it, however, logical empiricism was fast evolving in the direction of American pragmatism. Kuhn's list addressed that, too, in a way that Dewey and

William James would have approved: the human mind, he emphasized, plays an active role in creating these abstractions and generalizations. "Scientific generalization," that is, must be understood "as a creative imaginative process."

At this point, however, Kuhn may not have read *On Understanding Science*. For when he did, he'd see that Conant had broken from this view that scientific knowledge is built up into theories from perceptions or observations. Nor did Kuhn's outline mesh well with the kind of course Conant had described in that book. Kuhn used specialized, abstract terms such as "operational definitions," "successive abstractions," and "dynamic as well as static unities" that seem ill-suited to reach the students that Conant designed the course to serve, mainly, future lawyers, teachers, politicians, and civic leaders.

Why should citizens such as these need to understand science at all? Conant gave three reasons. One, the atomic bomb had created misunderstandings about what modern science could, and could not, achieve. In the wake of Hiroshima and Nagasaki, some had come to fear science, while others naively concluded that modern scientists could achieve almost anything at all. Both attitudes were wrong, Conant knew. But modern science was not going to disappear, and the problem would grow only more pressing. What we need, he wrote, is

> a widespread understanding of science in this country, for only thus can science be assimilated into our secular cultural pattern. When that has been achieved, we shall be one step nearer the goal which we now desire so earnestly, a unified, coherent culture suitable for our American democracy in this new age of machines and experts.

A second reason was our "benignly chaotic system of political democracy"—a system in which nearly anyone can be "almost accidentally thrown into positions of temporary power" and will likely face problems that involve science or at least require a basic familiarity with how science works. As he wrote this, Conant likely had in mind the urgent concerns among Washingtonians surrounding Roosevelt's death in 1945, particularly whether his vice president Harry Truman, who had never completed a college education and was known as "Haberdasher Harry" for his career selling men's clothes in Missouri, was qualified to lead the nation in war—much less this "new age of machines and experts."[12] And third, recalling his participation at the Unity of Science conference when the war first broke, Conant nodded to issues about whether and how the sciences can work and grow together. If innovations in physics or chemistry spark developments in the social sciences, for example, this could be "of supreme importance to the future of the free people." Experts and ordinary citizens alike could only benefit from a better understanding of science, one grounded in the rich details of historical case studies.[13]

What those case histories showed, Conant explained, was that the heart of scientific research is never a matter of facts or theories alone. Progress always hinges on their *interaction* at the hands of scientists working toward goals that are ultimately practical and connected to human purposes. Yet this was not the impression one gained from most science courses and most science textbooks. Typically, they ask students to accept claims about nature and scientific laws "on faith," as if they were dogmas "handed down by a high priest."[14] Of course, Conant understood that textbook writers had neither the expertise nor the number of pages necessary to cover adequately any field's history. But the consequences of this shortcut were enormous. It led to widespread misunderstanding of science at a time in the nation's history—the world's history—when science played enormous roles in the lives of individuals, nations, and armies.

Conant's exhibit A was the philosopher Karl Pearson and his influential book *The Grammar of Science*, first published in 1892. Pearson takes modern science to be merely "the classification of facts," Conant wrote. And from this, he announced, "I dissent entirely." Pearson misled his readers, Conant charged, by idealizing research and ignoring that science is often messy, confusing, filled with wrong turns, mistaken hunches, and excessive confidence in the status quo. If science were as simple as empiricists such as Pearson suggested, then why had it taken humanity thousands of years to develop successful theories in physics or biology? In truth, science was slow and difficult. Progress came without the aid of a reliable guide or map of the territory, and it was often fascinating and dramatic: "The stumbling way in which even the ablest of early the scientists had to fight through thickets of erroneous observations, misleading generalizations, inadequate formulations, and unconscious prejudice," Conant insisted, "is the story which it seems to me needs telling."[15]

At the center of Conant's story were "conceptual schemes." These are sets of ideas and theories in some area of science, and they can take different forms and sizes. A conceptual scheme "on a grand scale" might contain nearly the totality of knowledge of an age—such as Aristotle's theories about nature in antiquity—while others concern narrow areas of experience or certain techniques. In either case, science involves testing, refining, and developing conceptual schemes through experimentation and analysis. The process Conant described was dialectical (though he did not use that word): "New concepts arise from experiments and observations, and the new concepts in turn lead to further experiments and observations." Conceptual schemes provide a starting point for research, while experiments tell scientists when their conceptual schemes need to be modified or, in times of scientific revolution, abandoned. If there is an essence to modern science, Conant explained, it was this constant, self-correcting interplay between theories and experiments—it lay not in the *content* of scientific beliefs or their truth; but in this relentless forward movement and growth of knowledge.[16] That was the lesson students should draw from the three case studies Conant presented: early studies of atmospheric pressure undertaken by Galileo and Torricelli in Italy and by Robert Boyle in England, research on electricity, and the discovery of oxygen and the modern understanding of combustion.

On this basis, scientific revolutions could be understood as the replacement of one conceptual scheme by another. Often, Conant wrote, they occur when an individual scientist makes an unseen or unsuspected connection between certain "unsolved riddles" and "a new discovery or a new technique" that points to a new, different, and fruitful conceptual scheme:

> [T]he keen-minded scientist, the real genius . . . the man who keeps in the forefront of his thoughts these unsolved riddles. He is then ready to relate a new discovery or a new technique to the unsolved problems. He is the pioneer, the revolutionist.

Usually, conceptual schemes expand and grow gradually. But when revolution looms, as it did when Torricelli's experiments introduced a new idea of atmospheric pressure, "a completely new idea" comes on the scene, captures scientific attention, and makes the original idea obsolete.[17]

Contrary to Marxists who believed that principles of dialectical materialism created an inner momentum and logic to historical change, this approach emphasized that human agency, manifest in what Conant called the "tactics and strategy" of science, pushes knowledge forward. It was crucial that Conant's pioneer revolutionists kept tabs on different conceptual schemes and remained open and curious about unsolved riddles. Albert Einstein, for example, had long wondered why gravitational and inertial mass were numerically equivalent—that is, why is the "m" that appears in Newton's law of motion $f=ma$ and the "m" in his law of gravity, $F=GMm/$

r² represents the same physical quantity? This puzzle helped lead Einstein to his new theory of relativity, a conceptual scheme in which the physics of gravitation and motion are intimately linked. In studying episodes like this, Conant's students would see that revolutions don't simply happen on their own as new facts become known.

Conant made no apologies for his military terminology—the "objectives" he and his committee aimed to reach, and the "tactics and strategy" of science he mentioned throughout *On Understanding Science*. Besides the case-study methods of Harvard's professional schools, the course he envisioned was inspired as well by the case studies familiar to military education. "The success of that educational procedure," he wrote, "is one reason why I venture to be hopeful about this new approach to understanding science."[18] Military language as well as military optimism came naturally to a leader who had recently helped create the atomic bomb and to a nation that had just won a war on two fronts. On the heels of those victories, Conant hoped that the redbook report and his *On Understanding Science* would together help direct the nation's momentum and postwar confidence toward the reforms he envisioned. The nation's next momentous victory, he hoped, would be educational.

Conceptual Schemes, Paradigms, and a World in Crisis

When Kuhn first read *On Understanding Science*, he would have clearly seen how conceptual schemes lay at the heart of Conant's understanding of science, and he was evidently impressed.[19] The book helped chart his future career, for Conant's book can readily be seen as a model for *The Structure of Scientific Revolutions*. Each book is short, readable at one sitting, and crafted to instruct general readers to issues in philosophy and historiography that are usually the province of specialists. If the essence of science was its relentless forward motion, the way "science moves ahead," *Structure* shows more precisely how and by what stages paradigm shifts occur. Along the way, Kuhn employed the same case histories to attack the same conventional wisdom that science was built up from observations and perceptions, to emphasize the role of ideas in the history of science, and to expose the false history presented by science textbooks. Like Conant's conceptual schemes, Kuhn's paradigms organize and punctuate eras in science's history. Their open, hard-to-pinpoint qualities allowed both Kuhn and Conant to emphasize that research is simultaneously intellectual and social. If a general education instructor is doing the job well, Conant wrote, "there can be left in the student's mind no doubt that science is indeed a social process."[20] There was no doubt about this for those who read *The Structure of Scientific Revolutions*. For scientific paradigms, Kuhn would later explain, exist at the heart of scientific communities and could not function as they do apart from the social conventions of scientific education and professional scientific life.

In other ways, however, *Structure* can be understood as a reply to the larger, political agenda to which Conant had dedicated the general education program and the roles of history of science within it. Just as Kuhn's collaboration with Conant began, however, that agenda evolved along with the dramatically changing contours of the postwar situation. As he originally conceived it, Conant chartered the general education program to reform American education in ways that would nourish and stabilize American democracy as it weathered the difficult transition from a wartime to a peacetime economy. Though Conant was never a fan of Soviet Communism, he then hoped and believed that peaceful international collaboration with the Soviets was possible and desirable. To the consternation of his more hawkish colleagues in the military and government, for example, he sided with the many leading atomic scientists who believed that given the nature of scientific research, there were no genuine "atomic secrets" for America to protect as her own. Progress comes not merely by obtaining a formula, an equation, or a drawing, but through the determination and cleverness of talented scientists—of which there were an abundance in both the United States and the Soviet Union. Conant and other internationalists knew that sooner or later Russia and other nations would develop their own atomic weapons. Instead of attempting to guard America's bomb-building skills as a secret, therefore, they hoped to cultivate future peace and trust among nations by creating international organizations to share and regulate atomic technology.[21]

But relations between the Soviets and the West deteriorated rapidly after the allied victories. At the Potsdam conference, where the United States, England, and Russia had met to address the administration of postwar Germany, Truman and Josef Stalin did not get along. Stalin seemed mistrustful and disingenuous, and he soon asserted military control over parts of Poland and the Baltics. Postwar Russia, it was becoming clear, was not going to be the friend of freedom and liberty that it seemed to be when Stalin's armies and tanks rolled into Berlin to end Hitler's tyranny. By 1948, Conant had given up on hopes for postwar cooperation and amity. The Western allies and the Soviets were instead locked in a "world-wide struggle" for influence that had "overriding significance for our national planning."

As Conant saw it, however, this struggle made his proposals to improve and expand public education only more important. It also imperiled his educational reforms, however, for the climate of geopolitical mistrust, he feared, could encourage public education's critics, such as conservatives who deplored Roosevelt's expanded, New Deal government and the nation's many Catholics who often preferred parochial education for their children. Fearing that these and other factions might seek to shrink or dismantle public education, Conant went on the offensive to argue that now, more than ever, the nation must reform and strengthen its public schools. Far from taking a step *toward* national or centralized, authoritarian control of education, he explained, the reforms he intended would strengthen the nation and its freedoms in its struggle against the Soviet threat.[22]

Conant made his case in *Education in a Divided World*, a book that praised America's public schools as nothing less than "the sinews of our society."

> They are the product of our special history, a concrete manifestation of our unique ideals, and the vehicle by which the American concept of democracy may be transmitted to our future citizens. The strength of this republic is therefore intimately connected with the success or failure of our system of public education.

These ideals included representative government, a free, competitive economy, social mobility and equality of opportunity—all of which depended crucially on how the nation's citizens were educated. Writing in an age of segregation, the Jim Crow South, racial quotas at universities, and discrimination throughout society, Conant knew that some would find his proposals idealistic, if not utopian. But those critics did not understand how the nation worked, he insisted. History showed that the ideal of a classless society and the principled disdain of hereditary privilege had helped make America what it is. Ideals, even if only partly or barely realized, he argued, are not thereby meaningless:

> As long as a national ideal—be it equality before the law, personal liberty, social justice or "in America there are no classes"—as long as an ideal represents a goal toward which a community of free men may move by concerted action, the phrase in question has real meaning.

Like science, that is, a healthy society always "moves ahead'" toward its unrealized ideals. Now, especially, this motion would be crucial for meeting the ideological challenge from the Soviets: "Never-ending efforts must be made," he wrote, "to move society forward."²³ A healthy, thriving, and democratic United States would show the world that

> there is in fact a strong and vigorous rival to the Soviet views. This demonstration must convince even the most doctrinaire followers of Marx and Lenin that at least their time scale must be wrong. One can hardly expect at the outset to do more than to persuade the Soviet Philosophers that the advent of world communism has been postponed.

"Our answer to the Soviet philosophy," in other words, must be the enduring vitality of the West—"not only prosperity but a movement forward consistent with our inherent idealism."²⁴

Conant envisioned nothing less than a Manhattan Project in public education. Experts in sociology, anthropology, and social psychology must collaborate as did the "physicists and chemists in the battle against the Axis powers." There would be important roles for leaders in business and labor, too. Done right, general education in the postwar struggle should be everyone's concern: "All citizens of the country," Conant wrote, should "look increasingly to our free schools for the effective demonstration of our answer to totalitarian ideologies in a divided world."²⁵

Geopolitics and General Education

These "alarmingly clear and grim" realities of the new cold war were on Conant's mind as he welcomed Kuhn to the general education program and as they worked together for the next few years. For Kuhn's part, on the other hand, his notes from the time suggest a different picture. In a typed document titled "The Sciences in the Harvard General Education Program," for example, he dryly described Conant's course as part of "an effort to provide the student with an understanding of the mental and manipulative tools by which scientific conceptualizations are evolved and selected or rejected." As it was in his review of the redbook a few years before, and in his early notes for teaching alongside Conant, Kuhn at twenty-five years old did not visibly engage the national and international dimensions of his new position at Conant's side.

For several reasons, this was not surprising. Kuhn was then joining Conant's world as a subordinate collaborator, not an architect of national and geopolitical strategy. His first job was to write and defend a thesis in solid state physics in order to obtain his PhD. And his next was to retrain and reorient himself as a different kind of intellectual—not a scientist but a historian of science—and to establish himself and his future career in this very different field. Kuhn was also reinventing himself in the wake of his wrenching undergraduate crisis during the intervention debates that he described to his professor Robert Gorham Davis. Discovering an "antinomy of pure reason" within himself refuted his beliefs that he could always think his way through "the problems of the world" and that human beings are fundamentally "amenable to reason" and rational conflict resolution. Beyond matters of "incidental day to day existance [sic]," he explained, "this was a blow to my scheme of thought and left no foundation for action" in the larger world. "Reason had failed as a source of objective clear-cut judgement. On what could I base decision?"[26]

Yet even if Kuhn had made a conscious decision to turn away from politics and devote himself exclusively to scholarly and professional pursuits, he remained working and studying with Conant within the program that Conant now saw as one of the nation's most powerful cultural tools for responding to the Soviet threat. On campus and at meetings and conferences around the nation dedicated to general education, Kuhn was a public face of that ambitious enterprise.[27] He also became close—at least closer than many—with Conant on a personal level. Not all of Conant's colleagues desired this, or even believed it was possible, for he was not infrequently regarded as "a dour, priggish, brusque, unimaginative bureaucrat." Conant even described himself as "a cold reserved New Englander." Decades later, Kuhn used the same word—he was "a quite reserved person," he said of Conant. But they nonetheless "developed a relatively warm relationship." Kuhn compared him to his biological father, Samuel, who had attended Harvard as an undergraduate alongside Conant and whom Kuhn praised as "the brightest person I had ever known"—except, that is, for Conant.[28]

After their first semester co-teaching *Natural Sciences 4*, Kuhn asked Conant to nominate him to Harvard's Society of Fellows. This was somewhat brazen, for nominations were usually initiated by faculty who spotted the right kind of talent for this honor. But Conant agreed and, in 1948, Kuhn joined the prestigious society for the usual three-year stint of independent research, discussion, debate, and camaraderie. Kuhn was now rubbing elbows with Harvard's best and brightest, such as the philosopher Willard V. O. Quine, whose own writings influenced Kuhn and stand today alongside Kuhn's as twentieth-century classics.[29] After Conant left Harvard in early 1953, he would continue to support Kuhn's career in helpful and sometimes public ways.

The Frustrations of Diplomacy

Kuhn's relationship with Conant during the dramatic and worrisome years of the early cold war set the stage on which Kuhn would begin to formulate his new image of science. As the chapters in Part III examine more closely, despite the narrow and scholarly approach to analyzing science and theorizing manifest in Kuhn's early notes, his outlook would expand to incorporate psychology, sociology, and the central pillar of postwar political analysis, ideology. By the time Kuhn completed *The Structure of Scientific Revolutions* in the early 1960s, readers could readily sense—and some would point out—that *Structure* had found in the history of science dramatic, world-making powers of ideology and propaganda.

One need not have read *Education in a Divided World* or other cold war writings by Conant to sense these influences, however. Conant's analysis of the world situation in 1948 reflected the emerging understanding of the postwar situation that became a foundation for United States policy and American culture until the end of the cold war. One of the original and now iconic documents of this understanding is the famous "long telegram" written by the diplomat George Kennan two years earlier in 1946. Amid growing worries and puzzlement about Stalin's behavior and intentions, Kennan cabled his superiors in the State Department with seventeen pages of clear and convincing analysis. He explained that to understand anything about postwar Russia, one had to understand its history and geography, the Kremlin's deeply insecure and "neurotic view of world affairs," and the way that Marx's theories wound tightly throughout Russian thinking. Marxism provided Russia's leaders with a theoretical justification for "their instinctive fear of outside world, for the dictatorship without which they do not know how to rule, for cruelties they did not dare not to inflict, for sacrifices they felt bound to demand." As Conant would emphasize in *Education in a Divided World*, Kennan urged his fellow diplomats never to "underrate importance of dogma in Soviet affairs."[30]

The telegram made such an impression in Washington that the editor of *Foreign Affairs* commissioned Kennan to explain his thoughts to the American people, which he did under the pseudonym "Mr. X." Here Kennan reiterated his view that the struggle taking shape hinged on the power of ideas and mental conditioning to control human thought and behavior. Not Pavlov's dog, but the canine logo of the American Radio Corporation of America (RCA) came to Kennan's mind as he explained how Soviet operatives function. They are

> unamenable to argument or reason which comes to them from outside sources. Their whole training has taught them to mistrust and discount the glib persuasiveness of the outside world. Like the white dog before the phonograph, they hear only the "master's voice." And if they are to be called off from the purposes last dictated to them, it is the master who must call them off. Thus the foreign representative cannot hope that his words will make any impression on them. The most that he can hope is that they will be transmitted to those at the top, who are capable of changing the party line. But even those are not likely to be swayed by any normal logic in the words of the bourgeois representative. Since there can be no appeal to common purposes, there can be no appeal to common mental approaches.[31]

Ideologically cleaved from the West, the inhabitants of the Kremlin lived in a world that made sense to them, given their ideological and philosophical commitments. But lacking those commitments, outsiders often could not make sense of it. This incompatibility promised only frustration and mutual incomprehension for diplomats unaware of the ideological forces in play.

In *Education in a Divided World*, Conant explained that the problem went beyond "normal logic" having no purchase on the Soviet mind. It extended to the meanings of words used by both sides. Could it be, he asked, that Soviets and their advocates are simply lying when they describe their totalitarian system as "democratic" and point their fingers at the capitalist West for threatening human freedom? Or, as seemed more likely, had liberalism in the West and Soviet philosophy in the East fostered two competing "universes of discourse" in which common words "like 'freedom,' 'democracy,' even 'truth' and 'beauty' have entirely different overtones for the two groups"?[32] Conant found this situation especially alarming because he saw no pathway by which tensions might be reduced. He had hoped that international collaboration with the Russians to control and regulate atomic energy had "a real chance of uniting the world sufficiently to achieve a major degree of disarmament and the abolition of the present nightmare of a global war."³³ But as those hopes dissolved Conant saw the West facing an enemy captivated by ideology ("fanatic supporters of a philosophy based on Marx, Engels, and Lenin") who saw the world through Marxist lenses, recognized no validity in other points of view, and dogmatically believed that the wheels of history and the metaphysical forces of reality itself were "on their side."³⁴

Conant largely agreed with Kennan about what this situation required. Above all, both insisted, the nation must not panic and wage war on the Soviets. The tasks at hand were to refrain from exacerbating military tensions (both Kennan and Conant attempted to dissuade Truman from developing a new generation of more powerful atomic bombs), to recognize the powerful ideological dynamics in play, and then to educate Americans to understand and accept those civic responsibilities that could tip the geopolitical balance in the long run. Where Conant urged all Americans to get behind postwar reforms of public education and to make as great strides as possible in the direction of American democratic ideals, Kennan argued more broadly in his long telegram that the final outcome of the postwar conflict "depends on health and vigor of our own society":

> Every courageous and incisive measure to solve internal problems of our own society, to improve self confidence, discipline, morale and community spirit of our own people, is a diplomatic victory over Moscow worth a thousand diplomatic notes and joint communiqués.

These several points of alliance were not coincidental. Besides being fascinated in similar ways by the puzzles at hand of ideology and diplomacy, Conant and Kennan moved in the same diplomatic circles in 1946 and 1947. Conant quoted Kennan's celebrated "Mr. X" article when he toured Harvard alumni clubs to drum up support for Truman's Marshall Plan for rebuilding Europe, for example. And he consulted with Paul Nitze and other authors of NSC-68, the cold war blueprint calling for the "containment" of Soviet influence that was itself inspired partly by Kennan's long telegram.³⁵

Even commentators who disagreed sharply with Conant's and Kennan's recommendations accepted this picture of the Iron Curtain as an ideological and psychological border between two different worlds. In his books *The Struggle for the World* of 1947 and *The Coming Defeat of Communism* of 1950, the philosopher James Burnham argued that instead of attempting to contain and outlast Soviet Communism on the world stage, the United States and its allies must actively hasten its demise, even if that risked the cold war turning hot. Yet as he argued for a foreign policy very different from Kennan and Conant, Burnham similarly described Soviets and Westerners as if they lived in disconnected and irreconcilable mental worlds. When commenting on the first, carnival-like meetings with the Soviets at the fledgling United Nations, for instance, Burnham agreed that Western diplomats had trouble understanding the Soviet ambassador Andrei Gromyko because "these totalitarian movements are a species totally different from what we are accustomed to think of as 'political parties,'" As a result,

> When Byrnes and Cadogan and the others sit with Gromyko at the sessions of the Security Council, they are constantly puzzled by Gromyko's behavior; they find it "incomprehensible." It is, however, far more rational than their own. They are not aware that Gromyko sits there not because he has the slightest interest in solving fruitfully any problems of peace or prosperity, but precisely to aggravate these problems.

The fact that Russian motives were "incommensurate with the motives of a non-communist," meant that savvy diplomats such as Gromyko could deliberately and effectively flummox his counterparts, the American secretary of state James Byrnes and the British U.N. representative Alexander Cadogan. Instead of trying to figure the Russians out, Burnham proposed that the West turn the tables and take advantage of the ideological mechanisms in play: because communism "requires an absolute, a totalitarian hold on society, on minds and spirits as well as bodies and things," he argued, even a small propaganda victory engineered by Washington somewhere (almost anywhere) behind the Iron Curtain could have "a surprisingly extensive repercussion" and possibly begin to destabilize and destroy the whole Soviet system.[36]

Conant knew all about Gromyko's antics at the U.N. and alluded to them in *Education in a Divided World*:

> Without a better understanding of the way Russian rulers think—"how they are wired," as one American delegate who argued daily with them has said—without a better knowledge of Soviet philosophy and an accurate estimate of its hold on individuals, we are shadow-boxing in many areas.[37]

The delegate in question was likely General Frederick Osborn, whom Conant knew from both his war work and as a trustee of The Carnegie Corporation that had long supported Harvard. In 1947, Osborn headed the American delegation at the United Nations, observed the confounding behavior of the Soviets, and enlisted Conant as an outside expert to advise his delegation and hear his complaints about this puzzling behavior.[38] Osborn suggested to Conant that a "massive scholarly effort to probe Soviet behavior patterns" was required and Conant quite agreed. Through his Carnegie contacts, Osborn had money; as president of Harvard, Conant had an institution and intellectual talent. Together they created the Russian Research Center at Harvard to bring together experts in different areas—including psychology, anthropology, history, and sociology—for the task of better understanding the Soviets.

From its founding in 1948, fellows and associates at the center would closely analyze Russia in an effort to decipher the enigmatic Russian mind. They would study the "attitudes of Russians toward their homeland in relation to the rest of the world," their "attitudes toward Authority, Hierarchy, Suppression of Individual Freedom," and especially the distinctive mindset of that strange creature so vividly illustrated by Gromyko, "the Communist Bureaucrat."[39] A Carnegie Corporation document outlining future plans for the center cited Kennan's "Mr. X" article for inspiration and posed the kinds of questions Kennan and Conant had posed about how Soviet ideology affected language and communication.[40] What conditions must be met for the Soviets to appear intelligible and reasonable? What causes simple translations of language to fail, so that diplomats end up talking past each other?

Originally envisioned to exist for three years, the center became an enduring cold war institution. Scholars there and at Harvard's Department of Social Relations conducted research while government intelligence agencies—the CIA, in particular—welcomed the "new, high-powered source of experts and expertise to supplement and collaborate with their own efforts to understand their cold-war adversary." Coverage in *The Crimson* repeatedly cited director Clyde Kluckhohn's mission statement—to study "Russian institutions and behavior in an effort to determine the mainsprings of international actions and policy of the Soviet Union." Early publications of the center addressed aspects of the Soviet "national consciousness" and propaganda. A review of Alex Inkeles's book, *Public Opinion in Soviet Russia: A Study in Mass Persuasion*, appeared in the *New York Times Sunday Book Review* under the title "The Making of the Russian Mind." The accompanying photograph showed a handful Russian school children huddled around a composition book beneath a painting of Lenin teaching a schoolgirl to read. The caption read, "Russian school children study under the eye of Lenin."[41]

The new center embodied and institutionalized the emerging national consensus about the cold war and the roles scholars could play in addressing it intelligently and effectively. Conant had every reason to be enthused about it.[42] It joined his efforts to help diplomats and military planners to understand the nation's ideological enemy, his efforts to educate Americans about the nature of communism and the challenges it posed, and the new general education program in which science and scientific progress were seen as pillars of liberalism. For in teaching Americans how to understand the workings of science as a perpetual contest and succession of different conceptual schemes, he was teaching them how to understand precisely what had gone wrong in the Soviet Union: it had rejected the lessons of liberalism and pluralism by embracing a single, overarching philosophy that seemed to define and control the entire world of Soviet experience and understanding. The "intellectual humility" of history's most successful scientists, that is, was altogether missing in the Kremlin's fanatical devotion to its philosophy.

Especially for readers who lived through these geopolitical tensions, Kuhn's portrayal of different communities of scientists staring at each other across a revolutionary divide—their outlooks, perceptions, uses of language so different that they "will inevitably talk through each other when debating the relative merits of their respective paradigms"—was readily understandable through the geopolitical lenses that Conant, Kennan, and others had focused and polished. At times, Kuhn even used particular words and phrases common to this geopolitical discourse. Kennan remarked there were no "common purposes" and "common mental approaches" shared by the two sides and Burnham agreed by saying Russian motives were "incommensurate" to those in the West. Kuhn later wrote that advocates of different paradigms have "what we shall come to call their incommensurable ways of seeing the world and practicing science in it."[43] In a lecture of 1957, Kennan worried that conditions were growing worse as the mental world of Soviet leaders became more distant and peculiar:

> They view us as one might view the inhabitants of another planet through a very powerful telescope. Everything is visible; one sees in the greatest detail the strange beings of that other world going about their daily business; one can even discern the nature of their undertakings; but what one does not see and cannot see

is the motivation that drives them on their various pursuits. This remains concealed; and thus the entire image, clear and intelligible in detail, becomes incomprehensible in its totality.[44]

Kuhn used the same metaphor to describe the effects of paradigm shifts, writing "it is rather as if the professional community had been suddenly transported to another planet where familiar objects are seen a different light and are joined by unfamiliar ones as well." And when he wrote that revolutions "change the meaning of established and familiar concepts" and thus displace or reconfigure "the conceptual network through which scientists view the world,"[45] he responded in effect to Conant's and Osborn's call for scholars to better understand how leaders immersed within different ideologies "are wired" to think, speak, and behave by their training and education.

How did it come to be that a man who seemed to have turned away from the political interests of his youth, and who joined Conant's general education project as a scholar focused on historical and philosophical questions, wrote a book that so captured the geopolitical texture of its times? If there is a single, basic insight at the heart of *Structure*, it is that scientific and political revolutions can be understood in important ways as the same thing. This bridge connecting history and philosophy of science to Conant's geopolitics was supplied by Kuhn himself, who years after writing *Structure* would explain that the book was first sparked by a sudden, surprising, and revolutionary event in his own intellectual development. It raised for him a host of questions he would decide were ideological questions—not about the Russians and the mysteries of their peculiar mental world, but about Aristotle, one of history's most influential scientists. At that moment, on a warm summer day in 1947, Kuhn's narrow, scholarly interests in science and knowledge merged with this larger national conversation among Conant, Kennan, Burnham and many others about the power and fundamental importance of ideology in all human endeavors—not only geopolitics but the history of science.

Part II

"The Struggle for Men's Minds"

4

The Cold War Conversions of Thomas S. Kuhn and James Bryant Conant

On a hot summer day in the summer of 1947, Kuhn read Aristotle in his room in Kirkland Hall. Conant had asked him to read through texts in ancient physics with an eye for what problems, debates, or developments would make good case studies for the students in *Natural Sciences 4*. He cracked open a translation of Aristotle's *Physics* and began reading the great philosopher's dense and arcane theories of how nature works. He would have read passages such as,

> We must begin, then, as already said, with movement in general or progress from this to that. Now, some potentialities never exist apart, but always reveal themselves as actualized; others, while they are something actually, are capable of becoming something else than they are . . .

or,

> There must always be something that moves or changes, neither incidentally nor in the sense that some part of it moves, but in that it is in motion itself and directly.[1]

The text is called "physics," but it's nothing like the seventeenth-century physics of Isaac Newton taught in high schools and colleges, or the modern physics of Einstein. The only mathematics involved concerns proportionalities among speeds or other quantities, and there are no precise quantitative laws.

Even worse, the precise mathematical relationships one might extract from Aristotle's words are usually wrong. Where Newton's physics, for example, explains why heavy and light objects accelerate and move together under the influence of gravity, Aristotle believed that heavier objects will race ahead as they fall. This is because

> we see that of two bodies of similar formation the one that has the stronger trend downward by weight or upward by buoyancy, as the case may be, will be carried more quickly than the other through a given space in proportion to the greater strength of this trend.[2]

Aristotle was not wrong about everything. His biology and his logic, for example, remain instructive even today. But his physics was a bust.

Kuhn knew this, so he was not expecting any great enlightenment from Aristotle's text. As Conant's *On Understanding Science* suggests, there are good reasons why Newton's conceptual scheme replaced Aristotle's. It took almost two millennia for very smart people (such as Galileo) to diagnose these problems because science is not simply a matter of observing and measuring events, but doing so with the right ideas and concepts in mind. So it was that Aristotle's words seemed "full of egregious errors, both of logic and of observation," Kuhn later recalled. "Aristotle appeared not only ignorant of mechanics, but a dreadfully bad physical scientist as well."

And then, suddenly, everything changed:

> I was sitting at my desk with the text of Aristotle's Physics open in front of me with a four-color pencil in my hand. Looking up, I gazed abstractedly out the window of my room—the visual image is one I still retain. Suddenly the fragments in my head sorted themselves out in a new way, and fell into place together. My jaw dropped, for all at once Aristotle seemed a very good physicist indeed, but of a sort I'd never dreamed possible.

Aristotle's physics made sense after all, and it was no longer puzzling why his way of viewing the world held sway for centuries. "Statements that had previously seemed egregious mistakes, now seemed at worst near misses within a powerful and generally successful tradition."[3]

Was this some heat-induced reverie or daydream, Kuhn must have wondered? No, the sense and clarity he grasped did not disappear. New insights and connections among Aristotle's ideas kept unfolding. His concept of motion, for instance, meant more than change of place; it encompassed other kinds of change as well, like biological maturation and growth. And unlike Newton's physics in which masses and momentums attach to physical bodies independently of their sensuous qualities, the trees and falling rocks that Aristotle described possessed "forms" that could not be ignored by anyone seeking to understand nature.[4]

Were Aristotle's detractors just ignorant of the ways this made sense? Kuhn had to wonder what Conant would say about it. In *On Understanding Science*, Conant wrote confidently that scientific knowledge "has indeed advanced in many areas in the last three hundred years."[5] But his assurances were vague. What *kinds* of advancement? Relative *to what* have advances been made? Conant could not answer questions such as these abstractly because conceptual schemes change over time. One scientist's breakthrough, achieved within a moribund conceptual scheme, could be a puzzling failure to another. Conant knew these questions were complicated and required careful philosophical analysis. As a scientist, however, he had no doubt that scientific progress was real and objective. And he had a way to prove it:

> Bring back to life the great figures of the past who were identified with the subjects in question. Ask them to view the present scene and answer whether or not in their opinion there has been an advance. No one can doubt how Galileo, Newton, Harvey, or the pioneers in anthropology and archaeology would respond.

Science has *obviously* made progress and these resuscitated scientists would not hesitate to say so. For poets, artists, and philosophers, on the other hand, the outcome would be different. If the experiment involved zombie versions of artists such as Michelangelo or Rembrandt, writers such as Dante or Keats, or philosophers such as Spinoza or Kant, there would be no agreement. "We might argue all day," Conant wrote, about whether or not any of these figures would be impressed by the subsequent developments in their fields.[6]

This difference was crucial and Conant never hesitated to point it out: science improved human understanding by *adding* to and *building upon* existing knowledge. He was so confident of this, he presented this thought experiment over and over in his writings from the 1940s, '50s, and '60s.[7] And it holds pride of place in the redbook report that laid the groundwork for Kuhn's new career as a historian of science. Conant's "Objectives" committee for general education confidently noted "the accumulated fact and theory which science has inherited from the past" as one reason why students in America must understand how science works.[8]

Slack-jawed at his desk in Kirkland hall, however, Kuhn was not so sure. No, he had not brought Aristotle back from the dead to test Conant's claim. But he had instead gone back in time to see nature through Aristotle's eyes and think about it with his categories and presuppositions. From that point of view, he did not instantly recognize Newton's or Einstein's innovations as advances or accumulations of knowledge. They were different, but not necessarily larger or better. He had to wonder therefore whether Conant had it right. Would great scientists of the past quickly recognize today's science as progressive and fruitful? Would they even *understand* it? Or, like most historians of science—and Kuhn himself, up to that revelatory moment—would they find different conceptual schemes to be full of mistakes and misconceptions, possibly even nonsense?

"The Aristotle Experience" and the Conversions of James B. Conant

In publications and lectures from the 1950s, '60s, and '70s, Kuhn mentioned this encounter with Aristotle as a central, pivotal event in his intellectual development. He described it as a "shocking" experience, sometimes as a "revelation." In his notes from a 1957 lecture at Berkeley, he wrote, "Nothing in my physics education or my philosophy reading prepared me for the way science looks when viewed through the writings of dead scientists."[9]

Like his crisis and conversion from pacifist to interventionist, the Aristotle experience transformed him. He was now a young historian of science with a powerful insight about scientific knowledge that he would not fully articulate until he wrote *The Structure of Scientific Revolutions* nearly fifteen years later.[10] While the experience prepared him to reject Conant's understanding of scientific progress and develop his own, distinctive picture of science's history, the transformation wrought by reading Aristotle hardly alienated Kuhn from Conant and other intellectuals in these early years of the cold war. Crises and sudden conversions like these were familiar, recurring features of political and intellectual life in the United States.

For example, Kuhn had seen the convulsions among progressive educators and pacifists caused by the Nazi-Soviet nonaggression pact in 1939. Once he got to Harvard and paid close attention to Conant, he saw that

Conant's deepest convictions were sometimes upended and overturned, as well. Like Kuhn, Conant was first opposed to American intervention in the European war, so much so that he publicly denounced a poem written by his friend Archibald MacLeish. "Speech to the Scholars" appeared in the June 12, 1937 issue of *Saturday Review of Literature*. It began,

> O scholars schooled upon the books
> O skilled readers of the page
> Rise from your labor now! Enlist
> For warfare in this fighting age
> No longer may your learning wear
> The neutral truth's dispassionate peace[11]

The poem pressed Conant's liberal buttons in the wrong way. In private correspondence with MacLeish, and even in his baccalaureate sermon to the graduating class that year, Conant called the poem a "betrayal of what is important in the life of a university." No matter how ominous the threats, he believed, the advancement of knowledge requires scholars to resist and reject a warrior mentality. For if they take up arms they risk becoming like their enemies. It is altogether too easy, Conant told MacLeish, "to lose the very things one wishes to preserve by declaring war in favor of them." A scholar must instead stay above the fray and "let the weary world fight around him."[12]

Conant opposed intervention as late as 1938 when Hitler marched into Austria. That August, his sentiments were not too different from Neville Chamberlain's policy of "peace at any price." After *Kristallnacht*'s pogroms against Jews in Germany, he reluctantly agreed that Harvard should follow other institutions to create fellowships and scholarships for displaced Jewish scholars. But only after Hitler's invasion of France, the Netherlands, and Norway in the spring of 1940 did Conant revise his position and come to agree that Hitler must be actively opposed. That July he wrote in his diary, "We all live to change our minds in 1940."[13]

Conant changed his mind often. Early in the Manhattan project, for example, he was skeptical for scientific reasons that an atomic bomb could be created. But his doubts "evaporated" in mid-conversation with one of his colleagues in the project "and gave way to the 'rather rugged optimism' beginning to take hold among Allied scientists."[14] Once the bomb was dropped, Conant switched from supporting postwar world government—the only reasonable way to prevent misuse of the new atomic weapons, many believed—to opposing world government. After helping to develop the first generation atom bomb, he not only opposed but actively campaigned against calls for the United States to develop the new, more powerful hydrogen bomb.[15]

To his biographer James Hershberg, Conant's transformations are "a window on many of the revolutionary transformations in recent American history caused by World War II and the Cold War."[16] But Conant's—as well as Kuhn's—crises, conversions, and reversals also point to earlier centuries. Beginning with the Puritans and Pilgrims of the seventeenth century, waves of popular religious awakenings and the almost perpetual formation and dissolution of sects illustrate the nation's willingness to reject orthodoxy and embrace new forms of life and belief.[17] As a major American university, Conant's Harvard inherited this history and the proud, devotional sensibilities it illustrates; but Harvard also helped to rationalize these dynamics and to defend them as socially and intellectually productive. Radicals of almost any sort can find encouragement in Ralph Waldo Emerson's classic denouncement of conformity in his famous essay "On Self Reliance."

For understanding Kuhn's and Conant's twentieth-century conversions, however, a more immediate inspiration is William James and his pragmatic theories of truth and transformative mystical experiences. Kuhn's Aristotle Experience had each of the four defining characteristics James offered in his *Varieties of Religious Experience*.[18] It was *transient*, lasting only moments, but it was *noetic* and filled with lasting import and significance. It was *passive* insofar as it struck Kuhn unannounced; and finally it was *ineffable*—at least at first. As Kuhn said, nothing in his education had prepared him to understand this revelation, so he would have to educate himself in order to make sense of it.

From Scholar to Cold Warrior

Both Kuhn and Conant were drawn to these dynamics of personal conversion. As Kuhn had described it in *The Crimson*, Conant's militarization of the campus was a total "conversion" in educational methods and priorities. Conant used this word himself three years later when, at convocation in September 1945, he called for a second conversion to reverse the effects of the first. The "moral imperatives of the battlefield must be transformed into those of a free society which believes in the supreme significance of each individual man or woman," he announced. This was not only a "readjustment of our economic life," he explained, but a deep and fundamental "psychological conversion" that all undergrads would undergo.[19]

The conversion in Conant's life that perhaps comes closest to Kuhn's momentous encounter with Aristotle was his becoming a leading cold warrior in the late 1940s. Just as Kuhn's experience led him in a new professional and intellectual direction by abandoning his former understanding of science, Conant came to renounce principles and values he had long defended. In the 1930s, for example, he wrestled with the same reactionary currents and calls for loyalty oaths that bedeviled progressive educators like George S. Counts and Elizabeth Moos. Along with the presidents of Radcliffe, MIT, Amherst, and other colleges in Massachusetts, he stood against legislation proposed in 1935 that would require that all teachers in Massachusetts— even at private institutions like Harvard—to pledge loyalty to the nation. The legislation was driven by popular fears of Nazism in American schools that Conant and his colleagues saw as overblown and unrealistic. It seemed highly unlikely, they wrote jointly, that any "disloyal plotter or seditious conspirator" could be found "in our whole teaching force." If there were, they added, no mere oath would prevent any subversion or conspiracy.[20] The real issue for Conant was not how many subversives there might be; it was a matter of principle. Scholarship and inquiry must have the freedom necessary to move knowledge and society forward.

Conant's firm stance won praise and support from students and faculty. In *Atlantic Monthly*, he took the case to the nation in an article titled "Free Inquiry or Dogma" in which he bridled at being "pestered from the reactionaries to take a strong stand against 'the dangerous radicals' in the universities, and particularly at Harvard." These reactionaries had things just backward: as a crucible of liberal learning, a university *should* and *must* be "an arena for combat" among opposing arguments and perspectives:

> Our colleges and universities must not only guarantee the right of free inquiry, they must also see that the various points of view are represented so that a conflict of opinion really takes place. From such clashes fly the sparks that ignite the enthusiasm in the students which drives them seriously to examine the questions raised. We must have our share of thoughtful rebels on our faculties.[21]

By the fall of 1935, however, Conant lost this battle and the bill was approved. Hardly one to lick his wounds, a very different feature of Conant's personality came to the fore: his respect for unity and social consensus. He then sent letters to his faculty and to *The Crimson* admitting that the new law was an "unnecessary and unwise piece of legislation." But he refused those members of his faculty who called for the university to disobey for those very reasons. Conant took the noxious oath and expressed his hope that "all members of the various faculties of this University will do likewise."[22] In weighing the principles of free inquiry against the legal obligations of the university to the state in which it operated, Conant decided that this was a battle he would no longer fight.

The next time Conant wrestled with public concerns about the loyalty of Harvard's faculty, everything had changed. As 1950 approached, the ideological stalemate he described in *Education in a Divided World* became a military stalemate between superpowers armed not only with ideologies but bombs, armies, and clever spies. Again, Conant at first resisted. As late as 1948, when the first waves of anticommunist hysteria rippled across the nation, he remained skeptical about the growing fear of the red menace. He believed that State Department officer Alger Hiss, sensationally accused by Whittaker Chambers of being a former member of the communist underground, was most likely "an innocent victim of a vicious Red hunt." Conant especially feared what effects the headline-making case might have and what precedent it would set should Chambers's accusations turn Washington upside down. With so many officials having flirted with communism in their past, Conant seemed to reason, reactionaries like Richard Nixon could tear the government apart and, ironically, validate communist claims that American democracy was inherently fragile and unstable.[23]

One evening in September, Conant was visited by his friend, the lawyer William Marbury. Marbury knew Hiss personally and had represented him a month before when Hiss first answered Chambers's accusations before the House Un-American Activities Committee. Conant proposed to Marbury that Hiss should sue Chambers for libel. (That fall, William Remington would adopt the same tactic and sue his accuser, Elizabeth Bentley.) Marbury took Conant's proposal to Hiss and other attorneys, most of whom believed it was a very bad idea. Perhaps because Marbury thought the world of Conant's intellect and wisdom, Marbury and Hiss nevertheless proceeded to sue Chambers. Marbury sent Conant a copy of the complaint and acknowledged "a hard road ahead of us."[24]

Remington's suit against the Red Spy Queen worked. It helped him fend off his prosecutors for a year or two. But Hiss's suit backfired. It opened the door to depositions through which Nixon found inconsistencies in Hiss's statements. During these depositions, Chambers revealed his smoking gun: microfilm copies of State Department documents and notes that Hiss, he claimed, had given him ten years before. As a security precaution, Chambers said, he hid the film inside a pumpkin at his farmhouse in Westminster, Maryland. The "pumpkin papers" finally refuted Hiss's suspiciously evolving denials: first, that he did not know Chambers, then that he had known Chambers but under the name George Crosley, and finally that he knew him well enough to have once given him a car as a gift. By the end of the year, Hiss was indicted on two counts of perjury.

When Marbury saw the pumpkin papers and recognized Hiss's handwriting, "he suffered a shock that never quite wore off." Conant was shocked as well and later cited the Hiss case as a personal turning point. Suddenly, he realized, claims about spies or secret communists in government might be true. If Hiss was a communist spy, then there might be many others. And if that was true, the new cold war was likely to be much more—and more dangerous—than the ideological contest he described in *Education in a Divided World*.

Confirming evidence for this sinister worldview arrived rapidly. In September 1949, while traveling to San Francisco for speaking engagements, Conant received an urgent call from Washington. The message he heard on the other end was coded for just this situation, should it arise: "They have it." *They* were the Soviets, and *it* was a working atomic bomb. Again, Conant was shocked. He had just congratulated General Groves for winning a friendly bet about how long it would take the Russians to build the bomb—about twenty years, Groves had said. Until that moment, it looked like Groves would be right. Conant's most recent estimate had been five to fifteen years, and he staked a lot on it.[25] He had told President Truman that there would be ample time before the Russians had their own bomb. The time could be used to implement wise policies and to create necessary institutions to maintain stability and prevent the nightmare Conant feared most: "The age of the Superblitz," he called it, when nations possessed enough nuclear weapons to fight wars from the skies and target major cities. "They have it" scuttled Conant's hopes and plans.

Then bad news arrived from China, where Chiang Kai-shek and his nationalist forces fighting China's civil war retreated to Taiwan and left China in control of Mao Zedong and his fellow Communists. By the end of the year, China was "lost to communism" and at the end of January 1950, Truman decided to propel the arms race by developing the new "super" bomb. In a report he co-authored with Caltech president Lee Dubridge, Conant argued strenuously that such a bomb was in a category by itself and not merely a more powerful version of what was dropped on Hiroshima and Nagasaki. But Truman was unimpressed and took one step closer to the ominous age of the superblitz.[26]

In three days, on the heels of Truman's decision to pursue the superbomb, came this revelation from the *New York Times*:

> In London on Friday, one of the highest ranking British atomic physicists, Dr. Klaus Fuchs, made a strange and disturbing confession. He had betrayed important U.S. and British technical secrets about atomic energy to Soviet Russia. The knowledge he transmitted may have helped the Russians build their first fission bomb in a shorter time than the Western Governments had expected. The shock of the Fuchs confession was great.[27]

Conant knew Fuchs. He had worked as a physicist at Los Alamos under Conant's administration. When this telephone call came, a student in Conant's office watched him put the phone down and turn pale. "That man knew everything," Conant repeated, over and over, his mind clicking through the implications of what this betrayal had meant for the past and what it could mean for the future.[28]

The crescendo of alarming news heralded the era of Joseph McCarthy. A week later, the senator stole headlines with his speech at Wheeling. The Truman administration had been asleep at the helm, McCarthy bellowed, and the nation's capital was now crawling with subversives plotting a communist, atheist, collectivist, authoritarian, totalitarian revolution. Conant did not take McCarthy's claims at face value. They were hyperbolic and exaggerated for effect and, as he would observe later, after his face-to-face encounter, McCarthy was concerned less with the threat of communism than with what it could do for his political career. But he agreed, at least, that there might well be well-placed spies in Washington who could tip the balance in what was surely no longer merely an ideological contest. In June, Korean forces invaded the southern Korean peninsula, followed by Chinese forces in October. This drew the United States back into armed conflict only five years after the largest war in history had been concluded.

"The outbreak of the Korean war consummated Conant's conversion to cold warrior," Hershberg explains. What had been "a political and ideological struggle whose outcome would be determined by each system's economic and social vitality" became "a military confrontation whose outcome would depend on relative armed might and the willingness to use it."[29] The specter of the superblitz still had to be avoided. But the only way to do that now, Conant reasoned, was to rapidly build up the nation's military strength and readiness.

The Committee on the Present Danger

In August, Conant was summering near the White Mountains in New Hampshire. An evening walk took him to the home of his Harvard friend Richard Ammi Cutter who was then hosting the former undersecretary of the army Tracy Vorhees. In conversation, Vorhees insisted that the United States and especially Europe were susceptible to Soviet invasion, that Europe could easily be consumed by war like Korea. "I was all ears that night," Conant later recalled, as he learned from Vorhees and Cutter how easily Stalin could march across Europe to the English Channel. "If they were only half right," Conant wrote, "the nation was once again, as in June 1940, in extreme danger."[30]

Recalling how the White committee, some dozen years before, helped to persuade the nation to see Europe's interests as vital to its own, Conant suggested that Cutter and Vorhees adopt a similar plan:

> "Get a group of distinguished citizens together," I said, "draw up a program, put it before the public, get people to write to congress and, in general respond to the situation. From what I have just heard, I judge the country is asleep. You should wake it up."

"Would you be one of the leaders in such a Committee?" Vorhees asked. A case of peritonitis delayed Conant's answer, but he later agreed to help lead the newly formed committee and rally the nation toward an aggressive, anticommunist consensus backed by military force.[31]

The Committee on the Present Danger introduced itself at a press conference in Washington that December. It warned that unless the nation stepped up to meet the danger of immanent Soviet invasion,

> the time may soon come when all of Continental Europe can be forced into the Communist fold, and the British Isles placed

again in even graver peril than in 1940, at sacrifices of blood and wealth that the Kremlin would accept. No scruples of conscience will stand in their way.

The public was told that these warnings came from a group of concerned citizens. But Conant and his fellow committee members were well connected to military and government officers and privy to top secret information. In March, Conant was asked to read and comment on the report that would become NSC-68, the official blueprint for the State Department's policy of "containment" through the cold war. That policy, the authors of NSC-68 knew, would require public support and some had already considered Conant to be the perfect man to build public consensus.[32]

With Conant at the helm, a host of prominent Americans signed on to support the new committee—college presidents and trustees, former government officers, leaders of business and industry, journalists and movie stars (including Edward R. Murrow and the president of the Screen Actors Guild, Ronald Reagan). *The New York Times* praised it as a sane antidote to McCarthy's red-scare hysteria about communists in government and the media. The committee, that is, "puts the emphasis of its anticommunism where it primarily belongs—on the dangers of Russian and Communist imperialism."[33] For the next two years, Conant and his committee conducted a comprehensive publicity campaign. They made speeches, wrote articles, appeared before government committees, and worked with other organizations to promote greatly increased military spending, economic assistance to noncommunist nations, and to create outposts for U.S. troops and weapons around the world. To help meet the personnel requirement for all of this, Conant once again volunteered the nation's college-age students for duty by calling for mandatory military training and service.[34]

The Problem of Communist Teachers

Conant's cold war conversion also reversed his long-standing defense of absolute, exceptionless academic freedom. As late as 1947, for example, just before these tumultuous developments, Conant had once again stared down reactionary state legislators eager to purge the state of dangerous ideological influences. Attorney General Clarence Barnes sponsored a bill to ban from employment any person "who is a member of the communist party or who by speaking or writing advocates its doctrines."[35] Conant rallied friends and colleagues for an organized counterattack on the bill. In his year-end president's report, he insisted there could be "no compromise" on the matter of academic freedom, because any university is founded on a "charter of free inquiry."[36] More than a matter of principle, this was also proper strategy in the postwar ideological contest with communist ideology, Conant explained. As he soon put it in *Education in a Divided World*, public panic and hurried legislation only encouraged the communist agenda. Were he a communist at Stalin's side in the Kremlin, he wrote,

> I should eagerly await news that indicated that the American people had succumbed to panic, that they had lost confidence in those historical principles which had guided their development in the past.

Even full-blown Communists and party members should be allowed to teach in universities, Conant told *The Boston Globe*, if only because "you can't kill an idea by making martyrs of its disciples."[37]

Harvard prevailed against the Barnes bill, but public universities beholden to elected legislators had fewer defenses. In March 1949, the regents of the University of California decided that all faculty must take an oath that read:

> I hereby formally acknowledge my acceptance of the position and salary named, and also state that I am not a member of the Communist Party or any other organization which advocates the overthrow of the Government by force or violence, and that I have no commitments in conflict with my responsibilities with respect to impartial scholarship and free pursuit of truth. I understand that the foregoing statement is a condition of my employment and a consideration of payment of my salary.

The last sentence was a surprise. When faculty learned that the oath held their paychecks hostage, they concluded that the regents had gone too far. Many refused to sign and the ensuing battles among the faculty, administration, and state legislators lasted until 1952, when the courts finally determined that university policy was unconstitutional.

At first, Conant supported Robert Gordon Sproul, head of the California university system, and sent him an uplifting telegram. In the fight for academic freedom, he wrote, "you are the commanding general and we are looking to you for victory."³⁸ But Conant would soon reverse himself and agree that members of the Communist Party must not be allowed to teach in the nation's universities. His turnaround began when the National Education Association met in the fall of 1948 to assess the mounting controversy. Conant encouraged the Association's Education Policy Commission to make a tactical concession to the anticommunist mood, to "throw a bone to the anticommunists," as Hershberg describes it. It would shore up the teaching profession against the threat of state or federal takeover, Conant argued, and allow the profession to more effectively defend the majority of liberal, left-leaning scholars and teachers who did not belong to the party. "A lot of people would say that I sold myself down the road to the reactionaries," Conant later recalled; that he was betraying his principles and "crossing a dangerous line." But over the course of the meeting that day Conant prevailed—as he usually did—and the commission came to see his strategic wisdom and the helpful wiggle room embedded inside the verb *employ* ("I said not to employ, not anything about discharging" Communist faculty, he pointed out during their discussion).³⁹

The commission published its position in a pamphlet the following June. Conant himself went public at the same time. In a speech to alumni and administration, he couched his new position inside the familiar language he had used before to defend academic freedom:

> I maintain that a professor's political views, social philosophy or religion are of no concern in the University; nor are his activities within the law as a private citizen. I do not have to remind this audience that this is the traditional Harvard position and will be maintained in the face of whatever criticisms may come.

There was just one "single exception," he explained:

> In this period of a cold war, I do not believe the usual rules as to political parties apply to the Communist party. I am convinced that conspiracy and calculated deceit have been and are the characteristic pattern of behavior of regular Communists all over the world. For these reasons, as far as I am concerned, card holding members of the Communist party are out of bounds as members of the teaching profession. I should not want to be a party to the appointment of such a person to a teaching position with tenure in any educational institution.

Conant still opposed witch hunts, on-campus investigations, and loyalty oaths. He promised that "as long as I am President of the University, I can assure you there will be no policy of inquiry into the political views of the members of the staff and no watching over their activities as citizens."[40] But he had crossed a Rubicon by relaxing his principled stand for academic freedom to accommodate the nation's anticommunist consensus.

∼

Scientific revolutions are conversion experiences, Kuhn would later claim. Before and during his collaboration with Conant, he witnessed many. There was his own, over pacifism and intervention, and then his shocking experience reading Aristotle. There were also Conant's radical changes of mind and commitment—over intervention, over the very nature of the new cold war confrontation, and his momentous shift from defending academic freedom as an almost sacred principle to an official anticommunist policy for professors at the nation's leading university.

Conant had not simply adopted positions where he had none before; he converted from one position to another that was incompatible it, and he did so to increase national security and advance the causes of the West in the new cold war. With intellectual freedom and the progress of science fixed at the center of his struggle, it would have been easy for Kuhn to see two points that his *Structure of Scientific Revolutions* would later emphasize: progress occurs discontinuously in fits and starts, and it does so because of the ubiquitous power of ideology. The man teaching Kuhn how to understand the history of science was not just telling him and the rest of the nation that ideology can captivate and control the mind, or that it had already done so when it came to Communist faculty. He was teaching by example, updating William James's fascination with life-transforming conversions, and universalizing the transformative powers of ideology that drove the ongoing struggle for men's minds. It was not only Communists, in other words, whose beliefs and thinking were shaped by the revolutionary ideology they embraced, but intelligent liberal intellectuals—such as Conant, and Kuhn himself with his crisis over interventionism and his Aristotle experience—who grow and learn by embracing competing, even irreconcilable, points of view.

When he warned MacLeish that it was "easy to lose the very things one wishes to preserve by declaring war in favor of them," Conant could have been warning himself. He later came to regret how anticommunism blossomed into McCarthy's "big lie" and severely undermined the nation's ideals of intellectual and political freedom. In a way, his warning also applied to his relationship with Kuhn. Ten years later, when he first read Kuhn's new manuscript, *The Structure of Scientific Revolutions*, Conant recoiled upon seeing that these dynamics of psychological and political conversion had taken center stage in Kuhn's theorizing, that the power of ideology to unify groups of scientists had crowded out other, essential features of science and its history. Even in the scholarly realm of history of science, that is, some things Conant wished to preserve had been lost when he declared war in order to preserve them.

5

Sidney Hook and the Anticommunist Inquisition

On Friday March 25, 1949, the philosopher Sidney Hook was out to make news. Widely known as a professor at New York University and an authority on the dangers of communist ideology, Hook was ready to confront and expose the latest invasion of communism on the mainland United States. The crucial documents and phone numbers he would need were in his attaché. The reporter he had summoned from the *New York Herald Tribune* was by his side and ready for the scoop. It was nearing 6:30 in the evening. Everything was in place.

They walked briskly to 301 Park Avenue, but there was no need to check the address. The sidewalk in front was crowded with protestors and police. Hook's pulse quickened as he pushed through the crowd. He sympathized with the protestors, but he was taking his fight inside, directly to the man who orchestrated this invasion and was now commanding the enemy from within the opulent Waldorf Astoria Hotel.

They entered the lobby, took the up-elevator, and walked down the hallway. Hook scanned room numbers left and right as he got closer until . . . he could hear voices inside. He leaned in. A man's voice. That must be him, but there were others. Another man, it sounded like. And one, at least two, women. Wives? Probably. The clink of ice and glass? Russian vodka, of course. Hook looked at his watch. They would be on their way downstairs soon.

Hook nodded to his reporter, took a deep breath, and furiously rapped on the door. Too impatient to wait for an answer, he pushed it open and stepped in to confront Harvard professor of astronomy Harlow Shapley and his wife, dressed in formal attire. With them were Mrs. and Mr. Martin Popper, an attorney who had recently represented Hollywood screenwriters, the famous "Hollywood Ten," called before congress. Shapley jumped to his feet as Hook plopped down a letter he had written to Shapley on the coffee table. "There!" he said, presenting the proof, the final line in the argument he had been waging against Shapley for the past month: Shapley and his associates were up to no good. And Hook's reporter was there to tell the world all about it.

༄

For weeks before this moment, Hook had been writing to Shapley, demanding to speak at his Cultural and Scientific Conference for World Peace—the conference now set to commence in the ballroom downstairs. Shapley organized the event under the auspices of an organization called the National Council of the Arts, Sciences, and Professions. He invited luminaries in art, literature, and science from around the nation and the world—even a handful from Soviet controlled Eastern Europe and Russia itself (notably, composer Dmitri Shostakovich and poet and writer A. A. Fadayev)—to meet and discuss the world situation.

The politics of the conference were clear: Shapley and his conferees believed that American foreign policy after the war had become dangerously belligerent and mistrusting toward the Soviet Union. Truman was building more bombs, placing military bases around Europe, pushing for an anti-Soviet military alliance among nations (that would become NATO), and carrying on as if a final—almost certainly nuclear—showdown between America and the Soviet Union was inevitable. The conference would declare to the world that this aggressive posturing was not supported by everyone. These artists, scientists, philosophers, poets, and painters stood for mutual respect and above all peace between the new postwar superpowers.[1]

Hook was buying none of this. To him, Shapley's grand gesture was just another public-relations success for Moscow. Yes, it was wrapped up in liberal values and good, "peaceful" intentions. But behind the scenes he was sure it was being controlled by Communists, ultimately by the Kremlin itself, trying to soften up the people of the United States and weaken their defenses against communism. The conference had to be exposed and, if possible, subverted so that its message would backfire. As soon as he had heard about the event, Hook went into action—visiting, calling, and writing to participants to warn them that the conference "perpetuates a fraud upon the American people."[2]

Unfortunately for Hook, Shapley had a big head start. Not unlike Conant, he was a widely respected scientist, the head of the Harvard Observatory, who had been in the public eye for decades. He helped to found Science Service, the organization whose newsletters provided up-to-date scientific information to the public and the press, and just years before this conference he lobbied congress to create UNESCO, the United Nations Educational, Scientific, and Cultural Organization. With a Rolodex like Conant's, he had enlisted an impressive array of intellectuals and artists who would come to New York and public sponsors who would lend their names in support, such as Thomas Mann, Marlon Brando, Charlie Chaplin, Dashiell Hammett, Lillian Hellman, Arthur Miller, Eugene Ormandy, Ad Reinhardt, Paul Robeson, Henry Wallace, Frank Lloyd Wright and many others.

Hook counted some 650, most of whom he took to be true-believing communists or "hardened fellow travellers" who would probably not even take a phone call from an anticommunist like himself. But, some of these—about 150, he reasoned—seemed like they could be turned.[3] They were simply ignorant of the true nature of the event or were perhaps fooled by Shapley's overtures. Or, like an alarming number of Americans, Hook believed, they may have succumbed to the corrupting mental effects of communist ideology and had lost perspective and good judgment. Hook would help them understand what they had gotten themselves into.

First, he wrote to Shapley asking to participate. He wanted a prominent role, a slot at the plenary session of the conference to read a paper on "Science, Culture, and Peace." Hook's zeal and single-minded fixation on telling the world "that Stalin stinks" was well known and not always welcome, even among his fellow anticommunist intellectuals in New York.[4] So Shapley probably knew that were he allowed to speak, Hook would attack the conference itself, denounce the organizers as either malicious partisans or unwitting dupes of Stalin, and place the blame for the world's anxieties entirely on the international communist conspiracy.

Shapley didn't respond, and Hook grew more agitated as the days went by. Three weeks before the congress was to begin, he began writing to these 150 supporters to inform them that the event was a Communist front, and to implore them to turn on Shapley and take back their names and their honor. He asked a few to write to Shapley directly, to support Hook's request to address the conference, and thus help probe "the controlling group behind the conference." He asked Algernon Black, head of New York's Ethical Culture Society, to "write air-mail or wire Professor Harlow Shapley of Harvard demanding that a place be made for me on the program, and upon refusal resign with a ringing declaration."[5] Shapley's refusal, Hook figured, would vividly prove Hook's point.

But there was no mass defection from Shapley's event. Black stood firm and his name continued to appear on the official list of conference sponsors.[6] So did Kuhn's future editor, the University of Chicago philosopher Rudolf Carnap, who refused Hook's appeal because "it seems to me of the utmost importance at the present time that demonstrations of the will for peace are made."[7]

Finally, Hook heard back from the conference secretary that there was no extra room on the program and it was too late now to rearrange things for Hook.[8] Shapley was probably trying to avoid a direct confrontation and may or may not have been telling the truth. But by denying his demand to speak, Shapley was hardly censoring Hook or preventing his point of view from being heard. By this point, the news was out that Hook had organized a conference of his own, a counterconference to compete with Shapley's. It too was sponsored by a civic group, one called Americans for Intellectual Freedom that Hook formed for the occasion with his colleague George S. Counts—the ex-communist who, a dozen years before, had inspired Kuhn's teacher Elizabeth Moos to radicalize her school for pacifism and socialism. Like Hook, Counts was now a zealous convert to anticommunism and would help Hook steal Shapley's thunder by holding a one-day conference at Freedom House, just a few blocks away from the Waldorf.[9]

As if to provoke Shapley as much as possible, Hook rented a room in the Waldorf from which he planned his strategies, manned the telephones, and sent out invitations, press releases, and telegrams about his upcoming event. All the while, he lobbied for a place on Shapley's program. His break came when a few members of Shapley's organizing committee seemed sympathetic to his complaints. Dr. Herbert Davis, president of Smith College and Guy Emery Shipler, editor of *The Churchman* magazine, agreed to write to Shapley in support of Hook's request to speak. They sent Hook carbon copies of their letters, so if Shapley did not give in, Hook had all the evidence he needed to show that Shapley was so enmeshed in the Kremlin's conspiracy he was ignoring the members of his own program committee. Hook laid out this escalated charge to the University of Chicago physiologist A. J. Carlson, whose eminence and authority would pressure Shapley even more.[10]

When Carlson contacted Shapley about Hook's behind-the-scenes campaign, Shapley could no longer ignore it. On March 18, Shapley wrote to Hook:

> Several of the sponsors and participants in our coming Cultural and Scientific Conference for World Peace have sent me the let-

ters you have written them. In some of these communications I regret to see that you have made plain mis-statements of fact. For example, neither President Davis or Dr. Shipler have written me any instruction or request or communication whatever with regard to your demand that you be placed on the conference program.

Was Shapley telling the truth? In the rush of final preparations, had he forgotten about the letters Shipler and Davis had sent, had he never seen them, or did they get lost in the mail?[11] In any case, the battle escalated. The same day Hook received Shapley's note, he received a phone call from a reporter from the *New York Times* asking Hook to comment on Shapley's claim that neither Davis or Shipler had weighed in on Hook's behalf.[12] Shapley had gone public, that is, and now it looked to all the world that Sidney Hook, one of the nation's best-known professional philosophers, was scheming and lying.

March 21. Hook wrote to Shapley to demand an apology for impugning Hook's integrity. He received no response.

March 23. Another letter. Again, no response.

Now it was Friday, March 25. The conference was at hand and Hook felt he had no choice but to confront Shapley directly in his hotel room with the two letters from Davis and Shipler to prove to the world that *Shapley*—not Hook—had lied.

"There!" Hook slapped down his document. "Please excuse my visit," he said,

> but I have been unable to reach you in any other way since Monday. You have impugned my integrity, Dr. Shapley, by denying that my offer to read a paper at the conference was ever accepted by any members of the program committee. You have specifically claimed that President Davis and Dr. Shipler never wrote to you. Well here's the evidence.[13]

Hook's document quoted the letters Davis and Shipler had written to Shapley, and it scolded Shapley for thinking he could outwit Sidney Hook:

> What am I to think, Dr. Shapley? As an astronomer you know that the probability of three independently written letters,

mailed at different times about the same subject to the same person, being lost in the mails is indeterminably, if not infinitely, small! At the same time I am loath to conclude that you are deliberately telling an untruth although you have not scrupled to slander my integrity in charging me with plain and persistent misstatement of fact.

There is only one alternative explanation which plausibly accounts for your conduct, unexampled in scientific and academic circles. It is that you do not know what is going on within your own organization, that the inner Communist group are deceiving you to the point of suppressing letters sent to you which do not come to your attention, and that you have become the willing dupe of enemies of your own country as well as the ideals of moral decency and scientific truth you profess.[14]

Hook found no Communists hiding in Shapley's hotel room closet or under his bed, and none were later found by historians examining this episode. But Hook was sure nonetheless that Shapley and his colleagues in the National Council of Arts, Sciences, and Professions were controlled by some "inner Communist group."[15]

From Shapley's point of view, it was Hook who seemed fanatical and mentally possessed. He had an unblinking certainty that anyone publicly expressing pacifism or dissenting from the wisdom of U.S. foreign policy was in league with, or being manipulated by, communist agents or ideology. His determination to disrupt Shapley's conference made him seem unhinged and out of touch with his own behavior. Hook called it "my offer to read a paper," but it was plainly a demand ("Will you please let me know the time and place of the plenary session at which my address will be scheduled?" he wrote in one of his many letters to Shapley). Carnap, for his part, accused Hook of peddling "hysterical" and "gross exaggerations" about the threat of communism to the United States, and losing perspective about conferences of this sort. "No one has a 'right' to be allotted time on a program,"[16] Carnap scolded him.

The hotel room encounter had different endings, as well, depending on who was telling the story. Two days later, Hook called the *New York Times* to give his picture of what happened. He explained that the two letters "proved beyond doubt that my application had been received in time." Never missing a chance to portray Shapley as not in control of his operation, Hook mentioned that "possibly Dr. Shapley does not get all the mail addressed to him." "At any rate," Hook told the reporter, "I left the correspondence with him, and after a few more words, walked out of the room."[17]

But the paper had already covered this amusing and "unscheduled prelude" to the evening's proceedings rather differently. Quoting Shapley's friend Martin Popper, it described a dramatic, tense, and confusing encounter between "the toastmaster and an educator who wished to be included among the speakers at the dinner":

> According to Mr. Popper, he and Dr. and Mrs. Shapley were chatting about 6:30 P.M. when Dr. Hook and the fourth man pushed open the door, which was slightly ajar, and entered. The lawyer said Dr. Hook talked incoherently, but he gathered that the educator was charging Dr. Shapley with misstatements of fact concerning permission for Dr. Hook to speak. This, he said, Dr. Shapley denied.

Hook's reporter became "an unidentified man accompanying Dr. Hook" and Hook seemed unbalanced and hiding from the press: "Efforts to reach Dr. Hook last night and early this morning were unsuccessful."[18] If Hook was temporarily avoiding the press, it may have been because Shapley outmaneuvered him. Though he did "walk out of the room," as he told the *New York Times*, he recalled years later that Shapley lured him out, as if to talk privately in the hallway. Once Hook stepped out, however, Shapley dashed back inside and locked the door behind him. Now the voices behind the door were saying, "Call the Desk! Call Security!"[19]

Hook did not force his way onto Shapley's program, but he successfully stole the spotlight. The encounter and Shapley's dramatic escape became the talk of the conference. The New York rabbi Louis Newman later joked at the lectern, "The only thing red about this conference is the speakers' warning light." He added, "I was invited to speak here and didn't try to hook myself on the program."[20] *Life* magazine also recounted the "hilarious incident when Dr. Sidney Hook . . . burst into the hotel room of Dr. Harlow Shapley of Harvard" who "cleverly maneuvered Dr. Hook into the hallway and then quickly retreated back into his room, locking the door and refusing to come out again."[21] Still, despite the laughter at Hook's expense, Hook and his fellow anticommunists would have the last laugh in the coming days and weeks.

Heresy, Conspiracy, and the Cold War Consensus

Conant had known Harlow Shapley for years (see Figure 5.1). They were about the same age, both were prominent public intellectuals, and there had been talk some fifteen years before that Shapley, instead of Conant, might become the president of Harvard. But they took different political roads through the cold war. Conant left behind his hopes for peaceful ideological competition between the new superpowers and became a cold warrior who promoted military buildup in Europe. Shapley remained an unapologetic internationalist who believed that the progress of science in the nuclear age pointed to international cooperation and "the one-world ideal."[22] Along with George Kennan,[23] with whom Conant had also parted ways after his conversion to cold warrior, Shapley and his colleagues at this conference

Figure 5.1. Harlow Shapley. (Image courtesy of Harvard University Archives)

simply did not believe that the Soviet Union aimed to subvert democracy and control the entire world.

Conant knew Sidney Hook, as well. A year before the dustup at the Waldorf, Hook praised his book *Education in a Divided World* in the *New York Times*. Having studied philosophy at Columbia with John Dewey, Hook was a well-known expert in education whose notoriety rivaled Conant's and who shared Conant's position on the growing problem of communist teachers. Though they worked independently—Conant in Cambridge and Hook in New York—they helped established the cold war consensus about communist teachers and the power of ideology to compromise and control the human mind. While Conant proposed to the National Education Association that Communist Party membership be the "clear-cut" line marking the limit of the association's tolerance of communism in the teaching profession, Hook provided a basis and rationale for that policy. In his many writings and lectures on the subject, he argued that a prohibition against communist teachers was not a matter of politics—and especially nothing like Stalin's purges of intellectuals in Russia in the 1930. Instead, Hook insisted, it was a matter of professional ethics and standards to which no self-respecting educator could credibly object. Communist teachers who refused to renounce the Party and their former membership in it, Hook argued, were not really teachers at all. They were indoctrinators who had lost or surrendered their intellectual autonomy to the ideology of the Communist Party and the Kremlin that controlled it.

The crucial distinction, Hook insisted, lay between heresy and conspiracy. Like Conant, he understood that liberalism thrived on competing views and serious consideration of rival, even heretical, outlooks. Liberalism and freedom therefore had nothing to fear from communist ideas. Conant had even suggested in *Education in a Divided World* that students at the appropriate age should read communist texts to see for themselves what the ideological struggle was about. Conspiracy, however, was something else. Teachers or professors who belonged to the Communist Party, therefore, were out of bounds.

A month before Shapley's conference, Hook laid out his argument in the *New York Times Magazine*. Under a photograph of Conant's Harvard Yard, he argued that "there are no 'sleepers' or passive members of the Communist party." Each member is required to actively conspire against democracy and capitalism. Hook's evidence was Party literature from 1935, when Hitler's alarming rise to power and his vows to destroy communism led many communists to join liberal democrats in a united, "popular front" to oppose fascism. This literature had two broad aims, one to oppose Hitler, Franco, and other fascists in Europe, and another to fight for representation in American trade unions, most of which had long resisted Communist members and their goals. In the new postwar geopolitics, however, Hook deftly portrayed this literature as calling for the violent and conspiratorial overthrow of democracy in the United States. Communist teachers, in particular, would do their part in this conspiracy, he explained, as he quoted an article from *The Communist* magazine from 1937. "Teacher comrades," the article read, must be given a

> thorough education in the teaching of Marxism-Leninism. Only when teachers have really mastered Marxism-Leninism will they be able skillfully to inject it into their teaching at the least risk of exposure and at the same time conduct struggles around the schools in a truly Bolshevik manner.[24]

Having once been a devoted Marxist, Hook knew this reasoning well and he knew that Communists who may have read this and similar calls for action in the 1930s did not see themselves as trying to destroy the United States. Amid the relentless ravages of the depression, they believed instead that they struggled to save the United States and its democracy from the fascism that appeared to be engulfing Europe. Hook's colleague James Burnham put it well after an automobile tour of the Midwest in 1933. The Depression had so reduced regions of the country to class struggle, despair, and even starvation that either communism, fascism, or "complete breakdown" was in store for the nation. After that trip, Burnham became as convinced as Hook that some form of socialism was the answer. He spent the rest of the 1930s—often alongside Hook—fighting for a socialist future for America.[25]

As a postwar anticommunist, however, Hook used his knowledge of communism's history and its literature to press his case that communists were essentially anti-American and unable to think and reason independently of their conspiratorial ideology. "On the basis of its philosophy of dialectical materialism, a party line is laid down for every area of thought from art to zoology," he wrote in the *New York Times*. "To stay in the Communist party, they must believe and teach what the party line decrees."[26] In his book *Heresy, Yes—Conspiracy, No*, he insisted that teachers who remain members of the Party have surrendered their intellectual free will and therefore fall short of even minimal professional standards:

For their responsibility to help their students to mature intellectually and emotionally, for their responsibility to their colleagues in the quest for truth, and for their responsibility to the democratic community to develop free men, they have substituted a blind and partisan loyalty to the objectives of the Communist party.[27]

Those objectives were to advance Soviet communism around the globe and win the struggle for men's minds.

When Conant rejected Archibald MacLeish's call for scholars to take up arms, he did so because he feared it would lead them to destroy what they believed they were protecting. It seemed to many that Hook was blazing precisely this ironic trail. In an effort to oppose Stalin's influence in the United States, he envisioned every American educator thinking and acting like a devout Stalinist—informing on colleagues, accepting a logic of guilt by association with known Communists, and cultivating a repressive, cautionary atmosphere in schools and colleges. Hook loathed Stalin and naturally denied any such comparison as absurd.[28] All he aimed to do, he said, was to place this urgent matter in the wise and responsible hands of the profession itself. "There is no safer repository of the integrity of teaching and scholarship," he concluded his *New York Times* article, "than the dedicated men and women who constitute the faculties of our colleges and universities."[29]

The Ever-Changing Communist Party Line

Accusations of mental captivity and dangerous conspiracies had long been heard throughout American history. They had been directed at witches, religious sects, or at immigrant groups whose customs, languages, and appearance seemed subversive in the eyes of a Protestant majority. Catholicism was a frequent target not only because of historical discord with Protestantism but also because of Catholic fealty to the Vatican, its foreign central office. To their enemies, American Communists were easily compared to Catholics whose Vatican was the Kremlin.

As proof, anticommunists pointed to the way American Communists quickly accepted the new "party line" whenever Moscow changed its mind. During the years of the Popular Front, for example, many American socialists resolved to defeat Hitler and fascism by any means necessary. Some joined the Abraham Lincoln Brigade and went to fight in Spain. Once the Hitler-Stalin pact became known in 1939, however, leftist groups and organizations suddenly became pacifists. Then, in the summer of 1941 when Hitler invaded the Soviet Union, opinion reverted to supporting war against Hitler. In "Reds Here Shift in Stand on War," the *New York Times* charted these swift conversions within the Communist Party, the Young Communist League, the League of American Writers, and an organization called The American Peace Mobilization. This last group converted from pacifism to militarism and renamed itself "The American People's Mobilization."[30]

For Conant, these shifts of opinion underscored the need for educational reforms that would teach Americans to think for themselves and see through the false, utopian promises of any ideology or its fanatical supporters. When Kuhn came to Harvard in 1940, Conant's campus was itself coming to terms with this reputation of Communists for changing their beliefs, seemingly on command. In late 1939, editors of *The Crimson* commented on a recent meeting of the Harvard Student Union and the failed efforts of the Young Communist League to take over its governing board. After Stalin invaded Finland in a bid to secure a buffer around Leningrad, supposedly peace-loving Communists on campus were left scratching their heads and struggling to rationalize the invasion, perhaps as some part of Stalin's larger, wiser plans. "The communists took an apparent beating on the Russo-Finnish Question," *The Crimson* noted, "when their blanket support of the Soviet Union's foreign policy was snowed under by indignant liberals." Kuhn's predecessors on the editorial board were not staunchly anticommunist. But they grew weary of "the Communist attitude toward Stalin—the 'his country, right or wrong' attitude" and "red spots before the eyes, a malady which has reached epidemic proportions in all liberal groups."[31] Two months later, commenting on another YCL broadsheet on campus, the editors pointed to its "blind attachment to the Kremlin."[32] Kuhn's personal crisis over intervention occurred in these tumultuous years. It led him, as he put it in his essay "The War and My Crisis," to distance himself from the dogmatic pacifism of his teacher, Elizabeth Moos, and the student communists he observed on campus, "for whose minds I can have little respect."[33]

The Conversion of Sidney Hook

Hook's conversion occurred several years before Kuhn's, but it was just as dramatic and transformative. As a young, radical philosopher in the 1920s,

Hook published a two-part study of dialectical materialism that aimed to show "the power of Marx's ideas":

> I know of no more impressive evidence in recent times of the power and importance of the philosophic idea as such, than the political and social history of Russia from 1875 down to the present day. So strong has this influence been that those who have simplified the teachings of Marx and reduced them merely to a system of political economy have been at a loss to explain the power of Marx's ideas in terms of their own doctrine.

The next year, he traveled to Moscow to study at the Marx-Engels Institute and burnish his reputation as a leading American expert on Marxism and socialism. Along with Burnham, who became an inner-circle confidant of the exiled Bolshevik Leon Trotsky, Hook became a leading philosophical authority among those who embraced communism as a credible and ethically responsible antidote to the problems of Depression America. In 1933, Hook and Burnham crossed fully into political life by helping to organize the short-lived American Workers Party.[34]

By the mid-1930s, however, Hook began to see Marxist ideas differently—at least Stalin's appropriation of them. Dismayed by Stalin's dictatorial methods, he and others championed Leon Trotsky as socialism's last, best hope. Along with the literary critic Edmund Wilson, anthropologist Franz Boas, and theologian Reinhold Niebuhr, Hook created the American Committee to Defend Leon Trotsky to denounce Stalin and his incredible claims that Trotsky was a dangerous counterrevolutionary. Hook persuaded Dewey to lend his name and towering reputation to the cause by leading a special commission to interview Trotsky in Mexico City, where Trotsky had found refuge at the home of the muralist Diego Rivera. The Dewey commission would in effect put Stalin on trial by examining his charges against Trotsky and determining whether or not they were fair and accurate.

The Trotsky affair dominated New York intellectual life, further dividing Trotskyists from Stalinists and both factions from those who, for one reason or another, refused to take sides. Hook took it for granted that Dewey's final report—titled *Not Guilty: Report of the Commission of Inquiry Into the Charges Made Against Leon Trotsky in the Moscow Trials*—would unify and strengthen the American Left by exposing Stalin's misdeeds for all to see. But when the report was released in 1938, that did not happen and Hook was shocked and dismayed. "More startling than the behavior of the

Soviet regime," he later wrote, "was the response of predominant sections of American liberal opinion" to the Dewey commission's report. The editors of *The New Republic*, *The New Masses*, and *The New York Times* posed critical questions, such as, How credible can such an investigation be when no representative of Moscow was present, and when the sponsoring organization was overtly partisan to Trotsky? Did Hook's and Dewey's show trial really get at the truth when Stalin's did not? Hook became furious and charged these critics with intellectual myopia and moral spinelessness.[35]

In the wake of the Dewey commission's report, Hook continued to defend Marxism and prospects for socialism under Trotsky, but not for long. By waging war on these and other liberal intellectuals who so questioned the report, he began to reconfigure his career as a critic of the American Left. Hook's incandescent hatred of Stalin and of those who would question his political views made him one of the era's most visible and influential anticommunists. In this and other political disputes, Hook never admitted defeat or let an argument go. As Irving Howe once put it, he "had an amusing need always to have the last word." In his autobiography of 1987, Hook's fury at the American Left over the Trotsky affair remained uncontained. He compared his former friend Corliss Lamont to a fanatical zealot and he charged that Walter Duranty, the *New York Times* reporter who attended the commission's hearing in Mexico, "was of more value to Stalin than the entire Communist Party of the United States."[36]

Like Kuhn, Hook was a Jew from New York, living in a nation only beginning to loosen restrictions against Jews in employment and housing. As the nation's postwar tolerance for dissent became smaller and smaller, Hook was keenly aware of the potential for violent backlash against perceived heretics and conspirators of any stripe. In many minds, that meant Jews. As a lawyer for the American Defamation League once put it, Jews were "automatically suspect" during the cold war. "The general view was that people felt if you scratch a Jew, you can find a Communist."[37] Alongside Hook's quest to keep regulators and politicians at bay by urging educators to police the halls of academia on their own, his determination to be more anticommunist than anyone else may have been a kind of self-preservation. He knew that postwar tensions could become dangerous for intellectuals who might be targeted by government thugs or angry mobs of citizens for their radical beliefs or writings. When urging him to withdraw his name from Shapley's conference at the Waldorf, Hook reminded Rudolf Carnap that the nation had clamped down on the pro-Nazi German-American Bund, whose several leaders were interned or deported several years before.[38] "I can very well understand," Hook wrote to Carnap, how honest people can disagree with American foreign policy now as in 1940. But what would you have thought of anyone who in 1940 lent his name to the German-American Bund who were then agitating for peace? People will draw the same inference about you if you are a sponsor of the Waldorf conference and remain a sponsor.[39]

In 1949, as a well-known former socialist organizer, a former defender of Leon Trotsky, and published authority on the logic and historic power of Marxism, Hook could face these dangers, too—unless, of course, he was even better known as a vehement anticommunist. "No wonder his early writings caused Hook later anxiety," his biographer observed. As a leading anticommunist, he forbade the reprinting of his early book, *Towards the Understanding of Karl Marx*, originally published in 1933 (eventually reissued in 2002). As a "savvy and scrappy political combatant across his life, Hook had no interest in passing ammunition across the barricades."[40]

The Conant-Hook Alliance

In his review of Conant's *Education in a Divided World*, published months before the Waldorf conference, Hook was thrilled to see that Conant understood the urgency of the Soviet threat, that the educational issues raised were nothing less than matters of survival. As Hook put it,

> None of the ambitious plans of America educators have evolved in recent years have the slightest chance of being realized unless the United States and its ideals, "equality of opportunity" and "social democracy" survive. To the extent, argues President Conant, that the public school system can make American citizens aware of the political world they live in and fortify their allegiance to democratic values, the chances of survival will improve.

Hook did not endorse everything Conant had written. But his quibbles, he emphasized, "do not detract in the slightest, however, from the wisdom of Mr. Conant's discussion of the political and educational problems of an armed truce."[41]

Just a few days after the Waldorf conference and controversy had concluded, Hook traveled to Boston for MIT's mid-century convocation and offered these compliments to Conant directly. Conant was there to introduce James Killian as MIT's new president and to join thousands in Boston Garden to hear Winston Churchill speak. Hook spoke with Conant at MIT and wrote a week later to say he hoped to continue their conversation in the future.[42] In the same letter, he included a copy of a lecture he had recently given at Dartmouth. Titled "International Communism—There Is Nothing Mysterious About Its Core of Belief or About What It Is Trying To Do," the lecture nodded to Churchill and his famous remark that Russia was "a riddle wrapped in a mystery inside an enigma." As Hook saw it, there was no mystery at all about Russia. The real mystery, he insisted, was why so many American liberals refused to condemn the Communist Party as a conspiracy intent on destroying the United States from within?

The Communist Party was no ordinary political association, Hook explained in this lecture. Just like Nazism, communism "impels those who hold this ideology to embark upon a program of world conflict and conquest." Members of the Party in countries around the world,

> must function as fifth columns, disorganizing them by propaganda, infiltration, intensification of class struggles, and finally strike for power if a favorable revolutionary situation develops. At the very least, once hostilities break out, they must sabotage to the death the cause of their own country for the sake of the workers' true fatherland, the Soviet Union.

Hook based these claims on his expertise as a Marxist scholar, the logic of dialectical materialism, and by selecting quotations—often strategically edited—from the communist literature he knew so well. All this evidence proved, Hook insisted, that beneath the surface of communist paeans to democracy and peace lay a corrupt and controlling ideology.[43] One strategy these Communists follow, Hook explained, is "to subvert within our minds our attachments to, and understanding of, the nature of democracy, by taking over all the symbols and the terms of the democratic tradition and pretending that communism is a species of democracy." This is why American diplomats and "those who must negotiate agreements with Soviet representatives, if unacquainted with Soviet ideology, live a life of continual surprises at what words can be made to mean."

Armed with these techniques of semantic warfare, Hook explained, Communists easily fool or "dupe" American liberals. It worked like this:

> Communists begin by taking some slogan like peace, or democracy, or civil rights, or help for anti-fascist refugees, or even free

milk for babies, and then appeal to all well-intentioned people to join these organizations on the basis of some "big" names that have been acquired in various dubious ways. The purpose is to consolidate public opinion behind these organizations, not to achieve the declared objectives but [to create public support on] behalf of the political aims of the movement.[44]

Shapley's conference at the Waldorf—which took place months after Hook delivered this lecture at Dartmouth—seemed just such an event. In his eyes, Shapley and his hundreds of artists and intellectuals were staging the biggest and most elaborate Soviet front-event the United States had ever seen. They were trying to make "peace" synonymous with "let the Soviets rule the world."

Conant was apparently impressed by the transcription of Hook's lecture. He thanked him for sending it and said he would read it "with great interest."[45] Conant's own conversion as a cold warrior was triggered mainly by worrisome events unfolding on the world stage, especially the Alger Hiss controversy and the Korean conflict. But he received Hook's letter and his intriguing reprint between his policy work with the National Education Association in late 1948 and his public announcement that he could not support the employment of Communist faculty in June 1949. He then told the Harvard community, "I am convinced that conspiracy and calculated deceit have been and are the characteristic pattern of behavior of regular Communists all over the world,"[46] and it may well have been Hook's lecture that convinced him. The closing words of that lecture—the West must present the Soviets with "an overwhelming preponderance of force so that if the communist state undertakes a hostile act against the West, it will know it risks destruction"—expressed the logic and strategy of Conant's Committee on the Present Danger.

In subsequent years, Hook and Conant continued to exchange friendly and supportive letters. Hook sent a copy of his influential book, *Heresy, Yes—Conspiracy, No*, and alerted Conant to gossip or articles that he believed Conant would find important. As the years went by, Hook became increasingly forthright and collegial. In April 1952, Conant delivered a controversial speech in Boston to the regional convention of the American Association of School Administrators. Pushing back against advocates of private, mainly Catholic, education who did not like Conant's call to expand funding for public education, Conant argued that the nation should not support a "dual system of schools," some public and some private, for this would help only to create and maintain "group cleavages" among Americans. Many Catholics took the speech to attack parochial schools, if not Catholicism itself.[47]

"You are getting such a bad press on your speech concerning private education that I am moved to send you a word of support and encouragement," Hook wrote. He offered Conant some strategic advice (tell them, he proposed, "you have no objection to parochial or religious education after the hours devoted to public schooling") and referred Conant to his own book *Education for Modern Man*.[48] The controversy lasted months, erupted anew whenever Conant's speech was quoted or reprinted and, if the whispers Conant heard were correct, he worried that the controversy might scuttle his future appointment as High Commissioner to Germany. Conant refrained from defending himself with each dustup, choosing to respond to all of his critics in his forthcoming book *Education and Liberty*. When it was published, Hook reviewed it in the *New York Times* and ignited the controversy again—partly by writing an "uncritical review," as one reader fairly described it. The *Times* printed five letters about the book and Hook's review, all of them taking Conant's proposals as an insult to parochial education. If religion-hating Communists "wanted to put the clamps on the Catholic church in any one of a number of countries," one read, "the surest way to do it would be to prohibit the establishment of any more Catholic schools. . . . Such, in my opinion, is Dr. Conant's approach."[49]

Hook had the last word by writing his own letter aiming to refute the various complaints. He assured the public that Conant "is motivated by no hostility to any religious group" and his goal is not to attack private schools but rather to address "the improvement of our public schools." Second, any comparison of Conant's position to "the infamous Communist policy toward religion is grossly unfair," Hook charged. "It is comparable to the unfairness of arguing from the Kremlin's use of the Russian Orthodox Church or Franco's use of the Spanish Catholic church . . . to support the view that religion is always on the side of the status quo." By raising the specters of Stalin and Franco, Hook jabbed back at Conant's attacker at the same time that he pleaded for everyone to rise above uncivil and inflammatory rhetoric. After all, as Americans "we have infinitely more in common with each other than any of us has with totalitarianism."[50]

Though he was famous for being "Dewey's Bulldog," quick to bare his teeth and snarl at any who dared to criticize the great American philosopher, Hook was also protective of Conant and eager to serve as an advisor helping him to pick and conduct his battles. Their correspondence continued into the 1960s and in 1965 their names appeared together alongside a hundred other cold warriors calling for national resolve to oppose communism in North Vietnam.[51]

Brainwashed Academics and Fellow-Travelers

In 1948, when Hook first aligned himself and his anti-Stalinist crusade with Conant, he did so both to burnish his own reputation—if Conant's recognition of the Soviet ideological threat involves "extraordinary acumen," then so did Hook's—and to gain a powerful ally in his opposition to liberal intellectuals who disagreed with his views or his ironically Stalinist methods.[52] As Conant was the president of Harvard, the so-called Kremlin on the Charles, Hook's alliance brought him closer to some of his enemies. It may have emboldened him to attack Shapley, who worked at Harvard under Conant, and later in the 1950s to attack Erwin Griswold, dean of Harvard's law school, for his writings about the role of the Fifth Amendment in legal proceedings against alleged Communists.[53]

No matter how close Hook got to them, however, liberal intellectuals such as Shapley and Carnap who supported the cause of postwar peace posed a special problem for his crusade. Because they did not belong to the Party or embrace Marxist ideology on their own, Hook could not portray them as robot Communists, slavishly parroting each and every change in the Party line and conspiring to destroy the United States. Still, in his book *Heresy, Yes—Conspiracy, No*, he argued that these "fellow traveling" liberals were nonetheless mentally compromised and controlled by certain "political sympathies":

> Perhaps the most depressing feature of the habits of the fellow traveler is the completely unscrupulous character of his intellectual procedures as soon as he discusses a political question which concerns Communists or the Soviet Union. A man who would rather starve than misreport the evidence of an experiment, who would sooner sacrifice popularity and preferment than risk making a snap judgment about a manuscript, feels not the slightest compunction, once his political sympathies take on a communist tinge, about inventing his facts as he goes along, or refusing to investigate and verify evidence crucial to his [political] arguments.

These sympathies transformed the careful scholar into a fabulist, the scientist into a dogmatist who knows in advance what the evidence will be and who refuses "to be bound by consistent standards of judgement." This corruption is usually invisible to the fellow traveler in question. "I am not now referring to deliberate duplicity," Hook explained,

I am referring to the half-conscious belief, born of political euphoria, that everything goes because one knows in one's heart that it is all in a good cause, and that in the interests of human welfare it is not necessary to put too fine a point on truth.

"What explains this calamitous lapse from elementary justice on the part of men and women who in their own fields are so circumspect about intellectual consistency and moral decency?" As it was with fully fledged Communists, Hook believed, their puzzling behavior was caused by ideas lodged in their minds and powerful enough to suppress their native intelligence on these issues: "Most of it can be traced to an assumption deeply held, even when it is unspoken, that the Communist movement is an integral element in a wider movement of progress and enlightenment."[54] To the fellow traveler, in other words, ideals of progress and enlightenment justified themselves by forming a chain of mental associations that filled the mind and blocked out alternatives. To borrow from the banner that hung behind Hook when he took the podium at his Freedom House conference, communist ideology created a mental "frontier" that limited the mind (see Figure 5.2).

A few of Hook's philosophical colleagues publicly challenged his authority and his reasoning about the communist threat. One was Alexan-

Figure 5.2. Sidney Hook lecturing at the counterconference he organized to protest the Waldorf conference. (Image courtesy of British Pathé)

der Meiklejohn, the philosopher and former president of Amherst College, who saw the nation's growing intolerance of communism as extreme and simplistic. If we take seriously Hook's signature distinction between heresy and conspiracy, Meiklejohn argued in the *New York Times*, then how in good conscience can we condemn those scholars who arrive at communist beliefs and Communist Party membership not because they wish to conspire mindlessly against the United States but because of "a passionate determination to follow the truth where it seems to lead, no matter what may be the cost to themselves and their families"?[55] Communists who fit this description are not conspirators but heretics, whom Hook otherwise claims to respect and support, Meiklejohn argued. Another was the Johns Hopkins philosopher Victor Lowe, who seconded Meiklejohn's complaint in the pages of the *Journal of Philosophy*. Lowe charged that Hook and other anticommunist philosophers who endorsed Hook's fire-them-first policy made a philosophical mistake by rushing to the conclusion that each and every Party member must be a "perfect" and most effective conspirator. Pointing to William James's admonitions against "vicious intellectualism," Lowe wrote, "Those who predict a man's behavior from his -isms and his party forget the great fact of human variation."[56] As Lowe and Meiklejohn saw things, Hook simply had no evidence on which to insist that every Communist was, or would become, a dangerous, anti-American conspirator.

Hook replied to both of these scholars with his trademark barrage of arguments and insinuations that liberals like them were dangerously naive. A Communist teacher, he replied to Meiklejohn, is like a lobbyist who takes money from industry or government to defend its values and "champion its side on any issue." Surely Meiklejohn would not feel comfortable with a lobbyist teaching children, Hook argued, so why should he now defend Communists who will surely use the classroom to lobby for the Soviet Union? Replying to Lowe, he emphasized that Communist Party membership "is profoundly different from membership in most other political parties." Of course, William James was right, he conceded: people are usually different and their behavior is often unpredictable. But in an echo of Conant's "single exception," Hook said this was not true when it comes to Soviet communism. "If we are formulating a policy toward Communist Party members in 1951," Hook wrote, "then we must consider the psychological evidence which makes it much more probable that they are hardened conspirators prepared to carry out their assignments than might have been the case in some earlier year." That included not only "Communist directives for thought control" but the effects of the nation's anticommunist climate. With only hard-liners remaining in the Party, Hook argued, today's Communists would be the most dangerous, determined, and skilled conspirators.[57]

From the Waldorf to the Kremlin on the Charles

Hook's reputation as a philosopher lent intellectual respectability to the growing view that communism was a kind of mental sickness, "a disease of the body, of the mind, and of the spirit," as one pamphleteer put it.[58] By insisting that Communists were necessarily prone to "conspiracy" and "deceit," Hook, Counts, Conant, and Burnham laid a scholarly foundation for the anticommunist consensus that swept the nation in the late 1940s. More than any other event prior to McCarthy's crusade and the brainwashing sensation of the early 1950s, Hook's crusade against Shapley's conference in 1949 elevated anticommunism far above the pacifism and internationalism that Shapley and his colleagues aimed to defend.

On the day of the conference at Freedom House, an enormous banner reading "One World—*Or None*" hung across the stage to deny Shapley's premise that that peaceful coexistence between East and West was possible. Max Eastman, the former editor of the socialist magazine *The Masses* (and friend of Thomas Kuhn's father) pressed the issue of exiled or executed intellectuals in the Soviet Union by ceremoniously reading their names. Bryn Hovde, president of the New School for Social Research, ridiculed Shapley's refusal to admit that Hook's cause, "intellectual freedom," was paramount over Shapley's naive wishes for peace. If Shapley and his colleagues did not come around, Hovde remarked, then "we shall know how to classify the Americans at the Waldorf."[59]

As far as Shapley, Carnap, and other supporters of the conference were concerned, their public call for peace and postwar stability was credible and responsible given the geopolitical realities at hand. As Shapley explained in his letter of invitation,

> We do not think the question worthy of debate as to whether or not capitalism and socialism CAN exist together. Both DO exist. The only question worth discussing is how to restore the mutual acceptance of that fact, which brought victory in World War II, and which alone can avert World War III.[60]

But Hook understood that in the emotional recesses of the public mind, the victories over Germany and Japan had not really ended the war. Most

Americans still felt vulnerable and surrounded by enemies intent on destroying their values and institutions. With his letters to Shapley's conferees and his press releases, Hook baited Shapley, portrayed him as an agent of these enemies, and fed reporters from both of New York City's major newspapers stories and documents from which they often quoted at length. Coverage in the *Herald Tribune* frequently placed Shapley's overtures to "peace" in quotation marks, as if his pacifism were insincere or, as Hook insisted, a pretext for Soviet control. *The New York Times* approvingly described the overflow crowd that came to Freedom House as well as Hook and his stated ideal of universal freedom: "Addressing 450 inside the building and about 500 outside through loudspeakers, he keynoted the gathering by stressing its desire for freedom everywhere in the world."[61] Under the heading "Rival Group Taunts Shapley," the *Times* also quoted full paragraphs from an open letter from Hook and Counts challenging Shapley to openly condemn Stalin and his purges at the conference.[62]

Outside New York, the controversy captured national and international attention. The State Department followed Hook's lead, denounced the conference as a Communist front, and refused entry visas for some of the foreign participants Shapley had invited. The House Un-American Activities Committee called it "a 'lace curtain communist meeting' called to discredit the North Atlantic Security Pact" that would become NATO. A British Pathé newsreel titled "Battle of the Pickets" showed crowds carrying signs outside the Waldorf, outside Freedom House, and outside Madison Square Garden as Shostakovich performed inside.[63] Two weeks later, the controversy remained newsworthy to New Hampshire's *Nashua Telegraph*. "NY Rally Part of Red Move to Snarl Artists and Scientists," the paper reported as it denounced Stalin's "world drive for thought control."[64]

Life magazine did not cover Hook's event. But its multipart coverage of the Waldorf conference and the magazine's trademark photography pictured the event largely as Hook and other anticommunists saw it. In one photograph, a picketer's sign reading "Shostakovich! Jump Thru The Window" was explained by its caption: "A picket's sign refers to Mrs. Oksana Kasenkina who leaped from Russian consulate to escape in 1948." Under the title "Red Visitors Cause Rumpus," the main copy explained that Shapley's organization, The National Council of Arts, Science and Professions, was "dominated by intellectuals who fellow-travel the Communist line." Under "The Russians Get a Big Hand from U.S. Friends," the magazine compared the audience's applause for visiting Russians to symbolic treason.

Life's coverage was breezy, entertaining, and dismissive of the pacifist message Shapley hoped to send. However brilliant they may be in their respective fields, and however sincere they may be in their hopes to avoid a third world war, these pacifists were victims of the political ideas foisted on them by their Communist sponsors. They were "gentle souls, a little bewildered by the world outside their laboratories or textbooks" but they were far from harmless. In this magazine, at least, Hook's warning to Carnap that he must protect his reputation, that any sponsor of the conference could be "marked for life as a captive or fellow-traveler of the Communist Party,"[65] proved correct. Under "Dupes and Fellow Travelers Dress Up Communist Fronts," *Life* called the intellectuals who came to the Waldorf "weapons" of a Soviet offensive who "lend it glamor, prestige, or the respectability of American Liberalism." A two-page rogue's gallery of participants dismissed them and their beliefs with a light touch: "The Communist-front organizations have been exposed often enough, however, so that by now the perennial joiner whose friends try to excuse him because he is 'just a dupe' is clearly a superdupe."[66]

Shapley appeared twice through *Life*'s lenses, once at the podium next to Lillian Hellman, the blacklisted playwright, and in the concluding rogue's gallery. None of the coverage or notoriety helped his career as a public intellectual, already suffering because of his support and involvement with the Progressive Party and its spectacular failure in the presidential elections of 1948. In 1935, for comparison, Shapley was celebrated on the cover of *Time* magazine for his work in astronomy. After the Waldorf event, *Time* called him an astronomer who "gazed upward with red stars in his eyes." Even though Shapley was quoted as a critic of Soviet science policy (science "must be free and not warped to fit an irrelevant plan," he said), the magazine denounced him as "an ardent worker on numerous Communist front organizations" who had recently made a "major contribution to the cause" as chairman of the Waldorf conference.[67]

At Harvard, Shapley remained something of a headache for Conant. Alumni groups and politicians concerned for the reputation of the Kremlin on the Charles could simply point to Shapley (as McCarthy did in his Wheeling speech) and complain that the university should do more to eliminate subversives in its midst. To Shapley's great disappointment, Conant declined to bestow on him the prestigious title of University Professor. His colleagues in the astronomy department feared that his national reputation as a communist sympathizer would hamper their ability to receive government research grants.[68]

Aside from Shapley, however, Harvard was an obvious target for anticommunist politicians. When the House Un-American Activities Committee began its inquisition of Harvard faculty in early 1953, the first to testify was

Robert Gorham Davis—the professor to whom, twelve years before, Kuhn had submitted his essay "The War and My Crisis" as well as another political essay, "International Morale and a United States Declaration of War."[69] Davis was now a professor at Smith College, where he had been visited by a committee investigator and subpoenaed to testify about his years at Harvard. About a dozen former Harvard faculty members had been called. Like them, Davis had been a Communist in the 1930s. But he had turned away from the Party after the Soviet-Nazi pact, he told investigators, and now agreed to testify against what he now called a "dogmatic conspiratorial organization totally committed to the defense of the Soviet Union."[70]

Testifying for one morning and one afternoon, Davis described his regular meetings with fellow Communists in Cambridge, the reading group he belonged to, its fundraising activities, and its contacts with teachers' unions and other organizations. The main goals of this activity, he said, were to promote social justice, oppose anti-Semitism, and, above all, to help protect America from the fascism then engulfing Europe:

> MR. DAVIS. It was to a large extent a neighborhood organization and was intended to draw in people who were not primarily intellectuals, to educate them, to get them to take, as citizens, the political line which the Communist Party desired to see followed. It was not a tightly organized society.
>
> MR. TAVENNER. And that political line was to oppose Hitler?
>
> MR. DAVIS. But it also included defense of the Soviet Union, as they say.

Committee members repeatedly probed Davis's memory of faculty whom they presumed were "well indoctrinated in Marxian theory" and required to infiltrate teachers' unions and indoctrinate students. Tavenner read the same excerpts from the same article that Sidney Hook had quoted in the *New York Times* and asked Davis to elaborate on how his colleagues mastered Marxism and Leninism in order "to skillfully inject it into their teaching."

Davis defied HUAC's expectations, however. Each time the notion came up, Davis patiently explained that there had been no indoctrination at Harvard; that "there was never any direct attempt to influence the teaching in the classes," and that he then shared with his comrades "a lurking feeling that it wasn't quite good sportsmanship to try to influence young people—at least to make use of our position in the classroom to do this." He also emphasized that Harvard students seemed "politically conscious" on their own—sometimes more so than their professors—and that the campus hummed with political organizations and activities. Even if none of the faculty had been Communist during those years, Davis suggested, "I think student organizations would have taken very much the same character."

On the other hand, Davis agreed that indoctrination was rampant inside the Communist Party, whose leaders "dictate to its members how they should think." In the United States, however, this kind of mental control "had to take the form of persuasion, since they had no means physically of keeping a person from leaving the party. Therefore they had to do it through argument and pressure of all kinds." In the Soviet Union, he reassured these congressmen, things were much, much worse. Quoting from a book review he had written, Davis described how Soviet psychiatrists had learned to bring about the "extinction of the individual self" so that individual Communists came to perceive the world in specific, prescribed ways:

> [B]y torture, narcosis, hypnosis, and indoctrination in various combinations, the self's organic past can actually be negated, and . . . it can be made to "reflect" completely the party-partisan view of reality. Since those who do not come to reflect this reality are considered ultimate enemies of the people, there is no moral limit to the use which may be made of these psychological, physical, and medical means of extinguishing the self.

Davis's indignation proved to the committee that he had turned his back on communism. The excerpt he read documented "my present devotion to democracy in unqualified terms," as did other articles he had written for the magazines *Partisan Review*, *Commentary*, and *New Leader*—all of which "have for years fought Stalinism with informal intensity," he pointed out. Davis added that he "was one of the 88 intellectuals who signed the statement published in the New York Times on March 24, 1949, calling attention to the true nature of the Waldorf Scientific and Cultural Conference."[71]

In many ways a model witness for HUAC, Davis testified cooperatively, acknowledged the ominous tactics of mind control operative with the Communist Party, and performed the public ritual of naming names. Tavenner, the legal counsel for the committee, made it relatively easy on Davis to perform this demeaning ritual, however, by naming names himself and asking Davis to comment. For example,

Mr. Tavenner. Were you acquainted with Rubby Sherr—S-h-e-r-r?

Mr. Davis. Yes: he was a member of the group for a comparatively short time, but I think he was still a member when I left.

Mr. Tavenner. Or Wendell Furry—F-u-r-r-y?

Mr. Davis. I knew him very well. He was a member of the group.

The Ordeal of Wendell Furry

Wendell Furry he had been one of Kuhn's physics teachers at Harvard. He had been a student of J. Robert Oppenheimer and had been involved in Communist politics, but was no longer committed. Unlike Davis, however, Furry was uncooperative with the ongoing inquisition, refused to name names, and invoked his Fifth Amendment rights. The day after Davis's testimony, the committee asked Furry about his professional associations and whether the Communist Party had infiltrated them.

> Mr. Furry. Sir, this question obviously tends to inquire into my beliefs in associations. I do not believe that this committee has or any congressional committee has the right to make such inquiry, or that in America there can be any governmental body that has a right to inquire—
>
> Mr. Kearney (presiding). I am going to instruct the witness to answer the questions and not make a speech.
>
> Mr. Furry. Sir, on the grounds of my rights as a citizen under the first amendment, rights of free speech and assembly, and my rights and my privilege under the fifth amendment not to be a witness against myself, I decline to answer the question.

Two others from Harvard also resisted. They did so in Boston, where later that spring Senator William Jenner, chairman of the Senate subcommittee, arrived to interview selected faculty.[72]

Furry's ordeal began the day after Davis's testimony, when he appeared on the front page of the *Boston Globe*. Besides testifying once to HUAC and twice for McCarthy's Senate committee, he ran afoul of the Harvard Corporation because he admitted to once having misled FBI agents who had interviewed him about his political past. Along with others who resisted, Furry was questioned repeatedly by the corporation and managed to pay his attorney fees only because some of his colleagues in physics passed a hat to collect contributions.[73] In the fall of 1953, McCarthy called him to Washington again in connection with his and Roy Cohn's investigation of the Army's Fort Monmouth Signal Corps.[74] McCarthy questioned Furry privately and then went public as if he had unearthed yet another immense, diabolical conspiracy. He announced to *The New York Times* that Furry "refused to say whether he had indoctrinated students in Communist philosophy" and that he "refused to say if he ever had turned over to the Communists top secret material from 1943 to 1945" when he worked in military science at MIT. McCarthy could only conclude that Furry suffered from "a complete lack of academic freedom under the discipline of Russia," that "he is not a free agent." As for Harvard, McCarthy said the campus was "a smelly mess, and I cannot conceive of anyone sending their children anywhere where they might be open to indoctrination by Communist professors."

McCarthy's inquisition arrived at Harvard shortly after Conant resigned to take his new post in Germany and sparred with McCarthy in Washington over communist books and the budget for his high commission. In anticipation, the new president, Nathan Pusey, the Harvard Corporation and Overseers, faculty, alumni, and *Crimson* editors vigorously debated how to proceed, how to defend the university's reputation, and what responsibilities the administration had to faculty and students who might be in for trouble. Just as he had bullied Conant, McCarthy put Pusey personally on the spot by denouncing his institution and telling reporters he had telegrammed Pusey to demand some explanation of Furry's behavior.[75] The nation expected an answer.

Days later Pusey held a news conference to say he was unaware of any Communists on the faculty and to remind the nation that "Harvard was unalterably opposed to communism." Pusey seemed to miss the stern leadership that Conant would have provided during this difficult time, but he could at least invoke Conant's (and, tacitly, Sidney Hook's) authority in his response to McCarthy's telegram:

> I am in full agreement with the opinion publicly stated by my predecessor and the Harvard Corporation that a member of the Communist party is not fit to be on the faculty because he has not the necessary independence of thought and judgement.[76]

When Thomas Kuhn first arrived at Harvard in 1940, the age of ideology had just begun. As Nazism threatened to engulf all of Europe, Kuhn and Conant struggled along with the rest of the nation over whether and how to save Europe and how to protect the American traditions of liberalism and democracy. Thirteen years later, the problem of Nazism had been resolved. But now loomed the threat of Stalinism that was only more intense and urgent. If McCarthy, Hook, and other anticommunist authorities were correct, totalitarian ideology was poised to consume not only Europe but the United States and Kuhn's Harvard Yard. Kuhn's undergraduate education and now his new career as a historian of science remained surrounded by debates about an ideology so powerful it could grip and control the minds not only of ordinary Americans, but leading creative artists, writers, and Ivy League physicists such as Furry.

Kuhn probably dismissed McCarthy's sensational claim that Harvard was awash with subversive intellectuals indoctrinating students, just as Robert Gorham Davis had set the record straight about alleged indoctrination in his testimony to HUAC. Kuhn denied that any indoctrination had taken place at Hessian Hills, as well: "The ideas of my teachers were radical, but they were not forced upon me or any of us," he explained to Davis in "The War and My Crisis." But Kuhn found an important kernel of truth in this consensus that a political ideology could guide and limit a scholar's outlook and intellectual creativity. In his early attempts to decode his Aristotle experience, the idea would prove enormously helpful, and it would be the central focus of a new project Kuhn had recently taken on. The philosopher Charles Morris from the University of Chicago, a visiting lecturer on campus in 1952 in the Department of Social Relations, had asked Kuhn if he would write a monograph on the history of science for the *International Encyclopedia of Unified Science*.[77] Kuhn agreed and wrote to Morris the next July in the midst of the anticommunist inquisition. He had a new idea for the monograph, he explained: the book would analyze the several ways that scientific theories function as ideologies within scientific communities. Kuhn was very excited about this new approach, and he told Morris that he seriously considered including the word *ideology* in the title. He admitted, though, that "The Structure of Scientific Revolutions" was a better choice.

6

Brainwashing, or
The Structure of Philosophical Revolutions

The Russian Revolution meant different things to different Americans. For priests and ministers, it was an attack on theology. For businessmen, it was an assault on capitalism and free enterprise. For civil vigilantes such as Elizabeth Dilling, it threatened Protestant values and a cherished American "way of life." In whatever ways people feared the advance of communism, however, most agreed that it meant undergoing a philosophical revolution; a replacement of one system of ideas by another that was different and incompatible. Describing his trip into Russia through the Finland Station in 1919, Arthur Ransome, correspondent for the *Manchester Guardian*, put it this way: "Crossing that bridge we passed from one philosophy to another."[1]

In the wake of wars or revolutions, it usually takes years for economic and cultural institutions to settle into new, stable patterns. But seen from this philosophical angle, revolutions can be almost instantaneous. In John Reed's classic account, the Bolshevik revolution was *Ten Days that Shook the World*. In the pages of the *Philadelphia Evening Ledger Literary Digest*, communist revolution was compared to a short, violent explosion. In a cartoon titled "Too Slow for Me," a bearded, unkempt communist straddles a gunpowder rocket labeled "Bolshevism" that is pointed upward to the skies (see Figure 6.1 on page 112). As he reaches down with a burning match to light the rocket's fuse, we know it will not end well. But the true believer is convinced by their philosophy and must try.

Not long after "Too Slow for Me" was published, a real bomb exploded at the home of United States Attorney General A. Mitchell Palmer. Palmer and his family survived unharmed, but the explosion led Palmer to inaugurate an era of persecution, detention, and deportation of real or imagined political subversives. Thirty years later, the cold war and the McCarthy era

"TOO SLOW FOR ME!"
—Sykes in the Philadelphia *Evening Ledger*.

Figure 6.1. "Too Slow for Me," by Charles Henry Sykes, reproduced in *Literary Digest*, July 5, 1919, 27.

revived these popular fears of sudden, revolutionary change. But the image of the revolutionary conspiracy had changed. Cold war Communists in America were not generally violent. For one reason, there were not many Communists in America after the Nazi-Soviet pact of 1939, and their ranks were reduced again by the postwar thaw in Soviet-U.S. relations. By the time McCarthy commenced his crusade, membership in the Party had fallen to a small fraction of what it had been a decade before. For those that remained, the path to socialism they envisioned lay not in terrorism but in education, responsible politics, labor organization, and sometimes pacifism—as it was for Elizabeth Moos, her progressive Hessian Hills School, and her activism into the 1950s and '60s.

This posed a problem for anticommunist crusaders. After all, citizens who pledged allegiance to American democracy, to equality of economic and educational opportunity, international cooperation, and world peace could not plausibly be demonized, much less rounded up and deported as bomb-throwing subversives. They could, however, be stigmatized, marginalized, and publicly interrogated on the grounds that they had fallen prey to a most powerful revolutionary weapon—a seductive, tenacious, and controlling ideology or set of political ideas.

The Philosophical Side of Subversion

The literature of cold war anticommunism frequently addressed philosophy and the power of abstract ideas. What could it be, Herbert Philbrick asked in his book *I Led 3 Lives*, that drove those he observed as an undercover agent for the FBI?

> What was it that instilled such devotion into these fellow-travelling Communists, to labor so hard for such a distorted cause? What was it that got into them? Was it sincere believe in the prattle of communism? Was it a disillusionment and frustration that led them into these paths?

"It angered me to realize," Philbrick later concluded, "that the human mind could be so easily dominated, no matter what the cause."[2]

The invading philosophy was itself invisible, but Philbrick and other popular anticommunists styled themselves as experts who could detect its traces. There were the patterns of behavior that Philbrick recounted in his book, and there were verbal cues that exposed minds taken over. *LOOK* magazine ran "How to Identify an American Communist" in 1947, while the Army itself prepared a similar pamphlet in 1955. Both pointed to the use of certain words ("vanguard," "ruling class," "exploitation," for instance), a habit of speaking in long, complicated sentences, or as the Army pamphlet put it, a tendency to "stubbornly cling to Marxist ideals without being willing to question them."[3]

FBI director J. Edgar Hoover helped educate the public about philosophy and its place in the modern world. In 1944, when the nation was allied with Stalin to defeat Hitler's Germany, he criticized communism as an incomplete philosophy. Speaking at Holy Cross College in Massachusetts, he explained that while any particular philosophy can devolve into a set of dogmas, philosophy itself—understood as an activity and field of study—"must continually question all premises, conclusions, judgments, values and principles." The Soviet Union and Communists in China, however, had turned their backs on philosophical inquiry and placed themselves in a mental straightjacket. "In the strict sense of the word," therefore, communism is not even a philosophy. "It is, rather, an ideology," Hoover explained, "an interpretation of nature, history, and society which is developed with some logic from some premises which are demonstrably false but which are not open to question or criticism by its adherents."[4]

By the 1950s, Hoover could not even respect communism as a philosophy-gone-bad. In the wake of Sidney Hook's crusade against the Waldorf Congress and his campaign to educate Americans about the power and danger of communist ideology, Hoover described it similarly—as a conspiratorial mindset that turns believers into agents who deceive and manipulate their fellow Americans. "Something utterly new has taken root in America during the past generation," he warned in *Masters of Deceit*, his bestselling book from 1958, "a communist mentality representing a systematic, purposive, and conscious attempt to destroy Western civilization and roll history back to the age of barbaric cruelty and despotism, all in the name of 'progress.'"[5]

While Hoover's warnings were well known, the most visible promoter of this menacing picture of communist ideology in the early 1950s was McCarthy himself. Besides claiming to expose Communists in government and Communist Books in United States libraries in Europe, McCarthy reminded his fellow congressmen and the press that the main weapon trained on the United States was "the Communist philosophy." As chairman of his Senate investigative committee, he pronounced each demonic syllable with evident scorn as he interrogated his witnesses:

> THE CHAIRMAN. Today would you say that you feel sympathetic towards the Communist philosophy?
>
> MRS. LEWIS. Senator, that is a question now that is asking about my opinions and beliefs, is that right?
>
> THE CHAIRMAN. I think you understood the question. The question is: Are you now sympathetic to Communist philosophy?
>
> MRS. LEWIS. There are some things in the Communist philosophy that I am not particularly sympathetic with.

Brainwashing, or the Structure of Philosophical Revolutions | 115

THE CHAIRMAN. Can you tell us those things in the Communist philosophy you are not sympathetic with?

Mrs. Helen B. Lewis was a psychoanalyst who earned a PhD at Columbia University and testified in 1953. Her husband, Egyptologist Naphtali Lewis, came to McCarthy's attention when he was awarded a Fulbright Scholarship to study in Italy. When McCarthy came around to his wife's relationship to the Communist Party, Lewis repeatedly took the Fifth Amendment and suffered McCarthy's ire for doing so.[6]

With Harvard's Wendell Furry and others on the stand, McCarthy also probed their relationship to "the Communist philosophy":

THE CHAIRMAN. Did you ever attend Communist meetings with your students?

MR. FURRY. I refuse to answer that, sir.

THE CHAIRMAN. Did you ever try to indoctrinate your students in the Communist philosophy?

MR. FURRY. I refuse to answer that, sir, on the same grounds.

THE CHAIRMAN. Did you ever solicit any of your students to join the Communist party?

MR. FURRY. I refuse to answer that on the same grounds.

THE CHAIRMAN. Have you ever attempted to indoctrinate your students in the Communist philosophy?

MR. ROSENBAUM. I refuse to answer the question on the grounds of the Fifth Amendment.

THE CHAIRMAN. Were you ever instructed by the Communist party to indoctrinate your students in the philosophy of communism?

Mrs. Wolman. Fifth Amendment.

The Chairman. In your lectures, did you ever attempt to indoctrinate students in the Communist philosophy?

Mr. Grundfest. Sir, I only lecture on neurophysiology.

The Chairman. Answer the question.[7]

Almost surely, McCarthy had no intellectual interest in philosophy. But he understood well the menacing aura that "philosophy" could acquire in public discourse. The dangers he and other anticommunists ascribed to philosophy were a recent echo of the scandal that erupted nationwide in 1940 when the British philosopher Bertrand Russell was denied a position at the City University of New York. Russell was depicted by his critics not as a Communist but a nihilist whose libertine views on religion and sexuality threatened to corrupt his students as surely as Socrates in ancient Athens had subverted norms and made "the weaker speech the stronger."[8] In this new age of mind control, the fear was that "Communist philosophy" would convince weak proletarians that they were in fact stronger than their capitalist oppressors, as seemed to have been demonstrated by the Bolshevik Revolution. Philosophy's subversive reputation was only helped by the fact that Russia's revolutionary leaders, including Lenin, Alexander Bogdanov, and Leon Trotsky, were preoccupied with Marxist theory and wrote dense, theoretical treatises about epistemology and metaphysics. In the United States, too, some of the most ardent champions of socialism in the 1920s and '30s had been philosophers, including Sidney Hook and James Burnham. So it was, as McCarthy's concerns about "the communist philosophy" suggests, that academic philosophers were a frequent target of anticommunists during the McCarthy era—"six times as likely to be attacked by witch-hunters as English professors, and twice as likely as economists," John McCumber has calculated.[9]

Brainwashing, Mind Control, and Totalitarianism

The word *brainwashing* first appeared in the *Miami News* in a story about psychological techniques employed by Chinese communists to transform millions of peasants into model communists. Its author was Edward Hunter, a war reporter who in the 1930s covered the Spanish civil war and, in 1945, the Japanese invasion of Manchuria. Hunter also worked for the Office of Strategic Services and its successor organization, the CIA.

Hunter's scoop of 1950 brimmed with all the geopolitical significance of the conflict in Korea. Unlike the tales of bravery, determination, and victory that surrounded American soldiers during World War II, reports of American GIs in Korea worried the American public. Some 10 percent of those held captive collaborated with their captors in various ways—by writing statements favorable to communism, informing on fellow GIs, and (as immortalized in the novel and movie *The Manchurian Candidate*) even murdering each another at the behest of their captors.[10] Of the ten thousand Americans held captive at some point during the conflict, only about four thousand survived until the summer of 1953, when they were repatriated to the United States and studied closely by military psychologists. The Korean conflict did not go well for the United States and the nation needed to understand why.

Former prisoners testified to Congress about "beatings, solitary confinement, mock executions, and the tearing off of . . . fingernails and toenails." But the nation still could not understand why so many had sunk to collaboration with the enemy. Brainwashing provided an explanation: if they had acted like communists, the logic went, they must in some sense have mentally or psychologically *become* communists. Like the peasants in China—Hunter's "brainwash" was a literal translation from Chinese—their minds must have been wiped clean by Korean military psychiatrists and then repopulated with communist dogmas and values. Because the transformation was forced, the GIs could be seen as victims. Though the Army and Air Force scheduled dozens of courts martial, these stories of brutal captivity and the new phenomenon of brainwashing cultivated public sympathy, and all but fourteen were cancelled.[11]

The brainwashing sensation encapsulated and symbolized the geopolitical fears that possessed Americans during these years. While a military invasion of North America by Soviet or Chinese communists seemed unlikely, and was sure to be repulsed by the nation's war-tested military strength, the nation remained susceptible to persuasion and the power of ideas. Accordingly, news of brainwashing swept over American culture. In the popular media, *Time* featured an article in October 1951 and the anticommunist *New Leader* boasted that it had covered the story as early as 1950.[12] Hunter published two books, *Brainwashing in Red China: The Calculated Destruction of Men's Minds* and *Brainwashing: The Story of the Men Who Defied it*. Both were positively reviewed in the *New York Times* by writers who took

Hunter's reportage at face value. Overlooking or unaware of Hunter's role within the CIA, one described him as "a freelance foreign correspondent in the Far East." Another praised his "thorough anatomy of brainwashing, Korean brand," and noted that he "has traveled far and wide to interview the men who resisted."[13]

In 1955, the American secretary of defense created a special committee to study the experiences of the brainwashed POWs and to propose a code of conduct for any future soldiers taken captive. It consisted in six articles, summarized here by the *New York Times*:

> The first sentence of Article 1 is "I am an American fighting man," and the last sentence of Article VI is "I will trust in my God and in the United States of America." Between these two tenets, the American service man will pledge that he will never surrender of his own free will, and that he will endeavor to escape if caught. As a prisoner he will not betray his fellow prisoners. And he will resist brain-washing by refusing to give any information beyond his name, rank, serial number and date of birth.

The new code, Anthony Leviero reported, was based on fundamental "American principles." Implemented by an executive order from President Eisenhower, it concluded "a national debate over what should be done about Americans who were brainwashed in Korea and those who may receive barbarous treatment in any future war."[14] Outside of military circles, however, the debate had not concluded. In 1958, the story of American GIs in captivity was told again in *The New Yorker* by staff writer Eugene Kinkead. His article "The Study of Something New in History" was later expanded into a book, *In Every War But One*. According to its front cover, the book bravely explored "a failure in which we all share: why one out of every three American taken prisoner in Korea collaborated with the Communists, and why two out of five died, many at the hands of their fellow Americans."[15]

The claims being made about brainwashing were fantastic. But they were not more incredible than other axioms of cold war culture in the United States—that any of one's trusted friends, neighbors, coworkers, or teachers could be secretly working for Moscow; that powerful Soviet operatives had high-level positions in the federal government; or that an underground, backyard shelter supplied with canned food and breakfast cereal would allow a family to survive nuclear war. As historian Richard Hofstadter put it in his classic study of the "paranoid style in American politics," Americans have long perceived their enemies as having "some especially effective source of power," whether that be controlling the press, vast sources of wealth, or discovering "a new secret for influencing the mind (brainwashing)."[16] In a world awash with powerful, scientific tools of war—including the atomic bomb and the bacteriological weapons rumored to have been first deployed in Korea[17]—why shouldn't modern psychology have produced its own techniques for vanquishing one's enemies?

Brainwashing additionally helped make sense of the most troubling features of recent history, such as Nazism and the show trials in the 1930s that elicited false but seemingly sincere confessions of treason from Stalin's rivals. Columbia University psychology professor Joost Meerloo, who had been a psychiatrist in the Dutch Army in Nazi-occupied Holland, later explained in the *New York Times* how something like brainwashing lay behind the trials:

> It seemed nearly impossible to believe that these old-guard Communists had suddenly changed into traitors. When, one after another, every one of the accused confessed and beat his breast, we at first through that it was a great show of deception, intended for the international stage, until gradually it dawned upon us that a much worse tragedy was being enacted. Human beings were being systematically changed into puppets. They sang the tune their operator and stage director wanted them to sing.[18]

The Nazis had in fact discovered brainwashing before the Russians, Joost explained. The first time he observed how "the human mind can fall an easy prey to dictatorial powers" was the case of Marius Van der Lubbe, accused of setting fire to the Reichstag and setting the stage for Hitler's rise to power:

> When that pitiful creature—already known to Dutch psychiatrists as mentally unstable—came before his judges, the world wondered, "Can that little fellow be the heroic revolutionary who justified Hitler and his Brown Shirts in seizing power in Germany and taking justice into their own hands?"

Only later did it become clear, Meerloo explained, that "medical knowledge and psychiatric techniques had been misused in transforming the victim into a passive automaton, in the end merely replying 'yes' or 'no' to his questioners."[19]

The brainwashing sensation took root in fertile soil. Long receptive to beliefs about unseen religious forces capable of transforming individuals, Freud's theories of the mind had further prepared Americans for understanding that individuals could be controlled by complexes of ideas that are both powerful yet cloaked and unknown to the conscious mind. By helping to explain Nazism, Stalinism, and the rise of communist China, the idea of brainwashing complemented the concept of *totalitarianism* that guided political thought during the cold war. The word itself was not new—Conant spoke of "the intellectual tyranny of a totalitarian state" in his Baccalaureate Address of 1937 and in the prewar debates over the fate of the liberal arts he denounced "totalitarian ideas."[20] But alongside "brainwashing," this word gained currency as a generic name for the twin ideological curses that seemed to afflict the twentieth century. In 1949, for example, George Orwell's new novel *1984* spoke of "the totalitarians, as they were called . . . the German Nazis and the Russian Communists."[21] That same year, the philosopher Hannah Arendt finished her manuscript, *The Origins of Totalitarianism*, in which she too placed Nazism and communism under one conceptual umbrella. Neither Orwell nor Arendt mentioned "brainwashing," but each purported to show how totalitarian regimes maintain power through propaganda and mind control. Arendt explained that totalitarianism "has discovered a means of dominating and terrorizing human beings from within" by connecting rules to those they rule: "Hitler, who was fully aware of this interdependence, addressed it once in a speech to the SA: 'All that you are, you are through me; all that I am, I am through you alone.' "[22] In later editions of her book, Arendt dismissed popular theories about brainwashing as "magic."[23] But she nonetheless helped Americans and generations of scholars to understand the ideological dynamics behind "the struggle for men's minds."

Brainwashing in Politics, Popular Culture, and Academia

By the end of the 1950s, popular culture had largely accepted brainwashing as a regular feature of modern life. Movies and television programs, exposés of modern advertising techniques, and the self-help sermons of Norman Vincent Peale took for granted that the human mind could be shaped, if not precisely controlled.[24] Some remained skeptical, such as the military historian Samuel Marshall who reviewed Kinkead's *In Every War But One* in the *New York Times* and suggested that the hubbub over brainwashing revealed more about "that part of the American public which believes whatever it sees in print" than it did about serious military history.

Hunter responded to Marshall with an angry letter to the editor complaining that the eminent historian was trying to "brush brainwashing under the bed as non-existent."[25] Most Americans and their political leaders sided with Hunter: brainwashing was a real threat and it had to be understood. Hunter was therefore the star witness at a House Un-American Activities Committee hearing where he rehearsed his unlikely theory that techniques of brainwashing were rooted in Pavlovian psychology:

> MR. HUNTER. When the Communist hierarchy in Moscow discovered that it was unable to persuade people willingly to follow communism, when they found they could not create what they wanted, the "new Soviet man" in which human nature would be changed, they turned to Pavlov and his experiments. They considered people the same as animals anyways, and refused to recognize the role of reason or divinity in a human being. They took over the Pavlovian experiments on animals and extended them to people. . . . People, they anticipated, would react voluntarily under Pavlovian pressures in the way a dog does, to Communist orders, exactly as ants do in their collectivized society.
>
> MR. ARENS. What were the results of their application of these techniques to people?[26]

In the menacing picture Hunter presented to a sympathetic committee, the results were a long line of victories for international communism, from the Soviet show trials of the 1930s to the loss of China and the brainwashing of American GIs in North Korea.

Despite the almost supernatural powers popularly attributed it, some mainstream professors and psychiatrists paid close attention to the phenomenon. Four scholarly books about brainwashing appeared in 1960 and 1961. Theodore Chen, professor of Asiatic studies at the University of Southern California, published *Thought Reform of the Chinese Intellectuals*. Robert Jay Lifton interviewed victims and then published *Thought Reform and the Psychology of Totalism*. MIT professor of psychology Edgar Schein co-authored *Coercive Persuasion: A Psycho-Social Analysis of the 'Brainwashing' of American Civilian Prisoners by the Chinese Communists*. Albert Biderman and Herbert

Zimmer published a collection of essays by academic psychologists, psychiatrists, and medical experts, *The Manipulation of Human Behavior*, that would become an often-cited resource. Scholarly journals also took note of brainwashing and the new research it inspired.[27]

These authors acknowledged "the great shock," as Robert Waelder put it, caused by the stories of American GIs in captivity. But unlike the wide-eyed believers Hunter found in Congress and Hollywood producers found in cinemas, these intellectuals questioned the conventional wisdom and often disagreed with each other over what brainwashing was and how it worked. Schein, for example, disputed Hunter's and Meerloo's claims that the science behind brainwashing was Pavlovian. Chen claimed that others were exaggerating brainwashing's power. Meerloo, for example, had written of Chinese prisoners that "[n]obody is able to resist such brainwashing, whatever one's individual limit of endurance." But that was not always true of those Chen had studied: "Despite the attempt of the thought reform to probe the inner depths of their mind," he wrote, "they were able to keep inviolate the secret recesses of their being." Along with Schein, the social scientist Biderman was skeptical of Meerloo's sensational imagery of the immanent "totalitarianization" of society and he questioned Hunter's interviews that had started the brainwashing ball rolling. Many critics were quick to point out that the effects of brainwashing—whatever they were, exactly—proved temporary after victims were released from captivity.[28]

In taking aim at each other, these experts also took aim at the public and the press for sensationalizing a scientific phenomenon. Biderman noted that a "great deal of pseudo-scientific speculation" about it had saturated popular culture—almost certainly, he recognized, because of "the intensity of the political and psychological involvements of the moment." While Biderman blamed social scientists for feeding the hysteria, Raymond Bauer pointed to the public. Hysteria over the brainwashing of the POWs revealed only ideological insecurity and weakness in the United States. At a Chicago meeting of the American Psychological Association in 1956, he said, "We share in common with the Communists a lack of security in our own ideological beliefs, even though we both proclaim our confidence that we have the true way of life." Ironically, Bauer argued, the facts about the POWs demonstrated the strength and power of America's ideology: only a small handful of soldiers, after all, were converted to communism or collaboration.[29]

Despite their disagreements, these academics acknowledged the power of ideology to control human thought and behavior and they agreed with Hunter that whatever happened to the GIs in Korea, it involved "two processes, a softening up and an indoctrination process." As Hunter explained to Arens's committee, the GIs were softened up by being presented with alarming and incredible information—they had been betrayed by their families, their girlfriends, or their country. One of the most effective surprises for the American G.Is, Hunter told Congress, was evidence that some of their Americans were in fact pro-communist:

> One of the most corrosive publications was a magazine called *The China Monthly Review*, published by American pro-Communists in Shanghai. The men couldn't get over the shock of an American-edited magazine being put out in Red China. . . . In a number of cases, it was the decisive factor in their softening up.

At this point, the brainwashers

> presented themselves to [a victim] as his new friends, as "Big Brother," who would always stand by him through thick and thin, who would always love him. The cruelties they had perpetrated on him they now interpreted as the discipline of a kindly father.[30]

When the transformation was complete, these new communists possessed an entirely different moral character and outlook. Their healthy and patriotic grasp of "absolute moral standards" was replaced by dialectical materialism and its dogma that "everything changes, and that what is right or wrong, good or bad, changes as well." "They don't say this in so many words," Hunter explained, at least when speaking to outsiders. But they will use those words when speaking "to those who are already indoctrinated in communism."[31] For those who had already converted, it made perfect sense.

Academic accounts of brainwashing were similar, if only because any explanation had to take account of two phases, the before and the after, of ideological conversion. Some theorists, however, discerned a third, intermediate phase. Differences also emerged over the forces driving the transformation. Lifton believed the main engine of psychological conversion was "confession"—the honest and public realization that one's former beliefs had been in error. "Whatever its setting," he wrote,

> thought reform consists of two basic elements: *confession*, the exposure and renunciation of the past and present "evil"; and *re-education*, the remaking of a man in the communist image.

Along with the ritual of confession, Lifton spoke of a metaphorical resurrection. Two subjects he had interviewed, for example,

> took part in an agonizing drama of death and rebirth. In each case it was made clear that the "reactionary spy" who entered the prison must perish and that in his place must arise a "new man," resurrected in the Communist mold. Indeed, Dr. Vincent still used the phrase, "To die and be reborn"—words which he had heard more than once during his imprisonment.

It was "an appeal to inner enthusiasm through evangelistic exhortation," Lifton wrote, that "gave thought reform its emotional scope and power."[32]

The Normal Brain

While Lifton compared the dynamics of brainwashing to those of religion, Edgar Schein found a metaphor in basic chemistry. Schein described brainwashing as a process of "unfreezing," "changing," and finally "refreezing" an individual's beliefs in new, stable pattern. Schein hoped his icy metaphors would bring the debates about brainwashing back to observable, scientific matters. More than most investigators, he worried that much of the research about brainwashing was unscientific. In a review of Meerloo's *The Rape of the Mind* and Aldous Huxley's *Brave New World Revisited* (an update of his classic that includes a chapter devoted to brainwashing), Schein questioned Hunter's original reporting and doubted that prisoners of war who may have signed their names to a "peace petition" had really been converted to communism. After all, they might have signed in order to signal to their commanding officers that they had been captured and were healthy enough to sign their names.

As for the breakthrough in Pavlovian psychology that Hunter described, Schein was dismissive. He knew the current state of Soviet psychology and doubted "the conclusion that a Soviet corps of scientists have collaborated to evolve an irresistible tool for mental destruction." The techniques applied to prisoners of war in Korea are "not new, diabolical, or scientific." They are merely "the result of trial and error within the hierarchy of the secret police and the Communist Party and have been developed empirically."[33]

Still, Schein agreed that *something* remarkable had happened in Korea. As a young army officer with a PhD in psychology from Harvard, he was among the first to interview the American GIs after their release. He saw a pattern to the stories they told, but did not believe that Hunter, Meerloo, or others had found a scientific basis for understanding it.[34] Eschewing the word *brainwashing* and stereotypical imagery of drugged or hypnotized subjects, Schein proposed that the GIs had been subject to a process better called "coercive persuasion." "The essence of coercive persuasion," he wrote, "is to produce ideological and behavioral changes in a fully conscious, mentally intact individual."[35] What happened to the GIs, in other words, was not too different from what happens to those in psychotherapy, prisons, schools, Alcoholics Anonymous, religious orders, and monasteries.

Reviewing Schein's book *Coercive Persuasion*, Theodore Chen was ill at ease with this unorthodox approach. He thought Schein spent too little time discussing "the Chinese scene" and "Marxist ideology." And he predicted that "[s]ome readers may be disturbed by the failure to distinguish between Communist 'brainwashing' and indoctrination in non-Communist society." But he understood that Schein was a scientist, not a political commentator, and—at least to the extent of quoting it—he condoned Schein's disturbing conclusion:

> Chinese Communist coercive persuasion is not too different a process in its basic structure from coercive persuasion in institutions in our society which are in the process of changing fundamental beliefs and values.[36]

These two researchers, at least, had come to agree that whatever brainwashing is, it happens all the time in modern, even democratic societies.

Liberation by Philosophy

The Structure of Scientific Revolutions was published in 1962, amid this academic interest in the brainwashing phenomenon. There is no evidence that Kuhn followed these particular debates closely, but as the chapters in Part III will show, he shared these scholars' interests in psychology, the power of ideology, and the workings of language as they conspire to shape the very texture of experience and the contours of understanding. It is likely, moreover, that Kuhn's future readers would understand the provocative claims he made in *Structure* partly on the basis of this longstanding and widespread discussion about brainwashing.

His choice of the phrase "conversion experience," for example, evokes the dynamics of mind control as Lifton, Schein, and others discussed them. So understood, a revolution in some field of science involves similar phases and stages. They begin with the dogmatism and confidence characteristic of a normal scientific tradition. Then, as doubts emerge and grow, the community enters a period of "crisis" that concludes with the sudden, revolutionary transition to a new paradigm. What Kuhn described as "novelties" in normal science, like the puzzling observations that caught Roentgen's eye long before X-rays were understood—"his screen glowed when it should not"—function to "soften up" scientists and prepare them for the unthinkable prospect that their paradigm, despite their commitment to it, does not capture the truth of nature.[37] To borrow Lifton's terminology, those revolutionaries who reject the old and embrace the new paradigm will confess that their earlier scientific beliefs had been in error, that they had been blinkered.

Like Schein, Kuhn emphasized the power of social influence and persuasion. Given the incommensurability of different paradigms and their associated worldviews, paradigm debates must involve "techniques of mass persuasion" that lead scientists to reject one set of premises and to accept another. As he put it in his postscript to *Structure*, these debates are ultimately "about premises" and their "recourse is to persuasion as a prelude to the possibility of proof."[38] The resulting conversion can be sudden because once the mind is pried away from one scheme or ideology for understanding the world, it cannot function without *immediately* embracing a new one. Schein too had emphasized that the "normal" condition of the human mind is to be saturated or fully indoctrinated into some coherent system of beliefs and values. Brainwashers exploited this condition, for once they had succeed in weakening or "softening up" a subject's commitments, he or she quickly embraced the communist beliefs prepared for them—almost as if the human mind, abhorring an ideological vacuum, moved swiftly and naturally to a new, different kind of ideological commitment.

Conant had made this point, as well. In the scientific mind, he wrote, "a theory is only overthrown by a better theory, never merely by contradictory facts."[39] Kuhn developed this point within his sociology of normal scientific communities: "Once it has achieved the status of a paradigm," he wrote, "a scientific theory is declared invalid only if an alternate candidate is available to take its place." Scientists must either dogmatically cling to their current paradigm or let go of it in the face of mounting puzzles and anomalies so that a new way of seeing and understanding nature rushes in to fill the void. When this happens, Kuhn wrote, scientists will speak "of the 'scales falling from the eyes' or of the 'lightning flash' that 'inundates' a previously obscure puzzle."[40]

While Conant ascribed these dynamics to "conceptual schemes," Kuhn attached them to the paradigms at the heart of normal science and to this larger, cold war image of philosophy as a subversive enterprise that, by asking too many questions, threatens to overturn the status quo. What makes Kuhn's "normal science" normal, that is, is unquestioning commitment to a paradigm. It is "the activity in which most scientists inevitably spend almost all their time." To help maintain this normalcy, it instills a sense of confidence and dogmatism—"the assumption that the scientific community knows what the world is like."[41] Not unlike Hoover's communist ideologues who are caught in a system of ideas they do not question, Kuhn's normal scientists "have not generally needed or wanted to be philosophers." "Indeed," he explained, "normal science usually holds creative philosophy at arm's length, and probably for good reason." For creative philosophy excites the imagination, encourages exploration of alternative conceptual possibilities, and plants suspicions that paradigmatic truths about nature and how it is best studied may be incomplete or misleading. For the same reason, normal science makes "fundamental novelties" invisible to scientists. Since no theory can explain "all the facts with which it can be confronted," theories are naturally surrounded by subversive phenomena that, were they to capture scientific attention and become recognized anomalies, may inspire fundamental questions that derail normal research. "Normal science . . . often suppresses fundamental novelties because they are necessarily subversive of its basic commitments."[42]

Kuhn banished philosophy from normal science; but he welcomed its return during times of crisis. When the end of a paradigm's reign is nigh, and no amount of dogmatism and determination can save it, philosophy is welcomed back for the same reason that it was first banished: it helps unlock the mind from a now-faltering paradigm. It "loosens the stereotypes" and "prepare[s] the scientific mind to recognize experimental novelties for what they are." The philosophical "search for assumptions," he wrote, can be "an effective way to weaken the grip of a tradition upon the mind and to suggest the basis for a new one."[43]

Of course, any comparison of Kuhn's scientists to Edward Hunter's captured GIs goes only so far. The scientists Kuhn described are not forcibly captured, manipulated, tortured, or forced to accept a new paradigm during times of revolution. They are, however, educated rigorously and their professional careers are dependent upon their accepting the reigning paradigm that shapes their thought and the ways they perceive the world. When revolution looms, the realities of human psychology and the sociology of human groups have a productive effect similar to those processes Hunter described. By transporting them to new planets or worlds of experience and thought, scientific revolutions create something like the "new Soviet man" Hunter discussed in his congressional testimony.

The Black Queen of Diamonds

For his new theory to seem credible, Kuhn had to establish that, contrary to conventional images of open-mindedness and intellectual daring, professional scientists are in fact dogmatic and narrow-minded. The dogmatism of normal science—an "immense restriction of the scientist's vision" and a "considerable resistance to paradigm change"—plays an essential role in the emergence of novelties that first spark the revolutionary process:

> [N]ovelty emerges only for the man who, knowing *with precision* what he should expect, is able to recognize that something has gone wrong. Anomaly appears only against the background provided by the paradigm.[44]

To document how strong this confidence is, Kuhn appealed to research in human perception showing that a scientist's beliefs about nature will shape or bend perceptions to support or protect prevailing beliefs.

Normal scientists are like the subjects in the experiments undertaken by Jerome Bruner and Leo Postman in which "anomalous" playing cards (such as "a red six of spades and a black four of hearts") were briefly displayed to subjects who were then asked what they had seen. Kuhn explained,

> For the normal cards these identifications were usually correct, but the anomalous cards were almost always identified, without apparent hesitation or puzzlement, as normal. The black four of hearts, for example, might be identified as the four either of spades or hearts. Without any awareness of trouble, it was immediately fitted to one of the conceptual categories prepared by prior experience.

Only with increased display times would subjects realize that their perceptions did not match their "conceptual categories." Some, however, never recognized what was amiss, even given long exposure times, and became upset:

> One of them exclaimed, "I can't make the suit out, whatever it is. It didn't even look like a card that time. I don't know what color it is now or whether it's a spade or a heart. I'm not even sure what a spade looks like. My God!"

"We shall occasionally see scientists behaving this way too," Kuhn added.[45] The playing card experiments fit Kuhn's intellectual purposes at the same time that they called out to another, now-classic cliché about the power of brainwashing from Richard Condon's novel *The Manchurian Candidate* of 1959 and, more vividly, its film adaptation in 1962. Here the queen of diamonds, when flashed before the eyes of the brainwashed protagonist, opens his mind to instructions from his evil controllers.

If Kuhn's normal scientists are seen as brainwashed, what controls their thinking is not evil scientists or politicians but the sociological norms and conventions of professional scientific life. Among the most powerful factors at work in this regime are scientific textbooks that justify and celebrate currently accepted paradigms and cloak the revolutionary nature of scientific progress. "Once rewritten" in the wake of a revolution, Kuhn explained, textbooks "inevitably disguise not only the role but the very existence of the revolutions that produced them." Instead, they substitute

> just a bit of history, either in an introductory chapter or, more often, in scattered references to the great heroes of an earlier age. From such references both students and professionals come to feel like participants in a long-standing historical tradition. Yet the textbook-derived tradition in which scientists come to sense their participation is one that, in fact, never existed.[46]

Here Kuhn referred his readers to one icon of cold war interest in mind control, George Orwell's novel *1984*, in which the dominant party maintains its control over citizens by controlling how history is written and rewritten:

> Day by day and almost minute by minute the past was brought up to date. In this way every prediction made by the Party

could be shown by documentary evidence to have been correct; nor was any item of news, or any expression of opinion, which conflicted with the needs of the moment, ever allowed to remain on record. All history was a palimpsest, scraped clean and reinscribed exactly as often as was necessary.

Kuhn put it this way: "Scientific education makes use of no equivalent for the art museum or the library of classics, and the result is a sometimes drastic distortion in the scientist's perception of his discipline's past." He knew what many readers would think. "Inevitably those remarks suggest that the member of a mature scientific community is, like the typical character of Orwell's *1984*, the victim of a history rewritten by the powers that be."[47] "That suggestion is not altogether inappropriate," he added.

It was appropriate because Kuhn's theory of science in *Structure* would grow out his belief that scientific theories function "as ideologies" within scientific communities. Their essential function was to *restrict* thought and perception so that the welter of the world's complexity would not overwhelm scientific inquiry. This insight built upon something Kuhn may have learned from Conant: "A well-established concept may prove a barrier to the acceptance of a new one," Conant wrote in *On Understanding Science*, especially when it has been "long-intrenched in the minds of scientists." The history of science itself was a "fight through thickets of erroneous observations, misleading generalizations, inadequate formulations, and unconscious prejudice"—most spectacularly in the case of phlogiston chemistry (which Kuhn also discussed in detail in *Structure* as a crisis-bound paradigm). Conant marveled at how intelligent men such as Joseph Priestley, James Watt, Henry Cavendish, and "scores of others continued to cling to the phlogiston theory for a decade" when, in retrospect, there was powerful evidence telling them that they were wrong. Conant saw two reasons for their dogmatism: One, the experiments required to present the new conceptual scheme convincingly were difficult; and two, phlogiston chemistry "had acquired an almost paralyzing hold on their minds."[48] This "almost paralyzing hold" was more than a verbal flourish. In these pages Conant returns twice to this idea that phlogiston chemists were just not thinking properly—that they seemed to live in "an Alice-through-the-looking-glass world!" which they were mysteriously unable to escape. "How they twisted and squirmed to accommodate the quantitative facts of calcinations with the phlogiston theory" that had captivated their minds.[49]

Conant understood that dogmatism such as this was a genuine part of science; it was a part of being human. But it was an *impediment* to the growth of knowledge. It was viewed the same way by Kuhn's friend, Barnard College sociologist Bernard Barber, who, about a year before *Structure* was published, examined how scientists may "resist" discoveries that are in front of their noses, seemingly obvious and waiting to be made. Barber found this fascinating, but he too took it to be an ironic curiosity—a deviation from "the powerful norm of open-mindedness in science."[50]

Beginning with his Aristotle experience in 1947, however, Kuhn glimpsed that Barber, Conant, and others understood things backward: revolutionary advances in science occur not *in spite of* this dogmatism but *because* of it. Normal scientists are dogmatic and close-minded about the science they know, Kuhn would reason, because only then will a community push a paradigm to its epistemological limits so that there is no doubt that it cannot be made to solve the puzzles scientists wish to solve. In turn, only by meeting those limits will the members of that community begin to consider a possibility for which their textbooks never prepared them: the theory in which they have invested their careers just might be false. And only then, through revolution, can science liberate itself from its faltering paradigm and move into its future.

The insight was deep and beautifully ironic. Dogmatism, widely seen as science's bête noire, was in fact essential for science's success and historical progress. With the possible exception of the Polish sociologist Ludwik Fleck (who had called scientific knowledge a "harmony of collective illusions"), no scholar, historian, or philosopher had seen the scientific process in this light. Conant, of course, had plenty of scientific experience and he knew the history of science well, too. But Conant was a devoted champion of liberalism in science and society. So it was not surprising, Kuhn may have reasoned, that when Conant looked at the history of science he saw only one thing: the human mind's struggle to be free, independent, and autonomous.

7

The Necessary Dangers of Consensus and Unity

"My Dear Sir," James Conant wrote on the piece of stationery before him. He did not have a name, for he could not know who the president of Harvard would be fifty years in the future. But he assumed that the president would be male and that if anyone were to read these words at that time his worst fears had not been realized. The "prophets of doom" would have been wrong. The West and Harvard's tradition of learning had survived and now, in the wake of that victory, Conant wished his future counterpart to know what the early 1950s were like.

It was 1951 and Conant was extremely worried. He did not need to be a specialist in nerve gas during the first world war or an administrator of the atom bomb project in the second to know that the end of civilization might be near. With the superbomb project underway, destructive power would multiply and the age of the "superblitz" would be at hand. It was entirely possible that everything Conant knew and loved about Harvard University, Cambridge, Massachusetts, the nation, and the world might soon be gone. "We all wonder how the free world is going to get through the next fifty years," he wrote.

Simply imagining the letter being unsealed and read fifty years hence must have been a comfort. It meant that the Soviet assault on freedom, liberalism, and democracy had been defeated, perhaps by the massive buildup of arms and armies that Conant promoted through the Committee on the Present Danger. In that case, Conant could be sure that whoever was reading was "in charge of a more prosperous and significant institution than the one over which I have had the honor to preside."

Upon finishing his letter, he put down his pen, folded the paper into an envelope, and sealed it for the university archivist along with clear instructions: give this message to the current university president at the turn of the next century, fifty years hence, "and not before."[1]

⁓

Conant revealed these worries and doubts neither to his colleagues nor to the public. He knew that leaders must be self-assured, confident, even dogmatic about the beliefs on which their leadership and public mission depend. Were the public to learn of his doubts, they might fall into despair and cynicism, or they might panic and overreact. That is why his public speeches typically aimed for a calming balance between ominous circumstances soberly described and reassuring plans to keep things in perspective and in control. In his first radio broadcast as head of the Committee on the Present Danger, he "tried simultaneously to soothe and steel the nerves of the American people, to prepare for war yet not provoke it, to seek peace yet not make hasty concessions for it."[2]

He cultivated this style of leadership during World War I. "Senior chemists coached him on the art of leadership, urging him to remember to keep a firm voice and to conceal doubts in order to project confidence and keep him from undermining morale."[3] He applied the same strategy as a national leader in education, blending a respect for authority and consensus with calls for liberal, critical thinking, and reminders that any good and valuable consensus emerges from the rigors of open debate and uncensored intellectual creativity. The two styles may seem incompatible, but Conant joined them at different times during his career—sometimes to jarring effect. He would devote Harvard, a bastion of liberal learning, to military education. He would present his new way to teach the history of science as a course in "the tactics and strategy of science." He would claim in *Education in a Divided World* that a crucial and effective weapon in the West's arsenal was not a new kind of bomb or military technology but rather the reform and expansion of public education.

Conant could mix styles, and he knew when to emphasize one over others. He could speak credibly and sincerely (if not always persuasively) to rural farmers, urban civic leaders, educators, A-list scientists, and rank and file soldiers. To those in his inner circle, his intelligence and sensibilities could be almost intoxicating. For example, his friend William Marbury, who listened to Conant instead of a battery of contradicting lawyers in the Hiss case, believed Conant could not be wrong about anything. Kuhn's admiration seemed boundless, as well. Even decades after his collaboration he said Conant was "the brightest person I had ever known."[4]

Yet ten years later, when Conant first paged through a manuscript for *The Structure of Scientific Revolutions,* he would recoil at the extreme degree of unity and consensus reigning within Kuhn's communities of "normal scientists." These communities seemed regimented and uniform. Each member was trained according to the requirements and worldview of a single paradigm. They were taught to accept it as truth, as if it were a sacred scroll, immune to criticism and critical debate, or a commanding officer whose instructions must be followed without question. Conant knew that dogmatism and authoritarianism could be important in science; they came to the fore when consensus and agreement were vital. When Conant developed new poison gases, or when he organized scientists and engineers to create the atom bomb, he knew that scientists must get on with research and trust existing beliefs and theories. There was simply no time to let imaginations run free or to indulge in theoretical exploration for exploration's sake.

Crucially, however, this was not the whole explanation for science's historical progress. Kuhn's description of normal science left out skepticism, debate, and blistering criticism of current beliefs. Skepticism and doubt, as much as creativity and conviction, were the engines of learning and human progress as Conant understood them—even if there were times that doubts and worries had to be suppressed, or folded inside an envelope for only the future to see.

Thomas Kuhn and "The Crisis in Democracy"

The roots of this future disagreement between Kuhn and Conant wound through Conant's style of leadership and farther back in time to Kuhn's childhood interests and sensibilities. Before he arrived at Harvard in 1940, Kuhn was drawn to the ideal of unity and consensus that Conant so confidently cultivated as a public leader. "Disunity has a distinctly detrimental effect upon our country," he wrote in an essay about democracy in America. Displaying "Thomas Kuhn '40" in the upper right corner, the essay may have been a thesis he wrote at the Taft School, most likely when he was heading into the personal turmoil over pacifism and intervention he later analyzed in "The War and My Crisis." Whatever the setting, a teenage Thomas

Kuhn was thinking hard about the nation and how it should respond to the European situation.

That summer he sent a copy of the essay to Edward S. Greenbaum. Kuhn's father Samuel would have known Greenbaum from the Jewish Board of Guardians, a philanthropic social service in New York City on whose board of directors they served. Greenbaum was a prominent lawyer who, after serving in World War I, became a leader in court reform, an author, and a trustee of the Institute for Advanced Study in Princeton, New Jersey. Between the wars, Greenbaum's life mixed intellectualism and the kinds of progressive liberalism that would become professionally dangerous during the cold war. Morris L. Ernst, for example, Greenbaum's partner in his law firm, helped to found the American Civil Liberties Union as well as the National Lawyers Guild, first established as a liberal counterbalance to the conservative American Bar Association. In her *Red Channels*, Elizabeth Dilling refers to "Morris Ernst and other radicals" behind the A.C.L.U., the American Newspaper Guild, and various labor organizations. Though she misspelled his name, Dilling pointed to Greenbaum, too, as a partner of "Greenebaum, Wolff and Ernst, N.Y.C."[5]

Despite these potentially controversial associations, Greenbaum and Ernst remained far from public controversy. Ernst was in fact on friendly terms with J. Edgar Hoover, served as his personal attorney for various matters, and vigorously defended Hoover's FBI in the popular pages of *Reader's Digest*. Greenbaum's reputation as a veteran and a "legal perfectionist" (as the *New York Times* put it) positioned him far above suspicion. He returned to service in World War II as a brigadier general renowned for solving problems and brokering disputes between the Pentagon and its many suppliers and contractors. "Everybody felt he was a reasonable man," the *New York Times* reported, and everybody knew him from afar when he drove his "brown bomber"—a wheezing 1932 Chevrolet—to work every day at the Pentagon. After serving as executive officer to Robert Patterson, the nation's secretary of war, Greenbaum was awarded a Distinguished Service Medal in 1945.[6]

As the cold war commenced, Greenbaum became a cold warrior who joined Conant's Committee on the Present Danger. He helped to plan the new committee and his name appears with Conant's under the ominous official statement that the committee first released to the public: "The bitter fact," it concluded, "is that our country has again been thrust into a struggle in which our free existence is at stake, a struggle for survival."[7] Greenbaum's patriotism would be publicly acknowledged again in 1967, when his skills as a political go-between led George F. Kennan, his Princeton neighbor, to propose that Greenbaum handle the West's latest geopolitical triumph: the defection to the West of Svetlana Alliluyeva, Josef Stalin's daughter. The master Americanizer, now seventy-seven years old, met her in Switzerland, ushered her to the United States, and became a friend who helped her publish her best-selling book, *Twenty Letters to a Friend*.[8]

When Kuhn sent his essay to Greenbaum it was 1940, and Greenbaum belonged to the board of the Citizenship Education Service, an organization within the American Jewish Committee that sought to educate and Americanize newly arrived immigrants—especially to warn them away from subversives that might try to recruit them. The service published cartoon pamphlets such as "Footprints of the Trojan Horse" that warned of radicals seeking to "divide and conquer from within." In campaigns like this, Greenbaum's committee defined Americanism as Conant's objectives committee defined it—a unity rooted in the principled toleration of differences. After Pearl Harbor, Washington itself would adopt this approach to minimize ethnic rivalries within the ranks. An advertisement for war bonds from Rhode Island's Independent Textile Union announced that Hitler aimed to set "Negro and White against each other" and "native-born against naturalized citizen." But it was easy to foil Hitler's scheme, the poster proclaimed: "Be Big . . . be liberal . . . be tolerant . . . be American!"[9]

In sending "The Crisis in Democracy" to Greenbaum, Kuhn evidently believed his essay could perform a similar, unifying function, not just for recently arrived immigrants but Americans of all stripes. "Today, wherever we look," he wrote, "storm clouds hover. The European war threatens to engulf the entire civilized world." But "even if we should succeed in avoiding the conflagration," he explained, "we will still be faced by grave internal difficulties."[10] The remedy for those difficulties, he would explain, was a clear and strong national consensus about the power and supremacy of American democracy.

Kuhn was still a pacifist and his socialist values remained in place, as well. The internal difficulties he described included high unemployment, friction between labor and capital, and the nation's debt—all of which called for a solution that may require "a vast change in our economic structure." On the same pages, however, he voiced dislike and mistrust of Communists and revolutionaries seeking to change government. "Recently there was held in Madison Square Garden a mass meeting," he wrote, likely referring to the 1939 meeting of the Young Communist League. All citizens should be proud that meetings like this can take place, he wrote.

> But all of us should be frightened that there was no reaction among the American people, no reaction to show our youth that the ideals upheld were false and to inculcate in them a passion for democracy and an eagerness to live, to work, and to fight for it.

One need only listen to the soapbox orators in New York's Union Square, Kuhn wrote, to see an enormous difference between revolutionaries and defenders of democracy. The revolutionaries "are trained men with a complete knowledge of all the known devices used by a demagogue to sway a mob." The conservatives are sincere but mere "amateurs" unable to rally their audiences with comparable conviction and fervor. This was the root of the crisis at hand. "Today the star of the dictatorships is rising while that of the democracies wanes" because too many Americans are ignorant about their own political circumstances. They lack "a passion for democracy and an eagerness to live, to work, and to fight for it."[11]

Do I exaggerate? Kuhn asked. Not at all. The situation was so bad, he explained, that fellow students he interviewed about the European situation were apathetic, even nihilistic. "My view is of no importance," one told him. "I shall just have to sit here until the President or the Congress decides to send me to be shot." Having heard similar remarks frequently from his peers (supposedly "a particularly enlightened section of the populace"), Kuhn was sure that he was not exaggerating and assured his readers that the problem must be even worse in the streets and among the general public.

Parts of Kuhn's diagnosis anticipated the objectives report by Conant's general education committee. Citizens were apathetic and uninformed for many reasons, but one was the nation's rapid growth. Another was the growing tabloid press that failed to keep citizens informed and engaged as they had been "in the day of the town meeting." The result was an "impotency complex," Kuhn explained—"a natural product of our success and expansion; but today it has weakened us tremendously in our hour of need. Today it is the real enemy!"[12]

Kuhn's proposed solution, however, lay not in the kind of general education Conant had in mind but in general propaganda. He quoted approvingly Thomas Mann, the self-exiled German novelist then touring the country to persuade Americans that they must not merely brace themselves and hope to ride out Hitler's crusade. In his speech "The Problem of Freedom"—subtitled "The Crisis in Democracy"—Mann had similarly urged Americans to go on the ideological offensive: "A militant democracy is the need of the day, a democracy freed of all self-doubt, a democracy that knows what it wants, namely: victory—the victory of civilization over barbarism!"[13]

"The task before us was well stated by Thomas Mann," Kuhn wrote: democracy must "renew and rejuvenate itself by again becoming aware of itself." But one man could never rouse the nation to achieve this goal, Kuhn reasoned. It would require the power of mass propaganda, a power shown to be effective in Hitler's Germany. Instead of a force for ill, that is, propaganda could be used to *reduce* the ideological confusion and lethargy at hand. Kuhn explained,

> We must sort many confused ideas; we must find among the confused mass of present day ideologies the one which was once the prime-mover of American life; we must personalize our love of democracy, as we have personalized our hatred of dictatorship, so that it may once again weld all Americans together as a force for the right. The only means available to a democracy is education.

"Through the schools, the press, the movies, the radio, and every possible modern means," he preached, "we must begin to reeducate the people of America." They must be made to appreciate "the history and meaning of democracy." Popular study of "the referendum, the recall, the merit system, the citizens' committee, and the lobby, should be a great step in combating the impotency complex."

"No, we have no American equivalent of the Ministry of Propaganda!" Kuhn exclaimed. But the nation needed one now, he believed. It must be powerful and firmly joined to schools and "public and private committees for the origination and dissemination of positive propaganda." "So long as we do not overstep in any way the limits of democracy," he reassured his readers, American could and should "learn a lesson from the efficiency of our opponents" in Germany and utilize the powers of propaganda.[14]

"I think you have done a swell job," Greenbaum replied. He encouraged Kuhn to think of ways that his ideas about democracy "could be promoted in schools and colleges." But a public call for a United States Ministry of Propaganda was probably beyond the comfort zone of the American Jewish Committee. As moving and rousing as the essay was, Greenbaum wrote, "I do not know of any way in which it can be used by the committee and am therefore returning it to you."[15]

Military Science and Political Revolutions

In science as well as politics, the unity of unquestioned consensus played a central and recurring role in Kuhn's thinking. Though he called forcefully for consensus and shared conviction in "The Crisis in Democracy," he found that he personally disliked consensus and shared conviction in science. His recollections decades later about why he left physics for history of science suggest a personal aversion to professional routine and orthodoxy. At Harvard's Radio Research Lab, where he worked shortly after graduating in 1943, Kuhn took a scientific job "largely cooking standard formulas" to locate enemy radar transmissions so that they could be jammed. "I couldn't have begun to understand" where the equations themselves came from, he recalled, "or at least I didn't think I could, and I wasn't given time to find out." By 1944, he was transferred to England, first to Great Malvern and then to a base outside of London at the United States Strategic Air Force Headquarters. There, he said, he was "having trouble, sort of fitting in and getting interested in what I was being asked to do."[16]

So far, Kuhn had worked technically as a civilian. Only in France, touring the nation to inspect and gather intelligence from radar installations erected by invading Germans, did he wear a uniform (to lessen the chances of being accused of spying were he to be captured by remaining German forces). In Rheims, ironically, he found his work less regimented and more interesting. He worked outside the U.S. Army with experts "who applied mathematics and science to strategic and other problems in a rather unsystematic way"—a way that allowed scientists "to talk back to the general, as nobody within the army structure really could." Still, Kuhn was learning "that I was not all that interested in radar work." There were many factors involved in his subsequent transfer into history of science, including, of course, his opportunity to work with Conant. Another was that he found the graduate-level physics he studied after returning to Cambridge dull and uninteresting.[17]

Kuhn disliked the narrowness and specialization he saw in his future. "If I was going to go through graduate school successfully and on, I really had to focus my attention in one spot and give my full energies to it; and I found that hard to do." Studying solid-state physics with John Van Vleck, Kuhn missed the excitement of intellectual discovery and theoretical change that he described in his undergraduate essays, that he had read about in books by Bertrand Russell, Percy Bridgman, and other scientists and philosophers. Looking at a future spent refining the parameters of well-established theory (illustrated by his published papers in physics) Kuhn knew that he would not likely encounter that kind of intellectual excitement. In the language of *The Structure of Scientific Revolutions*, his ambitions lay in "revolutionary" science, not "normal science" and its "mopping up operations" (such as refining precise values of theoretical constants or applying known formulas to poorly understood phenomena).[18]

Unlike his memories of military science, however, Kuhn's memories of his military life in Europe are dramatic and exciting. In France, looking for a group of Americans he was to meet, he and a captain drove to Paris, passed by Chartres Cathedral and a slower-moving convoy, and finally settled in at the Petit Palais.

> We'd been there for about an hour, when this convoy started to come down the Champs Élysées. It was de Gaulle entering Paris! And suddenly there was rifle fire from somebody on the roof of the building across the street. Somebody fired, some member of the milice [the French militia] who was shot down. And there was still fighting out at Le Bourget, the other side of Paris. So, it was an exciting time.

Another unforgettable moment, Kuhn recalled, was seeing "the flattened city of Hamburg, I'll never forget it. I also saw Saint Lô"—a medieval walled town almost completely destroyed during the war—"and I'll never forget it."

"None of this," he said of Hamburg and Saint Lô, "has a lot to do with what happened later, except that as all of this was going on I increasingly realized that I was not all that interested in radar work." Yet Kuhn would later write a book comparing scientific revolutions to political revolutions, a book whose basic architecture—the categorical difference between two kinds of scientific experience, the dogmatic routines of "normal science" and the dramatic, world-changing consequences of "revolutionary science"—seems to come together in these recollections.[19]

The Search for Unity

When Kuhn joined Conant in his general education project, unity and social consensus remained central themes. Chapter 5 of the redbook, the "Objectives" report that Kuhn reviewed for the *Harvard Alumni Bulletin*, was called "The Search for Unity." The chapter surveyed very different proposals for what the unifying, guiding principles of modern education and modern life might be. Some Americans, for example, opted for the unifying umbrella of sectarian, religious orthodoxy—as implemented by Catholic schools and universities and by most American colleges until several decades before. Other institutions such as the University of Chicago and St. John's University took a great books approach and found unity "in the great writings of the European and American past." Progressive, "life adjustment" educators chose instead to ignore the past and focus on the present. For them, education "tries to organize knowledge around actual problems and questions" that students will encounter throughout modern life.

More promising, but still not acceptable to Conant and his committee, was the scientific view of the world represented by "pragmatism"—by which the committee referred as much to the pragmatic philosophers John Dewey and William James as the philosophers in the Vienna Circle and the unity of science movement. Conant liked this empiricist sensibility and attitude—"a habit of meeting problems in a detached, experimental, observing spirit," as the report put it—and he once characterized his own philosophy of science as a "mixture of William James's *Pragmatism* and the logical empiricism of the Vienna circle, with at least two jiggers of pure skepticism." But what worked for Conant or Harvard, he understood, would not necessarily work for the enormous and varied population of the United States. In a pious, religious nation, the view that only science counted as genuine knowledge and that other realms of culture such as art and religion veered into mere subjective expression or empty metaphysics would plainly not do. No unifying platform for general education could "omit as irrelevant the whole realm of belief and commitment by which, to all appearances, much of human activity seems in fact swayed."[20]

A larger, more flexible framework was required, the Objectives report maintained. No single source, be it religion, great books, or science and logic would ever capture the subtle balance between unity and diversity that Conant sought:

> This logic must be wide enough to embrace the actual richness and variegation of modern life. . . . It must also be strong enough to give goal and direction to this system. It is evidently to be looked for in the character of American society, a society not wholly of the new world since it came from the old, not wholly given to innovation since it acknowledges certain fixed

beliefs, not even wholly a law unto itself since there are principles above the state.

The unity could not be imposed or mandated; nor could it rely on a specific creed, doctrine, or theory. For in America one would never find widespread agreement about these "ultimates." It had to be forged instead from an ethics of toleration and pluralism. It would rest on

> certain intangibles of the American spirit, in particular, perhaps, the ideal of cooperation on the level of action irrespective of agreement on ultimates—which is to say, belief in the worth and meaning of the human spirit, however may one understand it. Such a belief rests on that hard but very great thing, tolerance not from absence of standards but through possession of them.[21]

This unity was complicated and filled with tensions, balances, and even outright contradictions. The United States remained united through its history, including a civil war, partly by learning to balance and adjust conflicting ideas and values against one another: "Democracy is a *community* of free men," the committee emphasized, and it must balance the freedom of individuality with the obligations of cooperation and "social living."[22] As he would soon tell the nation in *Education in a Divided World*, a program of general education based on this conception of American unity would be a powerful and surely effective response to the threat of Soviet totalitarianism.

The Ever-Present Danger of Totalitarianism

Supporting Conant as an editor of *The Crimson* before the war, and collaborating in the general education program after, Kuhn had many opportunities to know Conant as a leader who cultivated unity and consensus in all spheres of human endeavor and action. But Conant was also wary of unity and consensus as an obstacle to progress. It can become an end in itself instead of a means, a framework, for "moving ahead," as his signature definition of science put it. One of the first signs of this degeneration, he once explained, is the formation of rigid, like-minded groups that will neither tolerate nor consider competing views and opinions. Speaking to the graduating class of 1937, Conant himself addressed the "storm clouds" gathering in Europe that

Kuhn described in his essay and, for some measure of perspective, recalled how Americans responded to World War I, twenty years before:

> Separate worlds formed within this country and soon ceased to communicate with each other by rational speech: the pro-Ally camp, the pro-German group, the neutrals hated by both, then the pacifists and the militarists. All this while we were still only spectators of the conflagration which was consuming Europe.

Neither gifted poets (he would cite MacLeish's poem on this occasion) nor Harvard's professors were immune to intense pressures. On any issue, it seemed, there was "a potent blast of hostile criticism" from opposing groups. Probably anticipating yet another debate over intervention in Europe, Conant counseled these graduates to remain strong and independent and never to give in to some alluring ideology. He quoted St. Paul's epistle to the Romans—"Be not conformed to this world; but be ye transformed by the renewing of your mind"—and Ralph Waldo Emerson's praise for "the great man . . . who in the midst of the crowd keeps with perfect sweetness the independence of solitude."[23]

Conant described again the seductive allure of consensus and social unity in his valedictory speech in 1943, when a large segment of Kuhn's class graduated early, in January, to take up their new posts in the armed forces. (Kuhn was among a smaller group who remained to graduate in May, but he likely heard Conant speak in Sanders Theatre.) Conant addressed "the hazards to liberty when the war is won" and traced them to the contradiction between values and methods in times of war and times of peace. In order to achieve a speedy and effective victory, he explained, the nation was pulling together and devoting all of its resources to the war effort. At Harvard, Conant had already sacrificed liberal arts education. Across the nation, individuality, personal preferences, and even freedom of speech were taking a back seat to maximize morale and unity of purpose. But this unity and consensus could become a destructive and dangerous trap. "Indeed," he explained, all this unity and consensus required to win the war "could engender such conditions in our minds that we would be unable to preserve liberty when the time of peace had come." By fighting hard to defend those values in Europe, in other words, Americans may be transformed into citizens that no longer value liberty supremely. So "it will be your task as citizens of the United States," he told the graduates on the way to war, "to see to it that a totalitarian virus does not corrupt this nation."[24]

Yet again in September 1945, at the opening of the fall term, Conant returned to these worries as he discussed the "civil courage" with which citizens must reject the unified methods and collective mentality of war and once again embrace freedom, democracy, pluralism, and individuality. This postwar "period of psychological reconversion" was just as important and momentous as the reconversion of factories and workers to a peacetime economy and the return of soldiers to civilian life. It was a moral conversion, as well, Conant explained, since "the moral imperatives of the battlefield must be transformed into those of a free society which believes in the supreme significance of each individual man or woman."[25]

The United States had helped defeat Nazism, but that did not mean the threat of totalitarianism had been eliminated. The history of Germany from the rise of Hitler to the end of the war must always remain "a matter of study and reflection" for those Americans who value their liberty and freedom. He elaborated,

> Questions like these keep recurring in our minds: How could the Nazi doctrines gain such ascendency among a highly literate and apparently well-educated People? How could that nation breed such rulers and such brutal gangsters and allow them to terrorize the population? One answer to such questions has been given by a learned German professor in a private letter I saw not long ago. The writers philosophizing on the triumph of the Nazi party (which he had done nothing to prevent) explained the situation in these words: "One reason is that the education of the German people was carried out not only in frequently excellent schools but also on the military drilling grounds. Consequence: high mental and intellectual development, great military bravery, yet at the same time lack of civil courage, as many including Bismarck have said before. Lack of civil courage fosters political mass psychosis."

Conant did not say who the learned professor was, but he agreed with the diagnosis. It was worth asking, therefore, whether it might occur in the United States. Germany, after all, was not unique: "History records many examples of political mass psychosis" and—it seemed to Conant—"our highly industrialized and over-urbanized society provides a medium highly favorable for the development of this disease."[26]

But Conant remained optimistic. The danger he had pointed to three years before seemed to abate, for on all sides we see evidence that the nation recognizes fully the necessity for reorienting our sights. We know that the end no longer justifies the means. We know that the collective demands of a group—a ship's company, a regiment or a bombing crew—no longer have life or death hold over an individual. For military courage we must substitute civic courage.[27]

Still, he took pains and risks (it would appear in retrospect) to set a new tone for public discourse after the war. It was now not only possible but important, Conant seemed to say, to look squarely at the nation, to analyze and discuss its strength and weaknesses, and even to compare it in some respects to pre-Nazi Germany. That itself took civil courage and illustrated Conant's point: whether or not mass psychosis or the totalitarian virus took hold in the United States was not a roll of the dice. The nation's political future would depend on how Americans understood themselves and the world, the values they fought to defend, and how they acted and spoke to each other. That was why Conant did not keep this fundamental worry to himself. The best defense against domestic totalitarianism was widespread awareness of these difficult reconversions and their historic significance. "Every young American, particularly every soldier," he explained to Kuhn's class of 1943, must be aware. "For only by recognizing the dilemma may we hope that it will be solved."[28]

Reconversions Delayed

The cold war and the struggle for men's minds defied Conant's hopes. Instead of a postwar reconversion to liberalism and pluralism, a new wave of political conformity crested by the late 1940s, and Americans united and closed ranks behind fears and stereotypes of communist subversives in their midst. Within four years, Conant announced his policy opposing Communist faculty and both he and Paul Buck, one researcher later discovered, cultivated a secret and friendly relationship to J. Edgar Hoover's FBI.[29] Given Conant's and Harvard's national stature, these developments surely contributed to the repressive, conformist atmosphere that enveloped the nation. But even before Conant's conversion to cold warrior began, when he remained optimistic about international cooperation with the Soviet Union after the war, his writings about the history of science and general education remained saturated with military terminology and images of the military culture that Kuhn found constricting and dull.

True, the campus was no longer a naval training center, but the militaristic "tactics and strategy" of science as Conant understood them were front and center in the press proofs of *On Understanding Science* that Conant asked Kuhn to study closely in 1947. Military language also abounded in the Objectives report that Kuhn reviewed for the alumni magazine. Conant and his committee used this language self-consciously and with explicit qualifications, as if Conant's personal reconversion from military scientist to liberal educator was proceeding a little slower than he expected. As long as this incongruity was made plain, the text implied, there should be no confusion:

> The original title of this committee contained the word "objective." That is a term current in educational jargon but it almost belongs today to military science, and, though some of the implications of the comparison may be regrettable, it is clearly useful. We have been concerned with the strategy in the first place, to a less degree with the tactics, and in some measure with the logistics of an enterprise which is rightly to be regarded as a struggle. The struggle is as old as man himself. It may be looked upon as man's effort to become in actuality more nearly what he is in idea.[30]

Readers must not confuse this military style with the liberal, pluralistic ends Conant intended them to serve. If successful, the new general education program would teach Americans that liberalism and progress lay *not* in conformity, narrow-mindedness, or following the dictates of a single, widely accepted ideology. It lay rather in understanding and challenging other ideologies and other cultures on their own terms as sympathetic rivals. "Old fashioned isolationism," Conant would soon write in *Education in a Divided World*, "is dead."

Yet as cold war paranoia intensified, Conant himself eschewed this tolerant, ecumenical posture, joined the new anticommunist consensus, and as head of the Committee on the Present Danger became a nationally recognized spokesman for it. When he took to the airwaves on February 7, 1951, as head of the new committee, his first sentences signaled that national unity and like-mindedness were once again paramount, just as they had been during World War II:

> The United States is in danger. Few would be inclined to question this simple statement. The danger is clearly of a military nature. On this much we can all agree. The Congress of the United States has, with almost no dissenting voices, authorized a vast program of rearmament and mobilization.[31]

In his private letter to the future president of Harvard, he confided his doubts and imagined what Cambridge and the rest of the nation might look like if all the committee's efforts were for naught; or if they backfired, exacerbated tensions, and sparked a nuclear superblitz. But when speaking to the nation he projected confidence that unity and consensus were a path to survival. If all Americans were on board, and all these drastic and expensive steps were taken, he argued, there would be "a chance, a good chance, of avoiding World War III."[32]

In early 1953, when he chose to end his presidency of Harvard and embark on a new career as a diplomat in Germany, Conant returned again in his final presidential report to the theme of civil courage and the seductive dangers of unity and consensus. With Congress's inquisition headed to Cambridge that spring, he seemed to confess that the nation's obsession with ferreting out hidden communists at colleges and universities had gone too far. Projecting national unity and military strength abroad was one thing; but the nation's intense fear of communism now threatened to erode the ideals that anticommunism supposedly sought to protect.

He did not mention Sidney Hook by name, but Conant took aim at Hook's recommendation that educators police each other to keep the communist conspiracy at bay. On the contrary, "if there are members of the staff of any university who are in fact engaged in subversive activities," Conant wrote, "I hope the Government will ferret them out and prosecute them." This was a matter for law enforcement, in other words, and was not a legitimate concern for any academy that is truly free and liberal. For the nation's "tradition of dissent," long cultivated by its universities and colleges, was one of its greatest strengths and sources of creativity and progress. "The global struggle with communism turns on this very point," Conant explained as he once again pointed to "the dramatic examples of what occurred under the Nazi and Fascist regimes as well as what is now going on in totalitarian nations."[33]

For a smart and close observer of Conant such as Thomas Kuhn, already intrigued in his youth by the intellectual and cultural power of "positive propaganda," it is easy to imagine that these complexities and tensions within Conant's ideals of consensus and unity faded into the background—certainly so when Conant voiced his doubts and second thoughts only in private letters to the future or in his personal diary. Suppressing the liberal arts at Harvard during the war in order to defend and preserve them; projecting confidence and unity in public while keeping his fears of nuclear annihilation private; teaching the history of science in military language of "tactics and strategy;" and defending postwar liberalism by cultivating a national political consensus that, predictably, eroded "civil courage"—for Conant these were deviations from the ideals of liberalism he had defended all his life. They were steps taken either for urgent practical necessities, such as winning the war in Europe or reforming American education with military swiftness and efficiency. Or they were unfortunate consequences, such as the excesses of academic anticommunism he discussed in his final report.

Especially in times of national emergency and present danger, Conant would probably say, leaders and the institutions or nations they lead can be excused for courting these inevitable dangers. For most Americans, the years from 1947 to 1953 were extremely perilous and saturated with fears of atomic conflagration and domestic ideological subversion and revolution. For Thomas Kuhn, as it happened, those were the years in which he worked closely with Conant, reinvented himself as a historian of science under his guidance, and first conceived *The Structure of Scientific Revolutions*.

Part III

The Cold War Origins of
The Structure of Scientific Revolutions

8

The Language, Psychology, and Psychoanalysis of Scientific "Reorientations"

Thomas Kuhn always found learning easy. As he explained to Robert Gorham Davis in his essay "The War and My Crisis," he was proud of his intellectual ambitions and skills; they were part of his liberalism and his respect for reason. At Harvard, he reviewed poetry for *The Crimson*. He held his own discussing Proust over madeleines at dinners of the Signet Society. He worked through equations in solid-state physics with his dissertation advisor, John Van Vleck (who would later win a Nobel Prize). Intellectually, Kuhn could do almost anything, and he knew it.

But Aristotle he missed completely by trusting his preconceptions. When he first read *De Physica* closely, he agreed that Aristotle's methods, his observations, and conclusions seemed mistaken. Recounting the experience decades later, Kuhn first suspected the fault may have been his. "I could easily believe that Aristotle had stumbled," he wrote, "but not that, on entering physics, he had totally collapsed. Might not the fault be mine rather than Aristotle's?" Possibly a linguistic barrier had separated Kuhn from what Aristotle was really trying to say. "Perhaps," Kuhn put it, "his words had not always meant to him and his contemporaries quite what they meant to me and mine."[1]

Aristotle's vision of nature snapped into view for Kuhn when he allowed words he knew to take on different, more expansive meanings. "Motion," for Aristotle, did not mean only the movement of a body from one place to another, as it does within Newton's physics. It means that and much more, including growth (as of a plant or animal) and other sorts of qualitative and quantitative change. Objects are not just assemblages of physical matter that obey strict laws of movement through space, as Newton and before him René Descartes had presumed. For Aristotle, Kuhn learned, matter was "a neutral substrate" whose physical behavior was determined by the qualitative properties it took on. Heavy objects such as stones fall down toward the center of the earth because that is their natural place; while flames and embers, dominated by the lightness fire, flee up and away from the earth, toward the stars and planets we see at night, where such things naturally belong.

Aristotle's physics came alive with these linguistic shifts, Kuhn explained twenty years later.[2] At this point, he could better analyze and explain his Aristotle experience. It was as if he were able to travel back in science's history, he said, in order to compare paradigms and see the profound differences between them. ("The route I travelled backward with the aid of written texts was, I shall simply assert, nearly enough the same one that earlier scientists had traveled forward with no text but nature to guide them.")[3] But in the summer of 1947, when his jaw dropped and Aristotle's creative intelligence was first revealed to him, he was unsure what to make of it.

The Varieties of Scientific Experience

The underlying idea behind the Aristotle experience as Kuhn described it, that experience itself is plastic and requires shaping or filtering in order for knowledge to become possible, was an idea Kuhn had toyed with at least since his undergraduate days. William James, the nineteenth-century Harvard psychologist and philosopher, was a frequent touchstone in his thinking. In its final form, *Structure* turns to James to argue that paradigms are necessary for the mind to think scientifically. For it was an "undeniable fact," as James insisted, that our eyes and ears are not transparent windows on the world we experience. Instead, we must learn in infancy to divide and organize sensations. That is, James wrote,

> any number of impressions, from any number of sensory sources, falling simultaneously on a mind WHICH HAS NOT YET EXPERIENCED THEM SEPARATELY, will fuse into a single undivided object for the mind. . . . The baby, assailed by eyes, ears, nose, skin, and entrails at once, feels it all as one great blooming, buzzing confusion.[4]

James's *Principles of Pragmatism* appears in Kuhn's personal reading list from 1949.[5] But Kuhn had studied James years before. Shortly after graduating in 1943, he read *The Varieties of Religious Experience* and found it impressive and compelling from start to finish—"A fine and truly beautiful book," he wrote on a 3 x 5 notecard. "I must reread this book."

James explored religious beliefs and experiences pragmatically, with an eye for ways they functioned in actual believers' lives. For example, he discussed the mind-cure movement, a nineteenth-century blend of spirituality and medicine, whose devotees found they could lessen pain or illness by willing themselves toward restorative conversions. James admitted that hard-nosed scientists looked askance at such claims, but he sided with those whose reports and testimonials he quoted at length in his book.

> How conversions are thus made, and converts confirmed, is evident enough from the narrative which I have quoted.
>
> For the point I am driving at now, it makes no difference whether you consider the patients to be deluded victims of their imagination or not. That they seemed *themselves* to have been cured by the experiments tried was enough to make them converts to the system.

Kuhn was impressed by these testimonials as well as the metaphysical conclusion James drew from them: they "plainly show the universe to be a more many-sided affair than any sect, even the scientific sect, allows for," James wrote. These parts of James's discussion, Kuhn wrote in his notes, lent support to his own view that events in nature can be explained simultaneously and successfully by very different theoretical systems, that is, "in terms of different symbols & theoretical concepts."

As impressive as James's book was, Kuhn saw it more as a confirmation of his views ("my theory") than a source of new ideas. In fact, he believed that James had not gone far enough. "The ultimate Jamesian hypotheses" in the final section of the book—where James wrote,

> The whole drift of my education goes to persuade me that the world of our present consciousness is only one out of many worlds of consciousness that exist, and that those other worlds must contain experiences which have a meaning for our life.—

struck Kuhn as "weaker than necessary, and this is partly due to the restriction to 'religious' mysticism." The universe we experience is many-sided when it comes to religion, Kuhn agreed. But it is also many-sided when it comes to science and physics.[6]

Kuhn had earlier put this theory to his English professor, Robert Gorham Davis. In an essay titled "The Metaphysical Possibilities of Physics," Kuhn argued that concepts such as gravity, energy, and space and time are really *fictions* created by the human mind to organize the sensory phenomena. Describing the enormous changes in recent physics, he explained,

> [C]oncepts are constantly being changed to fit new data. . . . Gravity has disappeared; mass and energy have become basically the same; and all of this revolution has occurred at a time when quantum theory is necessitating a change in our concepts of cause and effect and of matter itself.[7]

James would have agreed that scientists can't simply read physical equations out of their sensations and perceptions; they must first organize and separate their sensations. This can be done, Kuhn added, in many ways—possibly "even an infinite number of entirely different concepts" can be made to fit the data of experience.

Writing while he was a sophomore at Harvard, however, Kuhn was not yet Kuhnian. The philosophical musings he presented to his English professor resembled more the slogans of logical positivism and "text book science" he would later criticize in *Structure*. Scientific revolutions could never usher in a truly different world from this point of view, for the "data" Kuhn spoke of remained constant and fixed throughout this possibility-filled adventure: "While the concepts of physics expand and the narrow fictions are replaced by broader ones," he wrote, "the structure of physics remains unshaken, for the basis of this structure is data, not concepts, and no change in concepts can invalidate the data."[8]

Linguistic Meaning in an Infinite Sea of Possibility

As an undergrad, Kuhn wrote less like William James and more like the French philosopher and mathematician Henri Poincaré, who emphasized the creativity within scientific theories: "The scientist must set in order. Science is built up with facts as a house is with stones," Poincaré wrote. "But a collection of facts is no more a science than a heap of stones is a house."[9] Several years later, his Aristotle experience would lead him past Poincaré to the insight that even these "stones," the raw data of experience, are themselves shaped by the human mind. They must be cut or carved out of the unorganized welter of experience before science can become possible. To make sense of the Aristotle experience, that is, Kuhn would later take back what he explained to Professor Davis. In fact, "data" are *not* stable and permanent. A change in concepts *can* "invalidate the data."

One striking aspect of the Aristotle experience was that when Kuhn "traveled backward" to see the logic and sense of ancient physics, he had to momentarily forget, or at least ignore, Newton's physics and the different ways that it cuts up and organizes experience. The two ways of organizing the physical world, Newton's and Aristotle's, that is, pushed his mind and his memory not only in two different directions, but two exclusive directions. Having been trained in modern physics and its very different way of organizing raw experience, were it not for the revelation he experienced, it would have been unlikely and nearly impossible for him to have seen and understood nature as Aristotle did. Like the gestalt or figure-ground illusions that he would present in *Structure*, when Aristotle's physics came into clear view, Newton's retreated to the background and became invisible, and vice versa. The "data" on which each system of physics rested and made good sense were in fact "invalidated" and hidden by the competing theory.

During his first years working in Conant's general education program, Kuhn explored the experience and what it said about the nature of science from a linguistic angle. In a sprawling, late-night document that he filed under "Incomplete Memos and Ideas, 1949," he struggled to figure out how language could function in ways that made sense of the experience. He had read the linguist Benjamin Lee Whorf, who explored similar ideas; but Kuhn thought things through on his own (and William James's) terms, beginning with the point that the world itself is overwhelmingly dense, complex, and overflowing with possibilities for organizing experience. We can manage this overflow only because our natural language "cuts" or simplifies experience for us:

> What I'm getting at is that natural language provides a finite means of mediating an infinitely complex universe. . . . Put more accurately—we in fact live in a world much more complex than our language admits. If we are to act in it, we must simplify it, and our choice of a particular manner of simplification (a cut) is pragmatically determined and is embodied in our language.[10]

"We can use language at all," he wrote, "only because we live in a world which can in practice be cut up into groups of sufficiently similar objects to pragmatically justify our attaching names to them." Any word's meaning, then, is some combination of its denotation, i.e., the range of objects to which the word can be applied or what it is attached to, and its connotation—the criteria or meanings that distinguish objects denoted from those that aren't. Denotation and connotation must work together, Kuhn figured, in order to make things such as the Aristotle experience possible, to shift in their mutual relations so that a new world of unseen objects and properties—a new constellation of semantic "boxes"—can come into view. This can happen, for instance, when we meet or discover something utterly unfamiliar and new.

> If in the future we encounter an object which does not fit into one and only one of our boxes we may either create a new box, or more often, alter our set of connotations so that the new object fits into one and only one box. In this structure we see connotation and denotation are mutually dependent and are actually capable of being changed together by experience.[11]

Aristotle and Galileo, he would later explain in *Structure*, had taken different, incompatible roads as they organized experience:

> To the Aristotelians, who believed that a heavy body is moved by its own nature from a higher position to a state of normal rest at a lower one, the swinging body was simply falling with difficulty. Constrained by the chain, it could achieve rest at its low point only after a tortuous motion and a considerable time. Galileo, on the other hand, looking at the swinging body, saw a pendulum, a body that almost succeeded in repeating the same motion over and over again ad infinitum.[12]

Given the infinite complexity of the world, Kuhn's notes read, "we have an infinity of possible different sets [of] connotations for each box." Aristotle and Galileo can therefore point to or denote the same thing, "a swinging stone," but recognize a different set of connotations precisely because "for a finite experienced world, fixing the denotation of words does not at all fix their connotations."[13]

There will be "fringe meanings," Kuhn theorized, that different people do not necessarily share, even though they seem to agree about the meaning of a given event or circumstance. Usually, these fringes remain in the background. Yet when faced with "an object which falls in the fringe for one but not for the other," there can be heated argument. In short, Kuhn summarized,

> a connotative (or meaning) system, sufficient to ensure agreement between two individuals about all the objects which they have experienced in the past, need not ensure agreement about the "boxing" of some object which they may meet in the future.

"Let me point out," he wrote in parentheses, "that the argument about the new experience lying in the 'fringe' may result in a totally new 'boxing' which separates items formally boxed together." This is how meanings (such as "motion" or "body" in physics) can change over time as new, unseen features of this "infinitely complex world" catch the attention of some scientists but not others. It seems necessary, Kuhn wrote, "that any language with the characteristics of natural language must be self-correcting to some extent—be capable of adjusting to new cuts as the old ones lose their pragmatic validity."[14]

The notes preview two arguments that philosophers would later find in *Structure*. One was directed against the logical empiricism that the *International Encyclopedia of Unified Science* represented: the view that beneath the trappings of natural language that scientists spoke to one another, a theory could be represented as a precise logical system, as if the axioms and postulates of Euclid's geometry could be adapted from points, lines, and planes to massive bodies in physics or cells and strands of DNA in biology. This crucial point about "fringe meanings," however, led Kuhn to see this as a hopeless possibility—just "doomed to failure." For logicizing scientific language, he reasoned, would strip away these all-important "fringes" of meaning that distinguish different scientists' understanding and give science the flexibility to change and evolve. A logical empiricist might be able to reconstruct Aristotle's physics, or Newton's physics; but that would not shed any light on the historical fact that Newton's very different "boxing" and "cutting" of the world historically emerged—somehow—out of Aristotle's. If science could be logicized, "any logicization which removed the fringe would fix our expectations for the future and would restrict science to the science of its past which it already embodied."[15] Science, quite simply, could never change.

The other argument points to Conant and his picture of science growing and "accumulating" insights, knowledge, and techniques through history. If the data on which Aristotle's physics made sense was "invalidated" by Newton's, and if the same held true for other areas of science, then science did not preserve the accomplishments of its past with each new era. Scientific knowledge lay in these networks of "cuts" and "boxes," which abstracted from and reduced the complexity of experience itself. A language that grew to capture the world in its infinite fullness, that could embrace at the same time multiple systems of knowledge, even *all* of them—"a language which would deal with the world of all possibles," he wrote[16]—would become useless as it grew to become just a linguistic duplicate of the world and its infinite complexity. Instead of growing larger over the course of history, these networks of cuts and boxes must therefore be replaced by others. Kuhn was not yet ready to declare Conant mistaken about progress—the announcement would come a decade later in the final version of *Structure*. But it was coming and Kuhn continued to amass evidence and analyses like this in his notes and notebooks to support his case.

The Semantics of Toy Automobiles

By April 1949, Kuhn had read Alfred Ayer's introduction to logical empiricism *Language, Truth, and Logic*. He took some five pages of notes and moved on to readings by Suzanne Langer (*Philosophy in a New Key*), Martha Ornstein's Columbia University dissertation on "The Role of Scientific Societies in the Seventeenth Century," and Robert Merton's *Science, Technology and Society in Seventeenth-Century England*. From the start, Kuhn found fewer clues in philosophy to make sense of the Aristotle experience than he did in sociology and psychology. He was particularly fascinated by what he found in the writings of Swiss psychologist Jean Piaget, who studied the development and use of concepts in children.

"The Piaget reading is useful primarily in shaping my own general view," he wrote that June. Piaget studied how children, some of them his own, came to understand basic concepts at different ages. Kuhn was deeply impressed by Piaget's paper "*Les notions de mouvement et de vitesse chez l'enfant*," published three years before, in which children observed toy automobiles, one red and one blue, moving from one place to another in different ways. During the trials, as Kuhn later summarized it,

> both cars were moved uniformly in a straight line. On some occasions both would cover the same distance but in different intervals of time. In other exposures the times required were the

same, but one car would cover a greater distance. Finally, there were a few experiments during which neither the distances nor the times were the same.[17]

Through experiments like this, which focused on concepts of speed and motion, Piaget mapped out four stages of human intellectual development, ranging from infants who slowly emerge from the perceptual welter of the world (at around eighteen months of age) to teenagers who find it natural to think and reason about concepts fully abstracted from perceptual experience. Kuhn was most intrigued by young children whose understanding of the concept "faster" mixed different, incompatible connotations. Sometimes, the "faster" car was the one that arrived at the goal first, while at other times it was the one that moved more rapidly. The two criteria could of course conflict, such as when the slower car reached its goal first (because the more rapid one began moving too late).

With Piaget in mind, Kuhn began to see the scientific process as a kind of development or maturation of human knowledge. "Piaget's kids," as he called them, illustrated how the mind carves up and organizes experience by attaching different kinds of meaning, including vague and contradictory "fringes" of meaning, to particular words. As Kuhn saw it, they formed what he called a "psychologically visible world" made from the "physically visible world," or "the pure raw flux, unorganized" of experience.

Over the course of a few paragraphs in his notes on Piaget, Kuhn's own language took a momentous turn: "visible" became "real" and he began to speak of "psychological *real* worlds" (or "psych. r. w.") carved from the visible flux of experience. He summarized his conclusion like this:

> In the broadest possible sense the scientific process consists of the attempt to minimize the verbal equipment implied by (or inherent in) the psych. r. w. . . . This is an attempt to achieve the most <u>ordered</u> & <u>adequate</u> psych. r. w. This process must be carried on consonant with logic and with the phys. r. w. Thus the evolution of concepts previously noted which is a change in the psych. r. w. Thus science changes the psych. r. w. Also, tho, science brings new features of phys. r. w. into psych. r. w. and ultimately expands (through machines & tools) the phys. r. w. itself. Thus the process is continuing.[18]

In this Kuhn combined and summarized several popular traditions in philosophy of science. From Ernst Mach's view of theories as economical descriptions of experience, he moved to the dialectics of Conant's conceptual schemes in which data from the physical world move into the psychological world of concepts and theories.[19] These theoretical reflections inspired by Piaget also point to the ideals of progressive education,[20] such as self-guided discovery and learning, and the kind of sudden illumination that Kuhn attached to his Aristotle experience. As he put it, "science changes the psych r. w." which in turn "ultimately expands" the physical world "itself."

In Piaget, Kuhn found a psychological key to understanding scientific change and the emergence of what he would later describe as new worlds of experience and understanding. These experiments were "particularly striking," Kuhn wrote, because they showed how children seemed to rehearse "Aristotelian and mediaeval notions on this subject" and to illuminate the "mechanisms of overthrow" involved.[21] "The growth of a sci[entific] conceptualization," Kuhn wrote in his notes, "represents a struggle between"

a) Psych[ological] reasonableness

b) Logical Consistency

c) Adequacy & applicability to phys[ical]. r. w.

The first force is perhaps the most important, tho it's commonly ignored.

Conant had not paid much attention to "psychological reasonableness." The scientists he presented in his case histories had neither the time nor the inclination to introspect about their inner psychological processes. That, too, was the failing of Ayer and the logical empiricists Kuhn had read, all of whom focused on the logical—not psychological—validity of concepts and theories in science. Of course, logic was important; Kuhn gave it second billing in his list. But by itself it shed no light on his Aristotle experience and what that promised to reveal about the history of science.

Piaget, though, was shedding very much light. Kuhn may have identified with Piaget's young subjects as they encountered contradictions in their thinking about motion and what it means to be "faster":

> That experience of paradox is the one generated by Piaget in the laboratory with occasionally striking results. Exposed to a single paradoxical experiment [in which both cars, by different

criteria, seem "faster" than the other] children will first say one body was "faster" and then immediately apply the same label to the other. Their answers become critically dependent upon minor differences in the experimental arrangement and in the wording of the question. Finally, as they become aware of the apparently arbitrary oscillation of their responses, those children who are either cleverest or best prepared will discover or invent the adult conception of "faster". . . . They had learned to avoid a significant conceptual error and thus to think more clearly.[22]

In his personal crisis over intervention in the war, as well as his Aristotle experience, Kuhn struggled to find order and rationality in or behind his changing views and values. He now saw "Piaget's kids" illustrating how history's scientists advanced knowledge—"I am supposing that the process Piaget sketches for children takes place at all ages when the range of conceptual thought is extended," Kuhn noted—as well as how everyone experiences the dynamics of intellectual growth:

> Everyone has occasionally had the experience of vehemently defending (with complete assurance) an idea or theory which he suddenly, in mid-course, finds totally contradictory. He is then shaken emotionally. Subsequently he can't understand how he could have thought this. Thus we get an illustration of the ease with which man embraces logical contradiction and the nature of the recasting of ideas upon its discovery.[23]

Writing in an age of conversion narratives, when intellectuals were swinging from the political Left to the Right, when American soldiers were thought to be forcibly "brainwashed" from the Right to the Left, and when his own conversion from pacifism to interventionism had caused an enormous and emotional personal crisis, Kuhn was on his way in 1949 to a theory of scientific revolutions: like Piaget's kids, science advanced when recalcitrant anomalies became noticed and forced a "recasting" of science's ideas.

Science and the Unconscious Mind

Three years before discovering Piaget, Kuhn explored a different branch of modern psychology that would leave its mark in *Structure*. His personal doubts about a future career in physics, his relationships (or lack of them) with women, and the fact that he "was clearly a neurotic, insecure young man" led him to Freudian psychoanalysis.[24] Freud theorized that personal development was a struggle between different compartments of the mind: the conscious Ego, the moralistic Superego, and the Id whose unacceptable desires and appetites remain unconscious or cloaked in symbols that emerge in dreams or unguarded ("Freudian") slips of the tongue. After Freud's American debut when he spoke in 1909 at Clark University in Worcester, about fifty miles from Boston, American psychologists and philosophers paid close attention to Freud, even if they did not accept all of his controversial ideas. As usual, William James was a trendsetter who enthused to Ernest Jones, the British psychoanalyst and biographer of Freud, "The future of psychology belongs to your work"—as if Freud were sure to revolutionize psychology. But James also harbored doubts and once remarked, "I strongly suspect Freud with his dream-theory of being a regular *hallucine*," a man entranced and captivated by his own fictions, "a man obsessed by fixed ideas."[25]

Freud's interests in human sexuality—at once scholarly, scientific, and therapeutic—could not be ignored for long in puritan America. After World War II, psychoanalysis flourished among elites and intellectuals who could afford to pay for weekly open-ended, intimate conversations with a therapist. By the 1960s, Freud was a familiar staple of popular culture, as illustrated by Philip Roth's best-seller *Portnoy's Complaint*, the entirety of which unfolds during one psychoanalytic session, and by the comedies of Woody Allen. Yet before Freud became so widely known, his theories were embraced by many progressive intellectuals, artists, and educators in New York City and Croton-on-Hudson. Elizabeth Moos prioritized the arts, including dance, plays, and music, in her curriculum because they allowed students to explore and develop their conscious and unconscious minds. Thomas Kuhn's mother, Minette, edited some published writings by the Germany psychoanalyst Karen Horney, who had relocated to the United States in the 1930s. In return, a grateful Horney publicly thanked "Minette Kuhn who helped me greatly toward a better organization of the material and a clearer formulation of my ideas."

It was Minette who encouraged her son to see an analyst and he did so for about two years. His quest for better self-understanding ended not because he had achieved any breakthrough in his mind, but because at least some of his concerns were resolved by his new relationship to Conant and the Harvard Society of Fellows. That, and the fact Kuhn's analyst "behaved extremely irresponsibly with me," he later recalled many years later. Woody

Allen was not yet a teenager but Kuhn observed that his therapist "used to fall asleep and then when I would catch him snoring he would act as though I had no business being at all angry or upset about it."[26] Kuhn remained angry with this therapist, but not about psychoanalysis and its intellectual goals. "The *technique* of understanding people and enabling them to understand themselves better—I'm not sure that it produces real therapy of any sort—but it sure as hell is interesting," he noted. His psychoanalytic interests and experiences, he suggested, may even have led him to his distinctive historiographical style, his ability "to climb into other people's heads" in order to examine history's scientific minds.[27]

Perhaps because of his family's longstanding interests in psychoanalysis, Kuhn also had social and scholarly relationships with prominent analysts in the 1940s and '50s. One was Sándor Radó, a Hungarian psychologist at Columbia University who had studied with Freud and wrote *Psychodynamics as a Basic Science*. In late 1949, most likely, Kuhn and Radó discussed the history of science from a psychodynamic perspective. In a long, undated letter, Kuhn referred Radó to his interest in Gestalt psychology, which shows "that our perceptions of the parts are affected by the manner in which we view the wholes, and that cultural and educational factors may alter our perceptual groupings."[28]

Kuhn also corresponded with Lawrence Kubie, a psychoanalyst who taught at Yale's school of medicine and counted Tennessee Williams among his list of high-profile clients. Kubie was a Harvard graduate who had a close relationship with the Conant family. Having frequent trouble raising their two high-strung sons, Conant and his wife Patty turned to Kubie in the 1930s for advice in understanding their sons' behavior, choosing where to send them for schooling, and managing the specter of mental illness that seemed to haunt Patty's side of the family.[29] When Kuhn corresponded with Kubie in the mid-1950s, Kubie's work also complemented Conant's political interests in American education. Kubie fully agreed with Conant that scientific progress in the West proved that liberalism was superior to totalitarianism, but he worried that the rigors of modern scientific education were ironically harming young scientists and cultivating crippling emotional neuroses.

Whether or not Kubie's essays about the prevalence of neuroses among science students had bearing on Kuhn's self-image as a "neurotic" young man working toward his PhD in physics, Kuhn also had a personal, familial relationship with Kubie. Kubie had graduated from Harvard alongside his father in the class of 1916 and remained a Kuhn family friend with whom Kuhn now shared a deep interest in the psychological, Freudian contours of scientific research. In 1955, for example, following up on discussion the two had begun at his father's birthday party, Kuhn wrote to elaborate the nonrational and unconscious dimensions of human thought—what he frequently described in his notes as the "non-formalizeable" aspects of the scientific process. All of the points he made in his long, six-page letter, Kuhn explained to Kubie, "are attempts to give the unconscious (or at least the non-rational and non-rationalizable) an essential role in the scientific creative process."[30]

Kuhn wrote almost as if science were itself a patient with a long, tortured history lying on Kubie's psychoanalytic couch—as if it had been told to ignore thoughts about the way things *are* and to dwell instead on how events are experienced and interpreted. Kuhn told Kubie that the concept "plastic" was central to understanding science's history:

> I can understand the history of science much better if I suppose that the world explored by the scientist possesses (or acts <u>as if</u> it possessed) considerable plasticity. The effect of any major conceptual innovation in the sciences is, on my analysis, to change the world which the scientist explores as well as the tools with which he does the exploring.

Was the world really plastic? Or did its overflowing, infinite complexity that Kuhn described in his notes on "cuts" and "boxes" make it *appear* plastic and changeable? Kuhn could go either way on this, he said. One can suppose that "the world itself is shaped by our beliefs" or, differently, "there is a fixed world, but it is too vast to be an object of knowledge or even perception" as a whole. In *Structure*, Kuhn adopted the second view, writing that "nature" can be experienced and understood only when simplified by paradigms. In this letter to Kubie, however, he wrote that he preferred the first way of seeing it—even though, "for most purposes the two statements are equivalent."

Kuhn also spoke tentatively of a collective or "group-unconscious" that belonged to any scientific community. Subtle but crucial interactions between the world and our understanding of it that seem to change the world itself—as he put it, "the processes by which a set of group beliefs act on the plastic real world"—had to remain invisible to the conscious scientific mind, he explained. Where Freud theorized that conscious awareness of the Id could overwhelm Victorian morality and propriety, Kuhn reasoned that awareness of these processes within science could overwhelm perception and scientific epistemology. They "must not be accessible to consciousness or they will not work," he explained, for once the mind is aware that it is seeing only a *part* of the world, it will become distracted. "When people know what they are not seeing," Kuhn explained, "they begin to see it." A similar point, confirmed by Kuhn's reading in political history, would later become one of the most striking and Orwellian features of *Structure*: scientific revolutions are and must be "invisible" to the normal scientists who work in their wake, lest knowledge of science's revolutionary past upset and overwhelm the values and beliefs of normal science.[31]

Kuhn did not elaborate his historical interests in Kubie's language of psychoanalysis as a courtesy or an abstract, theoretical challenge. His interests in unconscious aspects of scientific reasoning were central to his early goals as a scholar. He would cite Kubie in *The Structure of Scientific Revolutions*[32] and in 1951 he outlined these interests to Professor David Owen, then chairman of Conant's General Education Committee. Owen had asked Kuhn to describe his current research plans and Kuhn replied with the opening he would later use for *Structure*—that paying close attention to science's history "can produce an important reorientation" in our understanding of it. As things stood now, he explained to Owen, the sources of knowledge were broadly misunderstood: "For my starting point is that the implicit scientific injunction, 'Go ye forth and gather the fruits of objective observation,' is a meaningless one which no one could carry out."

It's impossible because of our primitive, Jamesian perceptual condition: the world allows us to collect "an infinity of independent observations; so that the process of scientific observation presupposes a choice of those aspects of experience which are to be deemed relevant." In his notes on cuts and boxes, Kuhn had similarly noted "our choice of a particular manner of simplification (a cut)" in the welter of experience. Writing now to Owen, however, he suggested that these choices are not deliberate and perhaps not even consciously undertaken. They are made, he explained, "on a largely unconscious basis" by what Kuhn now called "predispositions." Kuhn put it to Owen like this:

> Any particular set of observations in science (or everyday life) presupposes a predisposition toward a conceptual scheme of a corresponding sort: the "facts" of science already contain (in a psychological, not a metaphysical, sense) a portion of the theory from which they will ultimately be deduced. This "predisposition" . . . in its most trivial operation, leads the scientist to

ignore or discard certain portions of experience in formulating or verifying his theories.

In a real way, Kuhn believed, scientists were brainwashed by these predispositions. For they functioned not only positively to focus attention on particular theories, experiments, and observations. They also functioned negatively to prohibit scientists from paying attention to experiences and observations that lay outside the conceptual scheme in question. This was not just the "trivial" observation that scientists "ignore or discard" observations or areas of experience because they are irrelevant to their scientific goals. Kuhn's point was that predispositions function in a more "fundamental" way to keep vast portions of raw, unfiltered experience outside of scientific awareness: "The same 'predisposition,'" he continued, "exerts a far more fundamental influence in directing the scientist's attention to particular abstract aspects of experience and blocking his perception of alternate abstractions."

What exactly are these "predispositions"? Where do they come from and how do they work? What might science's history tell us about them? These were among the questions now central to Kuhn's current research project, he explained. In fact, he told Owen, they amounted to a new, interdisciplinary field of research "directed to the psychological, the experimental, and the social sources of these 'predispositions.'" This field would examine how predispositions operate in scientific research, and what happens to them during times of revolutionary change. Kuhn was keenly interested in "the mechanism by which particular observations or changes in scientific theory may cause one set of 'predispositions' to be replaced by another."[33]

Kuhn's unconscious predispositions were prototypes of *Structure*'s paradigms. Their source, he would explain in *Structure*, was the sociology of scientific communities and, in particular, the scientific education that inculcated the dominant paradigm of the day in young scientists. Kuhn would soon turn to sociology and sociology of science, but when he wrote to Owen in 1951—and emphatically when he corresponded with Lawrence Kubie four years later—he saw the dynamics of scientific revolutions through psychoanalytic lenses.

Structure would not be commissioned until 1952 or early 1953, but Kuhn knew as early as 1949 that he would write a book about these linguistic and psychological dynamics in science. "Consider the following outline for a book," he wrote in his personal notes on July 5, 1949. It would have three parts, one on "language and logic," a second on "The Scientific Function" that would explore "The Physical Real World" and "the Psych[ological] Real World," and a third titled "Science at Work," including "examples from the history of science."[34]

One of his outlines contained thirteen chapters—just like *The Structure of Scientific Revolutions*. But that was just a coincidence, for what Kuhn called THE BOOK in his notes from these years did not evolve into *Structure*. While *Structure* presented his theory of science as inspired by, and a way to understand, science's history, THE BOOK Kuhn envisioned here drew mainly on psychology and applied its insights to history of science at the end. Following his notes on Piaget and his interest in psychological reorientations, that is, this book would analyze the dynamics of science from a larger, psychological perspective in which scientific examples could sit side by side with "parallel processes from psychology, ethnology, and math problems, etc."[35]

Kuhn's mention of "ethnology," the study of different cultures around the globe, confirms the scope of his ambitions. Making sense of the Aristotle experience was going to illuminate not only ancient physics but the psychological foundations of human knowledge itself, wherever and however it takes shape around the globe and in human history. In these respects, THE BOOK and *Structure* were similar and similarly ambitious. But the book that made Kuhn famous was hardly in evidence at this time. As he began to write it, his emphasis would shift from psychology toward history and sociology; and the terms he used to articulate his post-Aristotle theory of science would evolve greatly. The *unconscious predisposition* he described to Owen as the basis of science would give way to *theory-as-ideology*, then to *consensus*, and finally *paradigm*. By building *Structure* around the concept of paradigms, Kuhn could move freely and easily among the several aspects of science that intrigued him—linguistic, psychological, psychoanalytical, historical, and sociological. Paradigms also helped him navigate the political realities that hovered around his research and his emerging theory of science. Treating these issues in the wrong ways—especially in public settings such as the Lowell Institute lectures that Kuhn was set to give in 1951—could be fatal to any young scholar's career.

9

"Attention Senator McCarthy"

The Perils of Methodology in Totalitarian Times

The advertisement read:

> **What are the Problems of Scientific Research Today?**
>
> The LOWELL INSTITUTE presents eight free lectures by Thomas S. Kuhn, Ph.D. of Harvard on "The Quest for Physical Theory"—Tuesdays and Fridays at 8 p.m., in the Boston Public Library's Lecture Hall, beginning March 2 (excepting March 23). For FREE TICKETS, write to Boston Public Library, Boston 17, Curator, Lowell Institute.

Kuhn put down his newspaper and telephoned Ralph Lowell, president of the Lowell Institute, to demand that he immediately cease publishing this advertisement. Lowell agreed and had his staff call the offices of the *Boston Globe* (as well as the *Boston Herald*, where the same ad was running). Except for the early-afternoon edition of the *Globe*, which was already printed, the local presses stopped, steamed, and creaked. The ad was never printed again.

Kuhn breathed a sigh of relief and thanked Lowell in a letter. Realizing that his reaction may have seemed extreme, he wrote, "I am sorry to disturb you with a situation which, except in its implications for the future, is essentially trivial." The future implications at stake concerned his career as a scholar—possibly also as a public intellectual. Kuhn was now a key part of Conant's campaign to elevate the public's understanding of science through general education. Delivering the prestigious Lowell Lectures would be a natural stepping stone to an even wider, public persona.

172 | The Politics of Paradigms

Three days later, however, Kuhn's anger and concern erupted again when he noticed that the Institute was circulating another advertisement, not in newspapers but in flyers posted around Boston and Cambridge (see Figure 9.1). He wrote to Lowell again: "When I wrote you last Saturday I was unaware that an expanded version of the 'blurb' to which I then called your attention had already been publicly distributed on the front of the Lowell Institute's flyer." This version, he made clear, added to his "already acute distress." It read,

What are the problems of  scientific research today

In a world in which science's quest for physical theory has already had results that promise to change the course of history, the fate of mankind may depend upon solving the problems of research . . .

The Lowell Institute

presents a course of eight free lectures on "The Quest for Physical Theory" by Thomas S. Kuhn, A.M., Ph.D. — Tuesdays and Fridays at 8 p.m., beginning Friday, March 2 (omitting Friday, March 23), at the Boston Public Library.

[over]

Figure 9.1. Lowell Institute full-page flyer advertising Kuhn's upcoming series of lectures, ca. February 1951.

> In a world in which science's quest for physical theory has already had results that promise to change the course of history, the fate of mankind may depend upon solving the problems of research . . .

The final, ominous ellipses were intended to intrigue Bostonians, to bring as many as possible to the public library for the lectures. But it appeared to Kuhn that the publicist who wrote the copy—a Mr. Stenbuck, he determined—was playing a dangerous game and putting Kuhn's future at risk.[1] As he had explained to Professor Owen, Kuhn was pioneering a new field of research. He wanted to debut on the rostrum of the Boston Public Library as a scholar with something new and original to offer. But Stenbuck's advertisements would give hundreds, perhaps thousands, the impression that Kuhn's upcoming lectures belonged to a familiar and suspicious kind of intellectual conduct—the genre of "scientific planning" that had already proved controversial in the new cold war.

Nineteen fifty-one was an especially tense year for intellectuals such as Kuhn. College campuses around the nation, the *New York Times* reported, were attacked by "a subtle, creeping paralysis of freedom of thought and speech." Faculty avoided speaking about controversial issues and students worried that using "the wrong word at the wrong time might jeopardize their futures."[2] Kuhn felt this fear as he saw his upcoming lectures described in print. He worried about how his iconoclastic thoughts about science would be received and whether, worst of all, the public might lump him together with stereotypical radicals who claim expertise about modern science and how it might be used to create a better—typically communistic—future for mankind. There was no shortage of intellectuals who had encountered trouble for similar reasons. Sidney Hook and McCarthy himself had helped the public see Harvard's Harlow Shapley as a pathetic "dupe" of communism. *The Crimson* had broken the story that Yale (and, for all anyone knew, Harvard, too) was awash with FBI informants and "caught in a mystifying web of 'cold war' security."[3] Kuhn's former teachers, Wendell Furry and Robert Gorham Davis, had not yet been called to Washington to account for their communist pasts, but Kuhn knew at this time that life was difficult for unregenerate leftists. Elizabeth Moos's reputation had crumbled, and her son-in-law was likely headed for jail. The Kuhn family's once beloved

Hessian Hills School was suffering for its radicalism and would soon close its doors for good.

Philosophers of science, in particular, seemed to be attracting anti-communist attention. William Malisoff, the founder of the Philosophy of Science Association and its journal *Philosophy of Science*, had been watched for years by Hoover's FBI agents in New York City, in Philadelphia, and in Croton-on-Hudson, where Malisoff had a summer home. Hoover knew that Malisoff was secretly an agent who regularly met and talked with Soviet contacts—which may explain why Hoover would come to believe a false rumor that Philipp Frank (who corresponded with Malisoff and taught courses about science alongside Kuhn at Harvard) was a high-level Communist working in the underground. Within a year, FBI agents and informers on campus would begin investigating Frank as a potential subversive.

As a Jew, there was a lot for Kuhn to lose. The progressive 1930s had indeed been progressive for Jews in America. The success of Kuhn, Loeb, and Co. and the banks founded by Lehmans, Goldmans, and Sachses had given New York City a class of Jewish elites powerful in finance, law, and often philanthropy and progressive social causes. Jewish intellectuals, such as those who founded *Partisan Review* and other magazines, gained public influence and credibility. Quotas for Jewish students and faculty at East Coast colleges and universities were gradually eliminated. The career of a philosopher like Sidney Hook proved that the educated public, at least, could accept Jews in the 1930s not only as public intellectuals but as defenders and promoters of progressive education, Marxism, and socialism.[4]

When Kuhn saw his name and his upcoming lectures announced in these advertisements, however, this progress was threatening to turn back. McCarthy targeted Communists strictly for their communism, never for their ethnicity or religion. But stereotypes of Jewish Bolsheviks, illustrated by Elizabeth Dilling's obsession with Albert Einstein, remained a cornerstone of academic anticommunism.[5] When the *American Legion Magazine* dedicated its cover to the ongoing controversy over Communist faculty in November, the illustration featured a teacher with stereotypical Jewish features gesturing confidently about "imperialism" and other controversial topics (see Figure 9.2). When citizens were arrested for passing atomic secrets to the Russians the year before, they were Jews—Julius and Ethel Rosenberg and Morton Sobell.

By the time he had graduated from Harvard, Kuhn had left behind his pacifism and his radical childhood. He had no memberships in Communist organizations to hide, and he was himself a veteran. But McCarthyism had a logic of its own that could easily find subversion in one's past associations,

Figure 9.2. *The American Legion Magazine*, November 1951.

activities, or publications. If, as Conant always suspected, McCarthy was out to make trouble for Conant, things could easily become difficult for Kuhn, too. Having nothing to hide was not enough. It was imperative for one's professional and social standing that no politician, none of Hoover's agents, and none of their informers sprinkled around the nation's campuses had reason to pay attention to one's politics or the political implications one might detect in one's lectures or writings.

As his Lowell Lectures approached, however, Kuhn was headed for the spotlight, and the new area of research he planned to discuss led him to some potentially controversial conclusions about the nature of science and the psychology of scientists. The professional scientific mind was not open and free; it was more like the stereotypical Russian mind scrutinized by the scholars at the new Russian Research Center and described by Kennan and Conant at the start of the cold war. Its thoughts and perceptions were confined and limited—necessarily so, Kuhn believed—in ways that could easily make Kuhn's research seem subversive, perhaps even "pro-Soviet." The political overtones of Kuhn's lectures might be hard to ignore, so he did not want preconceptions set or skewed by the institute's advertising copy.

The Perils of Planning

The philanthropist John Lowell founded the Lowell Institute in the nineteenth century. His great-grandson Abbott Lawrence Lowell preceded Conant as president of Harvard, but the institute was not a part of the university. Its charter was to enlighten the public about topics such as "The Development of Choral Music" and "The Migration of Birds." On the bill along with Kuhn in the program for 1950–51 were "Humor in Italian Literature" by a professor of Italian from California, "Japanese Culture Interpreted by its Art" by a former director of the Pennsylvania Museum of Art, and "Britain and the Challenge to Democracy" by an officer of the BBC. The institute often took advantage of its relationship to Harvard and its Society of Fellows, itself founded and endowed in 1933 by Abbot Lawrence upon his retirement, to recruit talented and creative lecturers.[6] William James delivered a famous series of lectures in 1906, two of which, "Pragmatism and Truth" and "The Many and the One," are among his most famous.

When he was first invited, Kuhn offered a title that pointed to the psychological and linguistic mechanisms that make knowledge possible: "The Creation of Scientific Objects." The title echoes *The Genesis and Development of a Scientific Fact*, the book by the Polish physiologist Ludwik Fleck, whose title grabbed Kuhn the moment he first encountered it. Along with the German sociologist Karl Mannheim, Fleck is now appreciated as a pioneer in the sociology of science and knowledge, a forerunner of Kuhn himself. But as much as Kuhn wanted a similar title for his lectures, he asked the institute for some time to think about it. For publicity purposes, however, the institute's curator William Lawrence told him they'd need a title by July 1950. Since Kuhn would be in Europe that summer, he wrote back in a week to say the series would be called "The Quest for Physical Theory—Problems in the Methodology of Scientific Research." With that he provided a list of eight titles for the individual lectures.[7]

By "methodology," Kuhn meant the path by which the scientific mind moves from the ignorance of primitive, unorganized experience to objective, scientific knowledge. The mind's "quest for physical theory" would treat the problems Kuhn had wrestled with since the Aristotle experience as well as the case studies he and Conant had been teaching in *Natural Sciences 4*. It was not a quest to develop new and better theories in contemporary science, as Stenbuck appeared to suppose; it was a quest to understand the human mind, how it works, and how it uses language and ideas to build physical knowledge.

Kuhn evidently did not anticipate that his title or the gloss on it formulated by Stenbuck could be controversial. But Engels famously said that the purpose of philosophy was not to understand the world, but to change it and make it better. One did not have to be a devout socialist or Marxist to agree that science was profoundly changing the postwar world. This was one of the main justifications for Conant's new general education project. But advocates of scientific planning took that logic a step farther, reasoning that science itself and contemporary research should be harnessed to change the world for the betterment of mankind. In the 1930s and '40s, experts in Europe and the United States argued that scientific research itself should be organized to meet these ends. Some were Communists, such as those who wrote in the pages of *The Communist*, the official intellectual journal of the Communist Party of the U.S.A. The founders of the journal *Science and Society*, on the other hand, professed sympathy with socialism, international cooperation in science and politics, and the theoretical insights of dialectical materialism—but as writers and editors they had no official relationship to Party communism or Stalin's Soviet Union. William Malisoff, who was devoted to Marxism and research to increase the human life span, was individually affiliated with communism and in regular contact with Communist agents. But the journal he founded, *Philosophy of Science*, and the professional *Philosophy of Science Association* he built around it professed no collective or official sympathy with Marxism, much less Stalin's Kremlin.[8]

Otto Neurath, the most prominent leader of the unity of science movement, understood his *International Encyclopedia of Unified Science* as an institution or forum to help facilitate collective, democratic planning of science. Scientists would guide research from the grass roots upward—without imposing any preconceived theoretical frame or ideology on the sciences from the top down, as Trofim Lysenko would do in Soviet genetics. "Our program is the following," Neurath had proclaimed in Paris, 1935 at the first official Congress for the Unity of Science: "No system from above, but systematization from below."[9] Neurath's American cousin, Waldemar Kaempffert, the science editor for the *New York Times*, faithfully championed the cause of organized science in the 1930s and '40s. "In the great industrial laboratories," Kaempffert wrote, researchers work together—"research is planned, organized, and competently directed. Scientists and engineers work as a team." This was quite unlike the unorganized research undertaken by "professorial prima donnas" who pay no attention to what other scientists are doing and create "a chaos of individualism." Kaempffert may have borrowed these formulations from Neurath himself—but if he did, he likely softened for his American audiences the overtly Marxist language his cousin liked to use. Making a similar point, Neurath had written that "[t]he Marxist always aims at informing himself about the main problem of the proletarian class struggle: how will the suffering of the capitalist order come to an end?" Bourgeois scientists very differently leave things to chance. To them, it doesn't much matter "whether a man thinks about some linguistic formations in Chinese or about a medieval legal text, about African beetles or about wind conditions at the North Pole."[10]

Controversy abounded over these calls to plan and guide modern science. Towering intellects such as the economist Friedrich Hayek and the philosopher Karl Popper, both from Austria, and the Hungarian chemist-turned-philosopher Michael Polanyi viscerally opposed planning and articulated libertarian, individualistic philosophies of science to justify their opposition. Because scientific progress depended upon individual genius, creativity, and talent—as opposed to collectivism and uniformity—any effort to plan the future of science, they believed, was as dangerous as attempts to plan and control the future of society. It would lead science and society alike toward a totalitarian nightmare via *The Road to Serfdom*, as Hayek's best-seller put it.[11]

Conant told Americans to keep an open mind about these issues. Any credible course in the tactics and strategy of science like the one he outlined in *On Understanding Science* should discuss the roles of scientific societies, national cultures in which scientists worked, and the fascinating stories of amateurs, without state or royal patronage, "by whose labors alone science advanced for more than two hundred years." In these ways, he noted, "there can be left in the student's mind no doubt that science is indeed a social process." But that did not necessarily mean that this social process could be reliably organized to produce more or better results. The workings of science were complex and too mysterious to be guided or planned—particularly when it came to "basic" or "pure" science. That was Conant's personal view, but he admitted that advocates of planning questioned the validity of this crucial distinction between "pure" and "applied" science. "I shall not attempt to prescribe how the instructor should balance the contending views," he wrote. But he emphasized that this debate was important and had to be treated responsibly by any general educator. If only because "echoes of this controversy find their way in to the daily press," students should be taught to see and understand the different sides of the controversy.[12]

Kallen Versus Neurath

The controversy Conant had in mind unfolded in 1945; but its roots lay in larger and older worries that the rise of fascism in Europe threatened not only political but intellectual freedoms around the world. Horace Kallen was a philosopher who put this charge to Neurath and his new *International of Unified Science*. Kallen taught at the New School for Social Research and, as a former student of William James, built his career by defending James's intellectual pluralism. Kallen and Neurath got to know each other as Neurath visited the United States in the 1930s and by their own accounts became fast friends.

But as Kallen's fellow New Yorkers Sidney Hook and James Burnham turned against Stalinism in the late 1930s, Kallen directed his wrath against his friend Neurath and Neurath's campaign for an organized and "unified science." In 1939, when Conant welcomed Neurath and his new encyclopedists to their conference at Harvard—news of Hitler's invasion of Poland, Neurath later recalled, came over the radio on the eve of the conference[13]—Kallen took the podium to declare war on Neurath's plans to unify the science and his "faith in synthesizing and systematizing observations, experiments and reasonings as a paramount condition for the progress of science." This faith was in fact harmful to science, Kallen insisted, for any synthesis or creation of a unified system meant excluding alternatives that did not seem to fit. Any single unifying framework or language of science—such as the language of "physicalism" that Neurath, Carnap, and others defended as a universal language of science—would require Neurath's encyclopedists to act like censors or dictators, he charged. They would have to "impose it by *force majeure*."[14]

Neurath was probably blindsided by Kallen's attack, not only because he knew Kallen as a friend or because he was a European perhaps unfamiliar with the bare-knuckle intellectual street fighting that Kallen, Hook, and other New Yorkers perfected. Neurath must have been additionally flummoxed by Kallen's charges because Neurath himself had emphasized that unified science was not a single, preestablished system of scientific knowledge. Physicalism did not mean scientists of the future would be lorded over by physicists, or that they would be forced to speak the language of physics; it meant that any credible science must remain connected to the physical world we actually experience; that is, connected to events in space and time—a proposal not unlike those found in William James's own "radical empiricism." In responding to his charge (and perhaps hoping to deflate it) Neurath even embraced Kallen's suggestion that the best metaphor for unified science was the *orchestration* of different sciences. On this view, chemistry and psychology, say, are as different as oboes and kettle drums; but they can be brought together with other sciences to achieve great things.

But Kallen pressed his charges and battled Neurath long after the Harvard conference in the pages of *Philosophy and Phenomenological Research*.[15] Nothing that Neurath said could assuage his worries, it seems, because the problem that so agitated him was larger than Otto Neurath and his new encyclopedia. It was the threat of totalitarianism, the demise of freedom and liberalism in Europe, and the plights of millions of Jews fleeing Hitler. "We are living in totalitarian times," Kallen said at the Harvard conference. He denounced "despotic countries" in which "not only are the lives and labors of the people 'unified,' their thoughts boilerplated; also the arts and sciences are 'unified' to the respective orthodoxies of the fascist, Nazi, communist, and clericalist dogmas."[16] Neurath's unity of science program and his new encyclopedia were simply inseparable from all this political repression, as far as Kallen could tell.

Kallen's view of science as a beacon of intellectual freedom helped to define the coming postwar struggle against totalitarianism and mind control. In a number of ways, it also prefigured the decline of Neurath's own reputation in the United States. Neurath's fellow *Encyclopedia* editor Charles Morris would later remark—to Kallen himself, ironically—that Neurath and his project were reputed to be "communistic." It was not the first time the charge had been hurled at Neurath, who was briefly imprisoned as a revolutionary in Germany after World War I and later fled his native Austria after being fingered as a communist.[17] But it surely hastened the demise of his encyclopedia project after the war and perhaps the demise of Neurath himself. Then living in England after fleeing Austria and then the Netherlands, Neurath died in late 1945 at the age of sixty-three in the midst of his debate with Kallen. His final missive was published posthumously, as if his demise were hastened by the stress of fleeing totalitarianism, wrestling with Kallen's ironic and frustrating accusations, and watching his cousin Kaempffert struggle to defend unified science. For the nation that had largely welcomed Neurath, his famous Vienna-Method of visual statistics or fact-pictures, and his new encyclopedia of science in the 1930s was by the mid-1940s worried that Kallen's alarming predictions about the creep of totalitarianism were correct.

The Postwar Future of American Science

The fears that Kallen expressed at the Harvard conference in 1939 took on new, national significance in the summer of 1945, when Vannevar Bush, head of Washington's Office of Scientific Research and Development (OSRD) published *Science, the Endless Frontier*—a report originally commissioned by Franklin Roosevelt to chart the future of science in postwar America. Bush concluded that the United States "can no longer count on ravaged Europe as a source of fundamental knowledge" or basic, pure science. It must therefore begin to cultivate its own and it should do so through "the creation of a National Research Foundation."[18]

Kaempffert championed Bush's proposal in the *New York Times* and reported on competing bills in Congress proposed to bring Bush's vision to life. Editorially, however, Kaempffert had his preference and flatly rejected the fundamental distinction between "basic" and "applied" science. "What is 'basic science'? Where does it end and where does practice begin? No line can be drawn," Kaempffert wrote. He believed that the Manhattan Project, which Bush and Conant had led within the OSRD, was a shining "example of what can be achieved rapidly by organizing and planning."[19] Kaempffert therefore endorsed a bill proposed by West Virginia senator Harley Kilgore for a national foundation that would not only pay for scientific research but provide oversight and planning by experts.

As Kaempffert saw it, the national debate over Bush's report was an opportunity for the nation as a whole to embrace the interdisciplinary ideals and the practical, social goals of his cousin's unity of science movement. And he was not alone. The day before he reviewed Bush's report, his paper printed an unsigned editorial making similar points with similar language. The influential *New York Times* itself, in other words, took Kaempffert's side against Bush and called for a national research foundation to plan and organize postwar science.[20]

At this point, Conant stepped in. In a letter to the editor, he strongly disagreed with Kaempffert and the editorial board. Noting clearly the geopolitical dimensions of the longstanding debate over whether science can be planned, Conant agreed that Kaempffert's position made good sense were the United States poised like postwar Britain to go "down the road of socialism" and to nationalize some or all of its industries. But given the outcome of recent national elections, Conant pointed out, that was not the case. Even worse, Conant added, Kaempffert seemed to misunderstand the OSRD's methods. Yes, it set goals for researchers to meet during the war; but if those goals were met it was precisely because the mysterious workings of scientific "genius" were given free rein:

> There is only one proved method of assisting the advancement of pure science—that of picking men of genius, backing them heavily and leaving them to direct themselves. There is only one proved method of getting results in applied science—picking men of genius, backing them heavily, and keeping their aim on the target chosen.

Kaempffert responded weeks later that this was all very confusing, and it was: Conant had admitted that so-called pure and applied science were similar; yet, for reasons that were "lost on us," Kaempffert wrote, Conant still defended the pure-versus-applied distinction as crucial to any future research organization.[21]

By the fall of 1945, Bush, Conant, and others had additional reasons to fear that Kaempffert's vision of science's future was ascendant. When he first commissioned Bush's report, Roosevelt himself took for granted the view that modern science—if properly cultivated and managed as it was during the war effort—could improve modern life by curing diseases, building the economy, and creating for all Americans "a fuller and more fruitful life." It now appeared that Roosevelt's successor was taking that view, as well. In his message to the Congress that September, Truman echoed Kaempffert's arguments that the OSRD itself had demonstrated the effectiveness of organized, planned research during the war, that there was no important distinction between pure science and technology, and that a "single agency" should oversee and coordinate publicly funded research in all fields of science in the new postwar age. Of course, Truman reassured the nation, scientists must not be "dictated to or regimented," for "science cannot progress unless founded on the free intelligence of the scientist." But that did not mean that the workings of "genius" should not be organized, supported, and actively cultivated in just those ways that Conant and Bush employed during the war. Possibly with the language of Conant's letter to the *New York Times* in mind, Truman wrote that the fruits of modern science "cannot depend alone upon brilliant inspiration or sudden flights of genius."[22]

Weeks later, it was reported that Truman was in contact with congressional leaders and proposing "straightline administration" of the new foundation—meaning that Truman and subsequent presidents would appoint its leaders. Fearing that this would mean nothing less than attempts by politicians to control and dictate the course of research, Conant and about forty leaders of foundations, businesses, and universities sprang into action. They formed an official organization, the "Committee Supporting the Bush Report," and wrote an open letter to Truman insisting that no administrator or board of directors be permitted to "control or coordinate all Government scientific activities." Kallen's menacing image of a scientific dictatorship hovered within their letter as they wrote,

> In our opinion it would be most unwise to subordinate the board to a single director appointed by the President. . . . No single person, however eminent or competent, could, except in a great emergency, command the confidence and support of all branches of science and of the many organizations and agencies, private and public, whose cooperation will be required.[23]

Still, as Conant later explained to readers of *Foreign Affairs*, freedom from scientific dictators did not settle important and difficult questions about which research projects should be funded with federal monies. Conant was quite serious, therefore, that once the new foundation was up and running, laypeople must be involved in these decisions—not only as representatives of the public but as disinterested and objective parties in complex and technical disputes. They would moderate and balance the influence of experts who, being human, "tend to be carried away with enthusiasm for their own ideas." Like the citizens Conant hoped to cultivate through courses such as *Natural Sciences 4*, these laypeople would know something about science, its history, and the often serendipitous, unpredictable ways that breakthroughs occur in pure and applied science.[24]

But not every signatory saw truth in both sides of the issue as Conant did. A single step taken to "unify" the sciences, Kallen had insisted, was a potentially disastrous capitulation to these "totalitarian times." That rhetorical baton was now picked up by Warren Weaver, an officer of the Rockefeller Foundation, which provided funds for research in natural science (and, for a while, the unity of science movement). Weaver angrily denounced Kaempffert and the *New York Times* for recommending that "all science should be mapped out and the gaps discovered, and that an all-high, all-embracing central organization, presumably set up by the Federal Government, should plan and direct scientific activities." It was foolish and dangerous, he insisted, for Kaempffert to ignore the crucial distinction between pure and applied science and to portray the atomic bomb as an illustration of effective planning. The "great bulk of scientific work during war," Weaver argued, rested squarely on "basic scientific knowledge." How had that knowledge been gained, and by whom?, he asked indignantly. "Why, gained by free scientists, following the free play of their imaginations, their curiosities, their hunches, their special prejudices, their undefended likes and dislikes"—all of which would be suppressed, if not eliminated, by some "group of scientific supermen who, seeing, hearing and knowing not only all of present science but also divining its mysterious future course, try to chart that course and tell the rest of us what to do."[25]

Kaempffert had earlier praised Neurath's unity of science movement in just these terms that now elevated Weaver's blood pressure. When the first monographs of Neurath's encyclopedia appeared, Kaempffert announced that unified science meant "bridging the gaps between the sciences" and helping scientists communicate with each other more effectively. "In the past," he explained, scientists

> have been independent navigators who paid little attention to the whistles, flags, semaphores, wireless and other means of intercommunication. The encyclopedia is intended to tell each navigator what other ships are doing and what he can learn from their signals, their movements, their errands. A heterogeneous collection of vessels on a common ocean, but each going its own

way, is to be converted into a homogenous fleet. There will be no admiral to give orders. Only this encyclopedia is to serve as chart and compass.[26]

In the same way, Kaempffert argued that the Kilgore bill would let scientists "map out the whole field of science to reveal gaps in knowledge and then contract with the best men to fill them cooperatively." But critics like Conant and especially Weaver could not be reassured. "For the nth time," Kaempffert pointed out in frustration, "we insist that no scientist would be compelled to join the organization and no director would tell anyone how to proceed in solving a problem." But the "totalitarian times" Kallen described in 1939 had only become more tense and anxious. Weaver had even resorted to alarming images of guns and thought police in his denouncement: "The earnings of science are not to be gained by organizing a super-control which holds guns at the heads of scientists and tells them what to do," he scolded Kaempffert. "The earnings of science are gained only by setting the scientists free."[27]

Saved by Methodology

Stenbuck's tagline, "What are the problems of scientific research today?" and the rest of his copy made it appear that Kuhn's lectures would engage these controversial discussions about what scientists should study, how they should go about it, and where that might lead the world. Even worse than Neurath or Kaempffert, who promoted interdisciplinary organization and communication among scientists to collectively chart the future, the tagline evoked the menacing image of the modern science dictator. What are the problems of scientific research today? Come hear Thomas Kuhn at the Boston Public Library and he will tell you.

Understandably, when Kuhn exploded in worry and distress over the tagline, the Lowell Institute believed he was overreacting. One officer gently pushed back to ask whether "you have seen the reverse side of these posters?"[28] The side that would be displayed when tacked on a bulletin board presented the title and description Kuhn himself had provided, along with the dates of the individual lectures. His complaint that Stenbuck's tagline "bears no relation" to the Kuhn's title, "problems in the methodology of scientific research," was itself an exaggeration. Stenbuck had taken three of its key words ("problems," "scientific," and "research") and crafted them into an eye-catching question. But by neglecting the crucial word *methodology*, Kuhn seemed to believe, his problems in methodology became problems in science itself. The poster's reference to "the course of history" only made things worse, for despite his deep interest in scientific revolutions, Kuhn was no revolutionary seeking to alter history's path. "It may help you to understand my dismay," he wrote to Lowell, "if I explain that the fascinating topic your copy writer has so clearly stated is one to which I believe no serious and responsible student of science would address himself." Those who study the methods of science, that is, have no business presuming to tell scientists how to proceed lest they ally themselves with planning advocates such as Neurath and Kaempffert and suffer the rebuke of leaders like Conant and Weaver. A month before, Kuhn had made the same point when Professor Owen asked him to describe the "long range values" of his research. Kuhn replied that "it can illuminate the nature of scientific activity" and "the nature of knowledge"; and "it can contribute to the scientist by indicating preferred technics for professional education"; but, "equally certainly it cannot ever venture to prescribe fruitful research procedures to the working scientist."[29]

The Quest Begins

Kuhn lectured on Tuesday and Friday nights for four weeks. The surviving drafts of his lectures show that his ongoing reading in psychology and linguistics was then leading him toward *The Structure of Scientific Revolutions* that we know today. The first night's audience heard a version of *Structure*'s opening proclamation that history could liberate us from "the image of science by which we are now possessed" when Kuhn promised to explain why Karl Pearson's influential book *The Grammar of Science* was mistaken. Pearson believed that the scientist "proceeds from objective experimental facts or meter readings to the unique laws which govern them." But this was false, as Kuhn would show by "turning to history as a source of data for the reconstruction of scientific method." His lectures would also draw from the epistemological and metaphysical positions he had worked out in his notes on language and in his letter to Professor Owen months before: "The impartial, dispassionate observation of nature is impossible,"[30] he explained to his audience, unless there is something that guides the scientific mind through the complexities of raw experience. The unconscious "predispositions" he described to Owen were now joined by unconscious "prejudices" that keep scientists secure in particular "scientific orientations" or "points of view":

> They suggest problems; they suggest the sorts of evidence relevant for the solutions of these problems; and they suggest the mode in which the answers to these problems must be cast. They are, if you will, predispositions to certain sorts of explanations . . . they are equally predispositions toward evidence, toward facts. They direct our attention to particular aspects of the phenomenal world, and they suppress other aspects.[31]

The main roles that paradigms play in *Structure*—as basic embodiments of scientific theories and as general worldviews—were covered not by paradigms (a word which makes only brief appearances in the lectures) but by an array of theoretical terms Kuhn utilized over the course of the lectures, including "orientations," "meaning systems," "metaphors," "points of view," and Kuhn's Piaget-inspired "behavioral worlds," all of which function to reduce and organize the welter of experience.

The lectures covered the experiments in perceptual psychology that Kuhn would later include in *Structure* and they elaborated *Structure*'s cyclical, historical pattern from "pre-paradigmatic" science to normal science, "crisis," and "revolution." In place of "normal science," however, Kuhn spoke of the "classical period" in the life of a science—it comes after vague and unsystematic views are refined into "a more or less precise scientific theory whose laws may be stated in scientific texts and verified by prescribed operation." Like *Structure*'s "normal science," this idea of "classical science" is marked by success and confidence, so much so that scientists may refuse to abandon theories when problems arise (such as when "new observations which were themselves suggested by the old orientation but which do not seem to fit with it"). But scientists cannot ignore these difficulties forever, Kuhn explained. The case study of phlogiston chemistry illustrated for his audience how desperate ad hoc hypotheses emerge to explain these difficulties away. "This is typical of what we may now call the crisis stage in the progress of a scientific orientation," he wrote.

At this point, fundamental and far-reaching change is not far away: "Some such crisis stage seems inevitably to precede the overthrow of a scientific theory and the orientation which has accompanied it," he wrote. As he would explain in *Structure*, he noted that theories are never simply discarded. They are instead immediately replaced by another, by "some alternate orientation which it is shown [will] resolve the difficulties." Because this alternative highlights different aspects or areas of experience, the two orientations will be very different. The new will be "totally disparate" from the old.[32]

As much as these views seem to anticipate *Structure*, Kuhn did not describe these revolutionary changes in terms of "incommensurable" theories or "different worlds" of experience and knowledge. Here, as in his opening criticism of Karl Pearson, Kuhn followed Conant by reassuring readers that he knew science well ("I have done research," he wrote, "and I am committed to the belief that scientific knowledge is good knowledge, that it is useful knowledge") and that he knew it to be a progressive, "cumulative" enterprise. Kuhn borrowed Conant's criterion of progress from *On Understanding Science* and explained that historical scientists themselves would see that progress had occurred. It can be "objectively described so that a seventeenth, eighteenth, or nineteenth century scientist, confronted with a textbook containing contemporary scientific theories would, without reluctance, admit that science had proceeded a long way since his own day."[33]

Despite these resemblances and affinities to Conant's writings, Kuhn left himself a path for his future iconoclasm about scientific progress. He did not ask whether a scientist such as Aristotle would agree "that science had proceeded a long way since his own day"—Aristotle lived long before the seventeenth century. And where Conant's test for progress stipulated that resurrected historical scientists need only "view the present scene" to see the reality of progress, Kuhn specifically imagined them reading "a textbook containing contemporary theories"—the very educational device (according to *Structure*) that inculcates a progressive, cumulative view of science's history. Kuhn agreed with Conant about the reality of scientific progress, it might seem, because he had rigged the test to obtain that outcome.

In his fifth lecture, however, Kuhn seemed to doubt the objective, perspective-independent reality of scientific progress. He explained that everything turned on one's point of view:

> I suggest then that it is only in retrospect that a scientific revolution can be viewed as a net addition to the sum of human knowledge, and that from the point of view of the man involved in it, it is always about as destructive as constructive. For it demands the abandonment of the perspective in which he has so far viewed natural phenomena, and the adoption of an alternate perspective, which to the extent that he has accustomed himself

to the old one, must always appear as a distortion of scientific vision. And this, I think, accounts for the battles which rage at the time of conceptual reorientation in science.[34]

In passages like these, Kuhn's Lowell Lectures are pregnant with the conclusion that this "distortion of scientific vision" is simply scientific vision itself; that observations are theory-laden, that scientific revolutions are "as destructive as constructive" of worldviews, and that our reigning image of scientific progress must itself be replaced by a revolutionary alternative.

The Misrepresentation

Kuhn handled the issue of progress carefully. In the wake of the problem Stenbuck had created with his advertising copy, he took equal care to manage his audience's perceptions of his scholarship and its place in the volatile political landscape. He told the opening-night audience that he would provide only "a description of a continuing research program of a rather unusual, though not unique, sort." This would introduce the audience to his new field of research, but should not be taken as "a report on the outcome of a completed study." Nothing that Kuhn would say, in other words, should be taken as a definite, much less controversial, conclusion.

"If any of you happens to have followed closely the advance notices of this series of lectures," he explained,

> you may have remarked that the topic just described bears very little relation to the one announced in some of the flyers prepared by the Lowell Institute's copy writer. That topic was, I believe, described under a banner head reading: "What are the Problems of Scientific Research Today?"

Kuhn hid his anger behind casual, offhand recollections of what the ad said and behind an insincere compliment to Stenbuck: "I can scarcely imagine a more fascinating question; I should gladly attend a series of lectures devoted to it." "Except," his lectures read,

> that I doubt whether any serious student of science or scientific method would consider himself equipped to address such a subject. Therefore with apologies for any confusion that the

misrepresentation may have created, I should like to announce that I do not intend to deal with any of the problems raised by that question at any point in this series of lectures.[35]

There was a line, Kuhn made clear, and he would not cross it in these lectures.

In part, the line was philosophical and it lay at the heart of Kuhn's research. He could not responsibly address methodological concerns relevant to the science of today because it was too difficult, perhaps impossible. If we seek to understand where theories and "finished conceptual schemes" come from, he explained, this

> is peculiarly hard to achieve in dealing with the science in which we happen to believe. For the theory in which we believe is necessarily and uniquely characterized by the apparent inevitability of its relation to the facts from which it arose. Our belief itself represents a commitment to the double position that only this theory will account for the facts which we know and that this theory will account for all the relevant facts. We may admit that in the future there will be other facts and other theories, but for the moment we cannot conceive these, so the appearance of the inevitable connection remains.[36]

Perhaps like patients unable to see the Freudian complexes that controlled their thinking, our personal involvement in the methodological dynamics of contemporary science blinds us to alternate theoretical possibilities and the different methodological steps knowledge might have taken in the past. They remain unconscious and hidden to us, our attention focused narrowly by our educations and our textbooks on the seemingly firm and "inevitable connection" between the world we experience and the theories we use to understand it.

Methodologically speaking, in other words, our knowledge holds us captive. We can only see the point of methodological study when we apply it to theories long ago superseded. Only then can we fruitfully ask, for example,

> "Why did this set of experimental findings lead to this theory, rather than to the alternate one we hold today?" or "Why didn't this 'fact' appear relevant to a test of the validity of that theory?" It is in answering such questions that we shall trace the progress of the mind in its pursuit of scientific theories, and it is in this

pursuit that whatever may properly be called scientific method is to be found.

That is one reason why Stenbuck's publicity was so misleading, Kuhn implied: "You will hear very little about twentieth century physics during the course of these lectures."[37]

At the same time, this restriction to science's past allowed Kuhn to steer clear of the issues surrounding scientific planning. His quest for physical theory concerned basic science, not its applications to social, economic, or military life:

> We shall be concerned with the sort of research that led to the Newtonian laws of motion, not with the manner in which these laws were applied in building new machines or instruments. We shall be concerned with the work of such men as Boyle and Dalton, in so far as this led to a new understanding and a new set of laws governing the formation of chemical compounds, but we shall not be concerned with the manner in which these laws, once arrived at and confirmed, were applied to the production of dyes, explosives, or plastics.[38]

By the end of his introductory lecture, therefore, Kuhn was doubly insulated from the debates over science that had animated Kaempffert, Roosevelt, Bush, Truman, and Conant himself. He would address the methodology of basic science, not its applications, and he was looking backward in time—not at the research of the future; not even, as Stenbuck had put it, at the "scientific research of today."

The Semantics of "Communist"

Still, Kuhn did not altogether ignore cold war politics and the vaunted power of Marxist ideas to guide and control science's progress. As for the sources of the "predispositions" or "orientations" that dominate the scientific mind, for example, he pointed to the same Marxist analyses of science that Conant had cited in his *On Understanding Science* and remarked that social and economic circumstances may trigger "crisis periods" that lead to new scientific ideas. Crises may be sparked by "a variety of sources many of which we have not touched upon this evening," he said.

They may, for example, be produced by social forces external to science. They may proceed from changes in economic structure which alter scientific motivation. ~~Attn. Sen. McCarthy~~.[39]

At some point Kuhn drew a line through the words "Attn. Sen. McCarthy" in his typescript, as shown here; so he may not have gone through with this joke about the senator then making almost daily headlines in his hunt for subversive communists. In either case, Kuhn's audience would likely have understood it, not only because students and professors were now prone to "an unusual amount of seriocomic joking about this or that official investigating committee 'getting you,'" as the *New York Times* article had observed. The planning debates had made it known that Marxist Communists took knowledge and science to be products of social and economic structures, and McCarthy, the nation's leading anticommunist vigilante, was publicly taking names of those indulging, or worse promoting, this and other doctrines within "the Communist philosophy."

McCarthyism popped up in Kuhn's final lecture, as well. This was a technical discussion of the semantics of scientific language that Kuhn introduced to cast doubt on the ideal of "a completely formalized language" for research. Though only a few philosophers of science had ever endorsed this ideal, one who did was Joseph Woodger, the British theoretician of biology who himself had written a monograph for Neurath's encyclopedia. Kuhn had read it and took care to explain here why Woodger was mistaken.[40]

First, Kuhn invoked his theory of language as a network of cuts and boxes and then applied it to a simple question Piaget might have asked: "How do we acquire our notion of the meaning of the word 'dog'?" It comes with our learning "to correlate the word with our perceptions," Kuhn answered. We may at first be clumsy and apply the word to cats or horses. But just as William James's infants learn to discriminate different sensations and perceptions, children "gradually learn to discriminate linguistically among these elements of our behavioral world." Crucially, while these discriminations are precise enough for everyday life, they are always surrounded by a penumbra of vagueness or a "vaguer fringe of meaning" that may or may not apply in people's experience (such as "most dogs can be domesticated" or "dogs are fur-bearing animals").[41] These fringes are important because they make scientific progress possible. Scientists either refine meanings more and more (in which case they illustrate "classical" science) or, in what Kuhn called "a more interesting case," attempts at refinement may lead to "the total destruction of the pre-existing meaning system" and its replacement by another—in other words, a revolution.⁴²

At the blackboard, Kuhn drew partly overlapping circles representing the core and fringe meanings of "dog," "bear," and "wolf." He did so to challenge Woodger and his stereotypical image of logical empiricism, as well as any other approach that aimed to clean and purify language of its ambiguities and fringe meanings. Kuhn mentioned, for example, the then-popular movement for "general semantics" that had been founded by Alfred Korzybski, a Polish scientist and polymath who came to North America in 1916 and later wrote the best seller *Science and Sanity*. In the '50s, however, the leaders of the movement also included S. I. Hayakawa, whose *Language in Action* was then a best-selling entry in the Book of the Month Club, and M.I.T. economist Stuart Chase, author of *The Tyranny of Words*.⁴³

Kuhn may have known Chase, for he had lived in Croton and until 1929 was married to Margaret Hatfield, co-founder with Elizabeth Moos of his beloved Hessian Hills School. In either case, Kuhn probably read Chase's introduction to *Language, Thought, and Reality*, Benjamin Lee Whorf's influential book that Kuhn later cited in *Structure*. (Kuhn's future editor, Charles Morris, had also taken a strong interest in popular semantics and had written a successful, semi-popular book, *Signs, Language, and Behavior* in 1946.) Kuhn was no doubt intrigued by the semantics movement, given his fascination with language and networks of meaning organized by his "cuts" and "boxes," and possibly by Korzybski's focus on Aristotle. When Korzybski founded the movement, his guiding insight was that our habits of speech commit us unconsciously to patterns of thought and scientific reasoning that made sense in Aristotle's day, but no longer in our own. Aficionados of general semantics were therefore self-proclaimed "non-Aristotelians" eager to develop a non-Aristotelian, scientific system of thinking.

As suggestive as Korzybski's "non-Aristotelianism" must have been for Kuhn, he was shrewd to keep the general semantics movement at arm's length in his lectures. With few exceptions, experts in this area were perceived by established scholars as mere amateurs who invented their own jargon (Korzybski's included "time binding" and the "structural differential") and who profited by selling their books to acolytes at their own, private institutes.⁴⁴ Kuhn did not reject the movement altogether, for he too was knee-deep in semantical theory and he too aspired to create a new field of original, powerful research, complete with its own special terminology.

He also agreed with Chase and the other semanticists that language was a powerful force in modern life—especially, Kuhn pointed out, when it came to "our political and social life":

> It is against these dangers that we have been so frequently warned recently in writings on the tyranny of words. Words like radical, reactionary, or communist are encumbered with all sorts of associations whose conjunction has little reference to our experience.

Perhaps thinking of William Remington, Julius Rosenberg, or the allegedly subversive professors fired from the University of Washington; or the semantic issues raised by Conant, Kennan, and other analysts of the new cold war geopolitics, Kuhn remarked that "words, as recent history shows, are weapons. We must exercise extreme caution in manipulating them."

> We lable [sic] a man a communist because of one relevant or irrelevant aspect of his behavior, and we then assume that he possesses all the other attributes which we associate with the name. And we do ourselves grave injury in this manner.[45]

The educated, cosmopolitan audience in the Boston Library was unlikely to disagree. A majority of Americans, especially intellectuals in New England, denounced McCarthy's smear tactics, at least in private.

But they might have been surprised by the right turn Kuhn took after denouncing these malicious and unfair political tactics: we ought not to worry too much about these linguistic confusions and their injurious consequences, he said. While an attempt to clean up and formalize our language may reduce the social harm of virulent anticommunism, he explained, it would be positively harmful for science. For both Woodger's vision of a clean, formal scientific language, and Korzybski's purified, non-Aristotelian language of everyday life would remove the all-important "fringe meanings" on which scientific methods depend. "I think this is an impossible demand to impose upon any language which is to serve us in everyday life, or for that matter in scientific research," he explained. "Beyond a certain point an increase in precision is actually harmful." Since "our research is always conducted within the area determined by these vaguer fringes," research would effectively stop and our science would apply "only to what we have already experienced."[46] Conant, Weaver, and especially Kallen were right, Kuhn implied: if attempts to organize science and advance its progress meant purifying its language, there was a great risk that science itself would suffer as a result.

∽

As he concluded his Lowell Lectures, Kuhn came down clearly on the side of Conant and other critics of scientific planning. But in doing so he rejected an assumption shared by critics and proponents alike—the almost universal assumption, embraced by communists, anticommunists, respectable philosophers, and self-proclaimed experts in general semantics—that what is good for society is also good for science. Conant firmly believed this was true. It was a foundation of his general education program and his career as a cold warrior. The equation of freedom in life with freedom in science linked his expertise as a chemist, a scientific administrator, his authoritative handling of the problem of communist teachers, and his leadership of the Committee on the Present Danger. His intellectual liberalism and his political liberalism, in other words, were united in the ideal of human freedom. But the more Kuhn disentangled the implications of his Aristotle experience and the more he understood the "predispositions" and "prejudices" that make science possible, the more his new field of research and Conant's liberal philosophy of science veered apart. ("I contend then that emptiness rather than openness characterizes the youthful mind" of the science student, he noted in his fifth lecture.)

Because of this subversion in the making, Kuhn moved carefully through the minefield of controversial issues that defined the cold war intellectual climate—so much so, it seems, that the quality of his Lowell Lectures may have suffered. Aside from the complications he addressed over Stenbuck's publicity, the lectures were erratic in tone and mood. Kuhn was sometimes casual and confident—even making a joke about Senator McCarthy; but sometimes stiff and formal—so much so it is difficult to imagine Kuhn's nonacademic audience staying the course through eight hour-long lectures. This passage from his last lecture is neither atypical nor ironic:

> The limitation of the alternate means of prescribing a meaning system are less apparent, and we will more profitably approach them by discussing the manner in which we actually arrive at meanings and then considering the manner in which the nature and function of language would be changed by an adherence to a more precise meaning system.[47]

Kuhn admitted decades later that the lecture series was difficult for him ("I had a dreadful time preparing it and I nearly cracked up"). He would not find his winning authorial voice until he wrote *The Copernican Revolution*, published six years later. The glowing reviews it received surely taught him to worry less about how his work was publicized and more about presenting clearly and directly the issues that engaged and fascinated him. At that point, however, with McCarthy dead and the worst excesses of academic McCarthyism past, Kuhn would have political problems of a sort he probably never anticipated: he was *too* successful among large, popular audiences to gain the kind of respectability he anticipated and sought in the eyes of other scholars. Despite the accolades that would pour in for *The Copernican Revolution* and the prestige of Conant's name and his generous foreword, Kuhn's tenure committee found the book neither sufficiently scholarly nor original to win Kuhn tenure at Harvard.[48]

Stenbuck's publicity was prophetic after all. As the success of *The Structure of Scientific Revolutions* showed, a very large public wanted want to hear what Kuhn had to say about science and its progress—not just about some distant, methodological past but about the exciting new worlds of science's future.

10

Ideology and Revolution in the *International Encyclopedia of Unified Science*

Even with the threat of nuclear superblitz hanging overhead, James Bryant Conant believed there was hope that Western liberalism could survive the threat of international communism. To do that, the communist East had to be better understood. This was the mission of Harvard's Russian Research Center at its founding in 1947. But while Conant helped to lead the nation intellectually and diplomatically against its cold war foe, many intellectuals reached for a more general and abstract way to understand the modern world—especially the causes of the two world wars and the ravages of fascism and totalitarianism. For them, the study of ideology, of systems of ideas and their complex relations to societies, economies, and individuals would be a road map for understanding what Conant once called "this troubled century."

None of these intellectuals was more influential than the Hungarian sociologist Karl Mannheim. In 1936, an English translation of Mannheim's classic work *Ideology and Utopia* was translated and edited by the sociologists Edward Shils and Louis Wirth at the University of Chicago. Shils would later become one of Kuhn's close colleagues to whom he would send the first draft of *The Structure of Scientific Revolutions* (and Shils would return the favor by inviting Kuhn to join the University of Chicago's Committee on Social Thought). In the early 1950s, however, Kuhn was a young scholar who shared with Shils and Wirth a keen interest in ideology. Years before he introduced "paradigms" into his evolving manuscript for *The Structure of Scientific Revolutions*, the centerpiece of Kuhn's developing theory of science was ideology.[1]

In his introduction to the new translation, Wirth could not overemphasize the importance of Mannheim's insights for the ominous world situation. The tragedy of World War I and the rise of fascism in Germany and Spain made it seem that rationality itself was exhausted and on the verge of collapse. This new edition of Mannheim's works, he wrote, could salvage whatever "prospects of rationality and common understanding" remain in the world.

> Whereas the intellectual world in earlier periods had at least a common frame of reference which offered a measure of certainty to the participants in that world and gave them a sense of mutual respect and trust, the contemporary intellectual world is no longer a cosmos but presents the spectacle of a battlefield of warring parties and conflicting doctrines. Not only does each of the conflicting factions have its own set of interests and purposes, but each has its picture of the world in which the same objects are accorded quite different meanings and values. . . . [S]ince the world is held together to a large extent by words, when these words have ceased to mean the same thing to those who use them, it follows that men will of necessity misunderstand and talk past one another.[2]

Wirth's diagnosis previewed the cold war impasse with the Russians that Kennan, Conant, and other cold warriors would describe ten years later. But it was bigger. To the modern ideologist, the landscape of mankind was a mosaic of "warring parties and conflicting doctrines" that threatened the very possibility of communication and understanding beyond one's party or tribe—anywhere, at any time.

From the start, ideology was understood through the lenses of war. The concept appeared first during the Napoleonic wars as a scheme for denouncing and vilifying one's enemy, Mannheim explained: "We begin to treat our adversary's views as ideologies only when we no longer consider them as calculated lies." After Marx, the concept grew into "total ideology" that applied not just to military or imperial conflicts but to class conflicts and their underlying economic realities. Ideology became ubiquitous—it could be found in every human being, in oneself, and of course in the work of the sociologist studying ideology: "At the present stage of our understanding," Mannheim wrote, it is impossible to avoid this "total conception of ideology, according to which the thought of all parties in all epochs is of an ideological character."[3]

Ideology and Education

If so, scientific thought and science's history must also involve ideology. Kuhn realized this evidently by early 1953, shortly after he was recruited by Charles Morris to write a monograph on the history of science for the *International Encyclopedia of Unified Science*. The insight was fitting, for the encyclopedia was itself an ideological response to the fragmentation of the modern world—not only the fragmentation of humanity that Wirth described, but the fragmentation and specialization of modern science. From the beginning of his collaboration with the famous Vienna Circle of philosophers, Otto Neurath insisted that the new scientific philosophy they promoted must address both fragmentation among the different sciences and a growing disconnection between science and public and political discourse. The two problems went together, for with increasing specialization—each science having its own journals, its own professional societies, and its own, specialized theoretical terminology—science lessened its power to address problems of modern life concerning economics, medicine, education, and other institutions of direct concern to the public. As Waldemar Kaempffert's metaphor had it, the sciences could be as powerful as a flotilla of naval ships—but only if they listened to each other, understood each other's research and discoveries, and worked together to address and bridge the gaps in human knowledge.

Neurath was a sociologist and educator, an expert in adult and public education. In prewar Vienna he founded the *Social and Economic Museum* that pioneered techniques of visual education. Before it became the international standard for signage directing travelers to baggage carousels, currency exchanges, and restrooms, Neurath and his colleagues created the International System of Typographic Picture Education or *ISOTYPE* for use in books and public exhibits. Even illiterates, he believed, and ordinary workers unfamiliar with scientific research and debate could nonetheless learn about the social and economic dynamics of modern life—about *Modern Man in the Making*, as his book on the *ISOTYPE* method was titled. In the scientific sphere, Neurath's new encyclopedists would write short booklets or monographs about one area of science, or one aspect of scientific methodology, and focus on the gaps and divisions that seemed important to fill. At the same time, Neurath hoped that his new encyclopedia of science would break down barriers so that ordinary people would better understand modern science. At one point he envisioned a future volume of the encyclopedia that used *ISOTYPE* to describe in pictures and diagrams how the sciences were becoming progressively more interconnected. The unity of science and the unity of society were twin concepts in Neurath's mind.[4]

Professionally speaking, Otto Neurath was no James Conant. This European sociologist, economist, and philosopher who never held a university title and the president of Harvard University had little in common. But they were both effective organizers seeking to improve the lives of broad masses of people through education and by cultivating unity and coherence in times of global conflict and irrationality. These interests and goals would converge repeatedly at Harvard, beginning with Conant's welcoming Neurath and others to Harvard on the eve of World War II (when Horace Kallen launched his attack on Neurath's project as "totalitarian"). In a few years, his general education committee would consider the unity of science movement in its "search for unity" in American public education. And again in 1950, after leadership of the movement had passed to Philipp Frank and his Institute for the Unity of Science, Conant again convened with the unity of science movement. At a conference dedicated to "Science and Man," Frank introduced Conant to the audience while offering a tolerant, big-tent picture of the postwar unity of science movement. Our understanding of science cannot stop with "logical and semantical analysis," Frank explained, for it must also adopt a pragmatic point of view that understands "science as a human enterprise by which man tries to adapt himself to the external world." The mission of the institute and the *International Encyclopedia of Unified Science*, he said, was to join these approaches. Frank's philosophical bar, in other words, was well stocked for Conant's favorite drink—his cocktail of pragmatism, logical empiricism, and "at least two jiggers of pure skepticism."[5]

"Both as a college president and a chemist I have personal reasons for being deeply interested in the unity of science," Conant told the group. One set of reasons concerned his general education project: "If we are going to give the lawyer, the statesman, the businessman, the news writer some understanding of science," he explained, "we educators ourselves must endeavor to understand the methods of science." As always, Conant chose his words carefully. By referring to the plural "methods" of science, he left his audience with the finishing taste of skepticism about whether there is just one path to knowledge. There may not be anything like final, definitive knowledge, at all. The unity of science that Conant recognized lay in its restless, forward motion:

> My own definition of science as part of accumulative knowledge is not in terms of any final goal but the nature of the process.

> The essence of the undertaking of the historian, the social scientist, the philosopher, the natural scientist to my mind lies in the fact that his efforts will be measured by their fruitfulness, not by their finality.

"Even physical science," he remarked, "to my mind is not a quest for certainty but rather a quest which is successful only to the degree that it is continuous" and keeps moving ahead.[6]

Conant's unity of science, in other words, was not something to be specified in a clear and simple formula; and certainly not one that might allow science to be controlled or planned. It was, rather, part of the larger, progressive unity of liberalism and human learning through time. In the year 1950, however, that tradition seemed increasingly threatened by postwar tensions with the Soviet Union. Within a year of this meeting, Conant would address the nation as head of the new Committee on the Present Danger and privately write to the future president of Harvard to confide his fears that the world might not survive nuclear war. With so much at stake, Conant gently but firmly pointed Frank's colleagues away from the mistaken and politically unpalatable view that there existed any simple and final way to understand science. But he earnestly embraced Frank's movement as a part of this grand, intellectual tradition now endangered by "this troubled century"—especially, as he concluded his remarks, if this meeting was itself no mere celebration of ideas at hand, but rather a stimulus "fruitful of ideas yet unthought."[7]

Science and its Existential Factors

Frank was deeply interested in ideology and its relations to science. In a bid to understand and defeat the ideological competition the unity of science movement faced, Frank wrote a long manuscript about the various ways by which political regimes favor and promote certain philosophies of life and knowledge (the resulting manuscript, *Science, Facts, and Values*, remained unpublished at the end of his life in 1966). He additionally accepted the Marxist idea (voiced as well by Kuhn in his Lowell Lectures) that social and economic forces could affect the direction of science and hence the very content of scientific beliefs. This could happen, for example, when science reached a fork in its road, a choice to pursue one of two very different theories, both of which are supported by the evidence and observations at hand. When evidence underdetermines theory, science's future is shaped by social or cultural factors, Frank explained, such as whether a theory accords well with common sense, with religious values, or whether it seems to support political goals of ruling leaders. Complex processes such as these remind us that science is "a human activity," as he put it when introducing Conant.[8]

In late 1952, Frank wrote to Kuhn about sociology of science and knowledge. At the conference that Conant addressed, Frank had spoken about his hopes to combine logical empiricism with sociological research into science. Now he was writing to say that his institute was ready to step in this direction. Frank had a small budget to fund researchers, but he needed to form a committee within his institute to choose projects and award research grants. So far, the committee included Frank himself and the philosopher Ernest Nagel at Columbia University. Would you like to join this new committee to help organize and promote this kind of research? Frank asked. There would be no others on the committee, for "we would like to keep it small."[9]

Indeed, Kuhn was interested. Frank and Nagel were established figures in philosophy of science, and this focus on sociology of science would complement his emerging ideas about meaning systems and "predispositions" in science. With Frank's letter in hand, he sat down and began typing a list of "possible research projects/Sociology of Science." Most of the topics had popped up in Kuhn's notes and lectures before. They included the "role of youth in creative research" and questions such as, How did different areas of science specialize and separate over history? Can sociological factors explain cases of simultaneous discovery, when scientists (such as Darwin and Alfred Russel Wallace) hit upon similar ideas at the same time without ever collaborating? What about "the resistance of scientists to new conceptual schemes"?

Along with the letter, Frank included a short position paper, titled "Research Project in the Sociology of Science." He said he had drawn it up with the help of Nagel and the sociologist Robert Merton, also of Columbia University. But their signatures at the end, evidently penned by Frank himself, suggest that Frank was the main author. In any case, the goal of the project was to better understand "the role which sociological factors have played in the acceptance of scientific laws and theories." These factors, it noted, "may be referred to as 'existential factors.'" The phrase "existential factors" suggest Frank's own interests in Mannheim's sociology of knowledge, for Mannheim used the phrase. Merton, too, was also a prolific commentator on Mannheim and his theories about "the existential determination (*Seinsverbundenheit*) of thought."[10]

Though still teaching in Conant's general education program, Kuhn was intellectually courted by senior scholars who were keenly interested in understanding science through sociological and ideological lenses. At least since his Aristotle experience, Kuhn was sure that nontheoretical factors—linguistic, psychological, perhaps Freudian—played a role in science. His earlier crisis over intervention, at least as he analyzed it for Robert Gorham Davis, led him to the ideological insight that "the basis of my belief went beyond reason," and that "all decision is composed of more than judgement." So Frank's suggestions were not wholly new or strange. Kuhn was probably at home with Frank's idea that nonrational "existential factors" involving "governments, churches, and other organized bodies, the pressure of public opinion, or the prospect of financial success or an improved social status, etc." can affect the course of science, sometimes crucially.[11]

But Kuhn was quick to notice something wrong. "It is generally possible," Frank's document read, "to distinguish quite clearly between acceptance of a theory because its consequences are in agreement with observed facts, and acceptance of a theory because of other reasons which may be referred to as 'existential factors.'" The logic of science, that is, and the ideology of science were different and independent of each other; as if ideology took over only when reason, calculation, and experiment failed to govern scientific choices. But the Aristotle experience had shown Kuhn that this could not possibly be true. Some kind of nonlogical "predisposition" to embrace a particular understanding and experience of the world (and to exclude possible alternatives) seemed to operate inside the scientific mind, even though its operation may be invisible to the scientists it affects. If these predispositions had anything to do with Frank's (and Mannheim's) "existential factors," then Frank and the others were not thinking about them the right way.

"Dear Professor Frank," Kuhn began to type.

> Thank you very much for your invitation to join the organizing committee of the Institute's project on "Sociology of Science." There is no group in which I should rather participate, and I look forward to hearing more from you about the committee's activities.
>
> Let me raise one question about the draft prospectus which you enclose. Would it not be appropriate to include in the committee's terms of reference an examination of those sociological factors which impinge upon an individual scientist not by virtue of his membership in a national community (say

the United States), but by virtue of his membership in a narrower professional group (say the American Physical Society)?

Kuhn agreed that scientists were normally surrounded by sociological, "existential" pressures and circumstances. And he agreed that these have

> an important bearing upon the problems which a scientist considers worth attacking, the experiments which he employs to resolve his problems, the abstract aspects of his experiments which he considers relevant, and the logical and experimental criteria which he demands of a "valid" argument.

But Frank, Nagel, and Merton had misidentified these factors. They were not like "religion, social status, economic organization" and other circumstances that exist normally outside of science. They included "all non-logical, non-operational factors influencing the choice between competing scientific theories." On that view, he explained, one must

> recognize the existence of sociological factors arising from the consensus of the scientific group (say the American Physical Society) about the problems of their science. Such factors are, I believe, intrinsic to science in the sense that no important "acceptance" or rejection occurs without them.

These factors are *always* at work, in other words, telling scientists every day that they had made the right choice, that their theories made good sense. They were the essential "predispositions" that Kuhn described to Owen and which carve the infinitely complex world into "scientific objects" for study and experimentation. The scientific procedures that Frank saw working in place of these "existential factors" actually worked along with them—normally, all the time. "Controlled experiments," Kuhn penciled in at one point in his letter, "play a central, possibly a decisive, role, but not in a sociological vacuum."[12]

Kuhn was eager to spell out his ideological insights. But he was not satisfied with his argument as it rolled up and out of his typewriter. His letter to Frank turned into a sketch pad in which he ran through the argument three different times (in as many pages) and then edited it by hand. His thoughts were in flux. In the Lowell Lectures two years before, for example, he had distinguished clearly between "internal" and "external" factors that might lead science into crisis; but after crossing out "Attention Senator McCarthy" he had also crossed out "internal factors" and replaced it with a reference to *other* factors, as if he was no longer sure of any robust internal-external distinction (as he was now trying to tell Frank).[13] Or perhaps, as his letter suggests at one point, the line between what is internal and external changes over time. "At this time and place," Kuhn wrote, the kinds of external factors "like Government, Church, etc." that Frank mentioned play the smallest of roles and "have relatively little impact upon decisions made by professional scientists." But that is because they changed their form and their location; they had migrated *inside* professional science as "socially conditioned, implicit, professional 'faiths,'" which now play the ideological roles that religion or other kinds of metaphysical belief had played centuries before.[14] Yesterday's externalities had become today's internal, professional realities.

Another reason for Kuhn's difficulties writing this letter was that Frank's thoughts were not as myopic about these ideological factors as Kuhn supposed. Frank acknowledged that these existential factors may operate "more subtly even in 'normal' social settings" and that it was not clear "whether, and to what extent, these existential factors are an obstacle to fundamental research." One of Frank's research topics was "the influence of social and economic factors on the actual conduct of scientific inquiry," involving "standards of accuracy" and "models used in construction of theories." Clearly, Frank understood that inherited traditions and preconceptions shaped scientific reasoning and behavior, so he was not blind to those functions Kuhn had ascribed to "predispositions" and "scientific orientations."

And then there was the tone of Kuhn's letter. After graciously accepting Frank's invitation, his arguments became forward and aggressive. The sociology of science that really matters, he remarked, at one point is "obscured by the present draft of the project." When he came back to read later what he had typed, he penciled "Tone down" in the margin. To the same effect, he replaced "one must, I think, recognize" with "I should find it helpful to recognize"; and "I do not really believe" with "I wonder whether." Kuhn toned it down because he did not want to insult or annoy Frank and his colleagues. Accusing them of "obscuring" the very field of inquiry they were undertaking to support was not too collegial a gesture (and, as it happened, Merton and Nagel would later become important colleagues and confidants for Kuhn).

When he wrote to Kuhn, Frank was not far from Cambridge. His letter came from the Institute for the Unity of Science, in Boston, and as a teacher in the general education program, he and Kuhn surely knew each other and may have had frequent, even regular contact. Yet Frank asked Kuhn "to please write me" if he was interested in the proposal so they could set up a meeting. That is what Kuhn set out to do, but it appears he may not have sent any written reply, after all. He may have sent a different letter of which no record remains, or he may have simply talked to Frank in person. But this document remains in Kuhn's personal papers with "not sent" written across the top in Kuhn's handwriting.

Ideology and the Structure of Scientific Revolutions

In a subsequent report to the Rockefeller Foundation, which funded Frank's institute, Frank mentioned this new committee and this collaboration with Kuhn (as well as the sociologist Bernard Barber, whom Frank evidently also invited to join). But it does not appear that the committee bore much scholarly fruit. Frank was at this time nearly seventy years old and his career and effectiveness as an organizer were waning. If Kuhn's "not sent" at the top of his document means that he decided not to accept Frank's invitation to collaborate, this may have been a reason. Another may have been that Frank was under investigation by Hoover's FBI. Four months before Frank wrote to Kuhn, Hoover decided to investigate reports (later determined to be spurious) that Frank had come to the United States "for the purpose of organizing high level Communist Party activities." By April 1953, four months after Frank wrote to Kuhn, Hoover's agents in Boston had interviewed informants and administrators on campus about Frank's background and activities. Frank's proposal therefore came to Kuhn during this first, active phase of the FBI's investigation in Cambridge, and it was widely known that Hoover's agents typically intercepted the mail of their targets (among the reasons Hoover ordered the investigation in the first place was that Frank was known to be associated with Harlow Shapley, of Waldorf peace conference infamy, whose mail and activities he also monitored).[15]

On the other hand, Kuhn did choose to collaborate with Frank as an author for the *International Encyclopedia of Unified Science*. After Neurath's death in late 1945, Frank joined Morris and Rudolf Carnap as editors of the encyclopedia and his institute became the official sponsor of the encyclopedia project after the war. Still, Kuhn appears to have had little contact with Frank once he agreed to participate. The surviving documents show that Kuhn corresponded mainly with Charles Morris in Chicago.

At that time, Kuhn was familiar with the encyclopedia. Along with Woodger's *Techniques of Theory Construction*, he had read Leonard Bloomfield's entry *Linguistic Aspects of Science*. Both monographs would have supported his view that philosophers in logical empiricism's orbit overrationalized the scientific process. They attended to its logical, formal properties at the expense of the "meaning fringes" and psychological "predispositions" he discussed in his Lowell Lectures. He also sensed that the encyclopedia had lately fallen on hard times. Neurath's energy and enthusiasm had been enormous and infectious. But after his death, neither Frank nor the other editors filled his enormous shoes. Neurath once envisioned a living, ever-changing encyclopedia that spanned dozens of volumes, each containing ten monographs and reaching from science, to law, to medicine, and education. But Morris and Carnap struggled simply to fulfill their editorial contract with the University of Chicago Press to deliver two preliminary volumes of twenty monographs. At the war's end they were about halfway there. Briefly, in 1949, things looked auspicious. Frank's institute was up and running and Morris and Carnap were buttonholing potential replacements for those authors no longer able to contribute. Neurath's encyclopedia project was set to "spring to life again."[16]

When Morris recruited Kuhn, his growing reputation as a historian of science did not signal any compromise or stopgap measure. From the beginning, Neurath and his co-editors insisted that some monographs should be historical. In an outline of the encyclopedia from 1936, Neurath sketched four: The history of logic, to be written by Polish logician Jan Łukasiewicz; the history of rationalism and empiricism, then slated for French philosopher Louis Rougier; a bibliography of the historical movement promoting the unification of the sciences—"only the last 100 years!" Neurath emphasized—and the history of science. For history of science, Neurath recruited the Italian mathematician Federique Enriques. By the late 1930s, however, the editors had replaced that monograph with one dedicated to sociology of science to be written by Louis Wirth. After Wirth's death in 1952, the monograph was rededicated to history of science and offered to I. Bernard Cohen, the Harvard historian who had helped Conant assemble the case histories he sketched in *On Understanding Science*. Cohen accepted the commission, but only temporarily. Perhaps through Cohen, through Conant, or through Frank (who had already eyed Kuhn as a recruit for his new committee on sociology), Morris learned of Kuhn. He was, in any case, close to a perfect choice from Morris's point of view. Kuhn was an interdisciplinary thinker, at home in science, philosophy, history, and literature. His early notes on "meaning systems" and semantics placed his interests close to Morris's own expertise in semiotic theory, while his sociological perspective—as his unsent letter to Frank shows—aligned his research with Frank's vision for the future of science studies. With the Lowell Lectures under his belt and his growing expertise teaching *Natural Sciences 4*, Kuhn had demonstrated that he could write for the wide, educated audience the encyclopedia hoped to reach.[17]

After their initial conversations, Morris wrote from Chicago to ask Kuhn for a title and a prospective outline for his monograph. Kuhn responded in July 1953 to say he was "reluctant to supply a title for an unwritten essay," but he understood that Morris needed one:

> How would you feel about "The Structure of Scientific Revolutions"? It is the best I have been able to contrive so far—all the alternatives make use of the word "ideology," which, at least in the title, I should like to avoid.

Morris's eyebrows must have gone up. "Ideology" was joining "international" and (thanks to Horace Kallen) "unity" as a controversial word in the new era of political inquisition and investigation on the nation's campuses. (As Kuhn wrote, his former teacher Wendell Furry was tangling with McCarthy and other anticommunists in Washington.) Kuhn was not the first to use the word for philosophical purposes. W. V. O. Quine had published an essay on "Ontology and Ideology" in early 1951, for example, in which ideology names "what ideas can be expressed" in a theory. Kuhn's intended meaning of the word exceeded Quine's, but in telling Morris that he "should like to avoid" the word in his title he joined Quine in recognizing its "unwanted connotations."[18]

Morris did not want those connotations either, for suspicions about the encyclopedia project and its editors were growing. A similar but older academic venture, the *Encyclopedia of the Social Sciences*, was then being scrutinized in Congress by the Reece Commission, charged with investigating alleged communist sympathies within tax-exempt foundations, such as the Rockefeller Foundation that funded Frank's Institute for the Unity of Science.[19] Morris was aware of Neurath's "communistic" reputation (as he had told Horace Kallen) and he would have been correct to worry whether Frank and Carnap might appear on Hoover's radar. In fact, Frank already had, and that investigation would soon lead Hoover's agents to Carnap—not, as Sidney Hook insisted might happen, because Carnap lent his name to support Shapley's peace conference four years earlier, but because he publicly called for Julius and Ethel Rosenberg to be spared their death sentence.[20]

In these perilous political times, therefore, Morris must have breathed a sigh of relief as he continued to read Kuhn's letter: by all means, let's go with "The Structure of Scientific Revolutions," and leave "ideology" out of it.[21] Judging by Kuhn's précis, however, ideology would remain central to the substantive claims the monograph would make. First, Kuhn explained,

> [F]or the professional group which employs it the content of a scientific theory is larger than the formal or formalizable content of a theory. Logical research may isolate the formal content of a theory, but it does so only by shearing away the functions of a theory as a profession[al] ideology for the practicing scientist.

As he had before, Kuhn planned to examine those aspects of science that exceeded what could be specified with logical analysis—the "cuts" and "boxes" of language, the "fringe meanings" he discussed in his Lowell Lectures, the unconscious "predispositions" he described to Professor Owen, and the "existential factors" within scientific communities that he described to Frank. In this ideological role,

> a theory serves to direct the scientist's attention to certain sorts of problems as "useful" and to certain sorts of measurements as "important"; it dictates preferred techniques of interpretation, and it sets standards of precision in experiment and of rigor in reasoning. Above all, the theory, as ideology, is a source simultaneously of essential direction and of disasterous [sic] inhibition to the creative imagination.

These nonformal aspects of theories and of scientific methodology "defy validation," Kuhn explained, because they are like the air scientists breathe—silently and inconspicuously structuring experience by guiding perceptions, arguments, and even the "creative imagination" itself within the professional scientific mind.

The focus of the monograph, Kuhn explained, would be scientific revolutions. From this point of view, a revolution—the replacement of one theory-as-ideology by another—is an eventual response to the sea of invisible "unsolved problems" that surround any successful theory:

> Theories preserve themselves by restricting the attention of the profession to problems which can in principal [*sic*] be solved within the theory and by inhibiting the recognition of important incongruities in the application of the theory to nature. In some sense every theoretical orientation excludes the existence of totally unsolved problems. A significant theoretical reorientation replaces a complete, but retrospectively exclusive, schema of knowledge with a more inclusive one. An innovator does not add new knowledge to old; he rather imposes a new set of categories on nature, destroying and replacing an older set, which, embedded in the profession by training and practice, die hard. So intellectual discovery is necessarily intellectual revolution.

Ideology ensured that there was "firm closure" in any area of science. But any area of research could be reopened by a new orientation, a new examination of previously ignored phenomena or "totally unsolved problems."[22] Ideology, in other words, made scientific revolutions possible.

∽

Generations of students have been taught that Kuhn's *Structure* almost singlehandedly revolutionized philosophy of science by closely examining the history of science—ironically, from within logical empiricism's encyclopedic stronghold. Kuhn had said history was key to understanding science in his Lowell Lectures, and he would say so again in *Structure*. But in the early 1950s his interest lay in ideology and his ideas were in flux. When he wrote to Morris in 1953 to announce his ideological focus, his terminology and even his view of scientific progress seemed unsettled. In one sentence, his revolutions seemed cumulative and progressive (they involve "a more inclusive" body of knowledge) and in the next they were not ("An innovator does not add new knowledge to old").

Kuhn apologized to Morris for the changes afoot. "I know this is not quite what you were after when you approached me about the volume," he said. But not everything had changed: "The point of view of the monograph is drawn from the history of science and the examples will be historical, so I am not quite so far off as the above discussion may indicate." But there was no doubt that Kuhn would offer no simple chronicles or textbook stories of science's past. He aimed to unearth the hidden ideological, and revolutionary dynamics of science by combining the methodological inquiries he explored in his Lowell lectures with his current interest in the sociology of scientific communities: "My basic problem is sociological," he explained to Morris, "since it arises because any theory which lasts must be embedded in the professional group in which it will be overthrown."[23]

Kuhn's philosophical relationship to Rudolf Carnap has also turned out to be more complex, and less antagonistic, than is usually supposed. Kuhn confessed in the last years of his life that he probably misunderstood Carnap's philosophy of science at the time. He never appreciated significant points of contact that were later pointed out, and he would likely have profited by knowing Carnap and his ideas better.[24] In some ways, *Structure* can be seen as friendly to Neurath's view of science, as well. Kuhn was an ideal contributor to the *Encyclopedia of Unified Science* because his theory of science was itself interdisciplinary and appealed to philosophy, psychology, sociology, linguistics, and semantic analysis. His broad, interdisciplinary, and reflexive approach illustrated precisely what the encyclopedia stood for: using science to better understand science itself. *Structure*'s view of progress as movement *from* the past instead of *toward* the future also complemented Neurath's encyclopedic vision of science's future as open, ever-evolving, and unpredictable in the long run. The new encyclopedia, he wrote, would be no "mausoleum of achievements of the past, but an instrument of most lively activity. . . . A hundred gateways are open."[25]

Neurath would have approved Kuhn's emphasis on the "fringe meanings" surrounding scientific words (Neurath had called them "clots" of indistinct meanings) as well as the scarcity of the word *truth* in *Structure*. He once infamously (and more than half-seriously) proposed that *truth* and other metaphysical words be placed on an *index verborum prohibitorum*, or list of prohibited words.[26] On this fundamental point Neurath joined Philipp Frank and Conant, who also de-emphasized the concept of truth in their writings about science. This anti-"truth" alliance points to the powerful influence of the American pragmatists Dewey and especially William James, who famously argued that truth is no "inert static relation" between our knowledge and reality itself. It is instead "something that happens to an idea" within the larger process of inquiry and research.[27] At *Structure*'s conclusion Kuhn joined this alliance, asking his readers to place aside the habitual belief "that there is some one full, objective, true account of nature" that scientific theories strive to capture. Again, *Structure* stands not as a rebellion against those of Kuhn's collaborators and philosophical predecessors, but as an expression of them—"a work of synthesis," as he once described it.[28]

Politically, however, Kuhn was headed in a different direction than Neurath and others who hoped and believed that understanding the ideological contours of thought would provide a tool for shaping science and for improving modern life. Given the widespread interest in ideology that surrounded the new encyclopedia, and its popular presentation (by Neurath's cousin Kaempffert at the *New York Times*) as a vehicle for unifying and advancing the whole of science, Kuhn was indeed an ideological mutineer within Neurath's encyclopedia. This mutiny, however, was less a matter of embracing different philosophical doctrines than matters of individual temperament and powerful political realities—the different ways that Neurath and Kuhn, countries and generations apart, responded to their times. Neurath the activist, the planner, and organizer could not resist the impulse to shape and guide the sciences and to educate the public to see the world more rationally and scientifically.

Kuhn had once been a child of this Enlightenment sensibility. His childhood "ideas on mankind and on the structure of a good society," he explained in "The War and My Crisis," were founded on "a complete trust in reason. My faith contained nothing mystic or divine; it was simple and complete."[29] But Kuhn's crisis over his pacifism before the war and the onset of the cold war turned him into a different kind of liberal intellectual. In America of the late '40s and early 1950s, the impulse to plan and organize scientific research became taboo, as Conant's and Warren Weaver's public attacks on Kaempffert made clear and vivid. Kuhn's distress over the Lowell Institute's publicity showed that he urgently feared being seen as engaged in these controversial debates. Still, *The Structure of Scientific Revolutions* was deeply shaped by this controversy as Kuhn positioned himself alongside Conant and Weaver and against Neurath and Kaempffert: the engine of scientific change and progress, *Structure* argues, is not organization and cooperation among different fields of science, as Neurath and other planners supposed. It is instead the very quality of modern science that Neurath and Kaempffert sought to reduce: specialization within scientific communities, the members of which share a single paradigm that leads them to see and understand the world in the same ways. The power of ideology makes progress possible, in fact; for without scientists' determination to deepen their paradigm's fit with nature, their ideological, dogmatic confidence that every puzzle it raises for them can be solved, anomalies would never be encountered and revolutions would never occur. As much as the future of science can be controlled by anyone, it is controlled by specialists *within* individual sciences—not by the scientists of the future Neurath and Kaempffert envisioned, seeking always to build bridges and reduce specialization.

By the time Kuhn finished writing it, *Structure* would include other themes and features of this politicized cold war culture, including the totalitarian historiography of Orwell's *1984*, an explicit comparison of scientific and political revolutions, and scientific "conversion experiences" that evoke widespread fascination with political brainwashing and mind control. In these and other political overtones, an air of revolutionary originality and subversion surrounded the yet-unwritten monograph Kuhn described to Charles Morris. His theory of revolutions as abrupt changes in *ideology* would surely be news to Conant, whose historical scientists normally moved from one conceptual scheme to another easily and without ideological restraint. It would be news to Frank, Merton, and Nagel whose position paper located the "existential" and sociological pressures of science mainly outside of the scientific machinery of logic, reason, evidence, and experiment—not at its core. Indeed the very "image of science by which we are now possessed," as Kuhn would put it in his introduction, stood to be overturned by the monograph—all of which reassured him that he was taking his exciting new theory of science to the heart of the enemy camp. When he reassured Morris about his new focus on ideology, he wrote: "Whether or not this rough and tentative description meets your original specifications, I hope that it raises problems appropriately discussed in the Encyclopedia. Certainly I know of no spot where I should rather discuss them."[30]

11

Progress, Ideology, and "Writing History Backwards"

Soon after Kuhn accepted Morris's invitation, he began sketching outlines. His "Notes Toward Unity of Science Monograph" imagine a short book in four parts:

I. The Problem of Scientific Revolutions

II. Functions of Commitment

III. The Structure of Scientific Revolutions

IV. Conclusions[1]

By the time Kuhn delivered his finished manuscript some nine years later, it had ballooned to thirteen sections or chapters, something much larger than he envisioned or Morris ever expected. As the years ticked by, it frustrated Kuhn that the project was taking so long and he became extremely apologetic in his letters to Morris. He broke his promises and missed the deadlines he gave himself.

In fact, Kuhn was very busy. Besides his teaching duties in the 1950s, he wrote and delivered talks, published essays, and wrote his first book, *The Copernican Revolution*, published in 1957. Once he arrived in 1956 at the University of California at Berkeley, where he belonged to both the departments of history and philosophy, he remained swamped with competing obligations. As a new faculty member, he told Morris, it was not easy to say "no" to administrative tasks.[2]

Yet none of these professional demands fully explains *Structure*'s slow gestation. Turning his ideas about "prejudices," "orientations," and "theories-as-ideologies" into a convincing, readable theory of scientific revolutions proved to be something of an intellectual odyssey filled with zigs and zags, reversals, and doubts. Perhaps to prove to his editor that he really could finish a writing project, one of Kuhn's letters of apology to Morris included manuscripts of two early publications and a preliminary draft of a chapter for the monograph.[3] These, he said,

> have cost me more intellectual struggle than I had thought possible. While writing them I again and again encountered problems that I did not know existed and that I needed a great deal of time in order to recognize and then to resolve. As I discovered and resolved them, I saw all sorts of new significance in my materials and all sorts of new ties between aspects of my story that had previously seemed quite isolated.[4]

Structure was not easy to write. Its famous first sentence—

> History, if viewed as a repository for more than anecdote or chronology, could produce a decisive transformation in the image of science by which we are now possessed—

suggests that the book is a dispatch from the archives, the result of historical research. But it is also the result of philosophical argumentation between Kuhn and himself, a struggle to make his ideas and contentions clear, convincing, and provocative. As he put it fifteen months before, again reassuring Morris that he would eventually deliver a monograph, "I am at last hard at work and at the typewriter. Though there are very few pages so far, the whole is beginning to take real shape in my mind.[5]

One reason that Kuhn's ideas resisted taking real shape was the sheer contradiction between his emerging revolutionary picture of scientific change and the traditional picture of scientific knowledge growing or increasing over time—the picture that Conant himself painted again and again. If it were true, as Kuhn explained at the end of his Lowell Lectures, that scientific revolutions are creative, imaginative leaps that "radically and destructively alter the behavioral worlds of professional scientists," then why was this news? Wouldn't scientists themselves routinely report such momentous and transformative events (perhaps in the same way that Kuhn was able to later describe his Aristotle experience)? How could it be that the intelligent and perceptive professionals who advance humanity's knowledge, who pay close attention to evidence and observations, fail to report these creative, dynamic, and especially disruptive events in their careers?[6]

Kuhn was in a position to solve this puzzle. For now, as a historian with his Aristotle experience behind him, he was privy to aspects of scientific methodology that he himself had been blind to as a military scientist and a doctoral student in physics. The dynamics of ideology and the phenomenon of brainwashing promised a solution: most scientists are typically blinkered and blinded to the disruptive dynamics of "creative science" by the false "textbook science" they have been taught to see and understand.

Ideology, in Outline

In section two of his preliminary outline, "The Functions of Commitment," Kuhn elaborated the dual function of "theory as ideology" that he described to Morris. It directs the scientific mind positively and that negatively inhibits "the creative imagination":

1. Types and degrees of commitment

2. Functions of commitment—positive:

    a) Guidance to unknown

    b) Explanation.

    c) Fruitfulness (?)

3. Functions of commitment—negative:

    a) Field closure

    b) blinders

    c) effect on data

4. The inextricable entanglement—nature of knowledge[7]

These positive functions, guiding scientists along some paths of inquiry and rewarding them with successful explanations and fruitful results, join with negative functions that close off areas of research and put scientists in "blinders." The two functions work together in an "inextricable entanglement"—one that denied Frank's view that logical and evidential factors can usually be separated from the "existential" factors determining the historical course of science.

Another outline appears to have been written later, perhaps after Kuhn's letter to Morris, for the title of Kuhn's third section, "The Structure of Scientific Revolutions," had been promoted to the title of the whole monograph. Here, the first chapter is "What is a Scientific Revolution?" while the second, formerly "Functions of Commitment," is renamed as "Theory as 'Ideology.'" Kuhn explained what he meant:

> really mean something closer to ideology transmitted by teachers, texts, & society than theory. But beware:
>
> a) "Theories" contain more of these extras than they should. Texts are full etc.
>
> b) so its at least ~~professional~~ a group ideology acquired by membership in profession.
>
> c) size of group & larger social environment relations vary greatly . . .

Kuhn also needed a new title for the third chapter. He chose,

> III. Ideological Breakdown—The Awareness of Incongruity.

What begins with an "incongruity"—what Kuhn would later call an "anomaly"—leads to "the process of creating a new Ideology."[8]

Ideology and Invisibility

In a draft of *Structure*'s first chapter, dating most likely to 1959 and titled "Discoveries as Revolutionary," Kuhn prepared a quick overview of "what I take to be the recurrent stages in the evolution of scientific revolutions":

> For periods that may, depending upon the problem and the science, be as short as a decade or as long as several centuries, scientists are raised within and base their research upon what I have previously called a stable textbook tradition. While they do so, they remain in close agreement about the validity of existing research theories, the problems remaining to be solved

with these theories, and the nature of acceptable solutions. They share, that is, what we now recognize as a single professional ideology, gained initially from popular writings and speeches about science, reinforced and elaborated by teachers and texts, and made very nearly rigid by the rigours of professional intercourse and publication.

Until this relatively late point in *Structure*'s development, therefore, Kuhn held that scientific revolutions were best understood as ideological revolutions. They were precipitated by crises in the form of "decreasing professional agreement about the meaning of the theories and the applications of the standards that the professional community agrees it holds in common." Revolution itself is the replacement of one ideology by another "that once again permits professionals to see the world 'scientifically' but which is in significant respects incommensurable with the 'scientific' view of the world that preceded it."[9]

While an ideology is in place, Kuhn wrote, "it governs both the activity and the evaluation of research, and science appears to proceed by accretion." Not unlike the way *Structure*'s paradigms guide perceptions and the understanding of nature, "theory as ideology" in this early draft guides the profession's perception and understanding of itself and its history. In *Structure*, Kuhn partially assigned this function to textbooks. They make scientific revolutions invisible by providing a false, rewritten history of how the paradigm came to be established: "They have to be rewritten in the aftermath of each scientific revolution," he wrote, "and, once rewritten, they inevitably disguise not only the role but the very existence of the revolutions that produced them."

> Partly by selection and party by distortion, the scientists of earlier ages are implicitly represented as having worked upon the same set of fixed problems and in accordance with the same set of fixed canons that the most recent revolution in scientific theory and method has made to seem scientific. No wonder that textbooks and the historical tradition they imply have to be rewritten after each revolution. And no wonder that, as they are rewritten, science once again comes to seem largely cumulative.

And no wonder that Aristotle, read by a graduate student in physics, would at first seem full of mistakes and silly ideas. Taken as an attempt to achieve what Newton later accomplished, Aristotle would of course appear to be deeply confused and mistaken.[10]

Ludwik Fleck and the Sociology of the *Denkkollektiv*

Kuhn knew that some readers would find his "ideological" picture of science politically disturbing. "Scientific education makes use of no equivalent for the art museum or the library of classics," he wrote in *Structure*, "and the result is sometimes a drastic distortion in the scientist's perception of his discipline's past." This may "suggest that the member of a mature scientific community is, like the typical character of Orwell's *1984*, the victim of a history rewritten by the powers that be."[11]

Kuhn may have first encountered this idea in Orwell's book, or he may have discovered a similar idea around the same time in the writings of Ludwik Fleck, the Polish physician and biologist. Kuhn chanced upon a reference to Fleck while researching "a revelation"—his Aristotle experience—"that had come to me two or three years before." In Hans Reichenbach's book *Experience and Prediction*, Kuhn was struck by a reference to Fleck's *Entstehung und Entwicklung einer wissenschaftlichen Tatsache—Genesis and Development of a Scientific Fact*. Naturally, Kuhn wondered if Fleck's book addressed these same processes of "creative science" that interested him.[12]

The book was first published in Poland in 1935, not long before Poland was overtaken—first by Hitler and then Stalin—and images of mental captivity began to feed and support Western anxieties about totalitarianism. Fleck's scientists, like Kuhn's, are deeply immersed in sociological dynamics. "Introduction to a field of knowledge is a kind of initiation that is performed by others," Fleck wrote; only by joining this social group, a *Denkkollektiv* or "thought-collective," could an individual begin to make scientific contributions. Traditional images of scientific discovery that celebrate individual genius are mythical, as far as Fleck was concerned: "[E]ven the simplest observation is conditioned by the thought style and is thus tied to a community of thought." Science "is a supremely social activity which cannot by any means be completely localized within the confines of the individual."[13]

In translation, Fleck's claims about the scientific mind being conditioned and modified by the thought-collective anticipate and evoke the brainwashing of Chinese peasants under Mao. Observations such as this—

> Cognition modifies the knower so as to adapt him harmoniously to his acquired knowledge. This situation ensures harmony within the dominant view about the origin of knowledge[14]—

were later echoed in Edgar Schein's accounts of how Chinese peasants, through study groups, social pressure, and being injected with "communist premises," adapted to the new realities of Mao's China.[15] Nothing less than a "harmony of illusions," Fleck said, kept the thought-style intact and maintained scientists' "tenacity"—their confidence in their observations, in the truth of theories, and their basic belief "in a reality existing independently of us."

Though he would later find Fleck's formulations somewhat disturbing, Kuhn originally found Fleck's approach reassuring and exciting. For "in 1950 and for some years thereafter I knew of no one else who saw in the history of science what I was myself finding there." In Fleck, he recognized "what had already been very much on my mind: changes in the gestalts in which nature presented itself, and the resulting difficulties in rendering 'fact' independent of 'point of view.'" And, he added, it encouraged him to accept that his interests "had a fundamentally sociological dimension."[16] The book appears in Kuhn's preliminary outlines for his Lowell Lectures, and Fleck's imagery of the scientific mind inhabiting an illusory world propped up by powerful sociological dynamics anticipated Kuhn's essay "The Function of Dogma in Scientific Research." In *Structure*, Kuhn notes that he was indebted to Fleck "in more ways than I can now reconstruct or evaluate."[17]

Ideology and Revolution

Fleck was likely as important as Mannheim (who himself appealed to the notion of a shared "thought style" or *Denkstil*)[18] for shaping and inspiring Kuhn's theorizing about "theory-as-ideology." As suggestive and helpful as the concept was, however, Fleck's book did not help Kuhn figure out how, when, and why scientific revolutions occur. Fleck's approach was more evolutionary than revolutionary—"Knowledge exists in the collective and is continually being revised," he wrote.[19] There was, however, no shortage of scholars at Harvard in the early 1950s interested in revolutions. The new Russian Research Center was dedicated to understanding the fruit of one particular revolution, while other scholars aimed for a more panoramic survey. Crane Brinton, whom Kuhn knew from the general education program and the Society of Fellows, had published *The Anatomy of Revolution* in 1938. Brinton compared and contrasted the English, American, French, and Russian revolutions, summed up their similarities and differences, and returned to these issues in the early '50s as he prepared a new edition. Brinton's approach was proudly scientific. He introduced his readers to "conceptual schemes" as a way to understand historical events. But he concluded that there really was no regular structure of political revolutions. The courses revolutions took, their underlying causes, and the kinds of people that tried to spark them were usually different. The popular view that the downtrodden workers of the world united for revolution seemed to have little to do with it.[20]

Kuhn knew at least some of Brinton's work. He took notes on his book *French Revolutionary Legislation on Illegitimacy 1789–1804*, and wrote that it is "a classic investigation of the relationship of ideology to act and administration. Nothing whatsoever to do with my field, substantively, but everything methodologically." Kuhn might also have consulted with Bernard Bailyn. He was finishing his doctoral studies in the early '50s and soon became an expert on the American Revolution. But Bailyn was nearly a contemporary of Kuhn's, and his classic work from 1967, *The Ideological Origins of the American Revolution*, with chapter titles such as "The Logic of Rebellion" and "Transformation," may owe more to Kuhn's *Structure* than the other way around.[21]

Kuhn's attention was captured, however, by Eugen Rosenstock-Huessy (1888–1973), the historian, sociologist, and philosopher of religion who taught at Harvard for two years in the 1930s after arriving from his native Germany. After a short career at Harvard, Rosenstock-Huessy taught for the rest of life at Dartmouth, where, as it happens, he taught William Remington, Elizabeth Moos's son-in-law. He was the only professor at Dartmouth to give the future Communist a "C."[22]

In a file of papers labeled "Notes for Monograph," Kuhn left a notecard filled with quotations from Rosenstock-Huessy's book *Die Europäischen Revolutionen, Volkscharaktere und Staatenbildung*. First published in 1931, the book offers a sweeping, dramatic picture of European history driven forward by successive political revolutions. Rosenstock-Huessy later presented his ideas in English in his book *Out of Revolution: Autobiography of Western Man*, writing there that "[a]ny real book conveys one idea and one idea only." If Kuhn's *Structure* conveys the idea of his Aristotle experience, Rosenstock-Huessy found his idea as a young man "when I was thrown into the turmoil of the great war." Despite the carnage and suffering he witnessed in the trenches, Rosenstock-Huessy glimpsed an ironic, positive function in the violence: each revolution in Europe's history, he argued, starting with the Gregorian Reformation and ending with the Bolshevik Revolution some nine hundred years later, was a crucible that created new, different, and progressively better types of humanity and modes of culture.[23]

These revolutions were like Conant's case histories. Far better than any abstract definition, the study of specific political revolutions, Rosenstock-Huessy believed, revealed history's creative and forward motion. Kuhn immediately saw the relevance of Rosenstock-Huessy's book, at least its first thirty pages, and took down quotations while adding his own paraphrases in English.[24] His selections emphasize the sharp break between the old regime and the new; that a revolution is total, transformative, and inaugurates a new "life-principle." Revolutions also introduce new languages—"in other words, the emergence of a different logic," his notes read. The last two quotes address how revolutions end—not, Rosenstock-Huessy points out, with the elimination of the old regime and its people, but rather with a new covenant or *bund* that embraces a new language and logic that belong to the future. The destructive, wrenching, and humiliating realities of war and revolution may appear senseless (*sinnloss*)—but in light of the larger, creative, revolutionary dynamics in play, they are senseless only in appearance.

Kuhn noted that "each revolutionary movement starts calendar anew and destroys the past." This offered another clue to the puzzle. Scientists are typically oblivious to the revolutionary changes of the past, Kuhn may have reasoned, because current knowledge, along with the way it carves up experience, cloaks the past in a congratulatory narrative about the current theory-as-ideology and the prosperous scientific future it promises. Through historiography, the new ideology reaches back into the past to depict earlier scientists (such as Aristotle) as trying but failing to develop comparable insights. The fact that their theories and beliefs were based on alternative, incommensurable ways of cutting and boxing experience therefore remains obscure, if not altogether unknown, to successive generations of scientists.

Rosenstock-Huessy's picture of political revolutions seemed to illuminate Kuhn's Aristotle experience and offered something that Brinton's research did not—a simple, schematic formula for political revolutions that could be applied to science's history. Kuhn aimed to make this connection clear in *Structure* by writing, for example, "Like the choice between competing political institutions, that between competing paradigms proves to be a choice between incompatible modes of community life."[25] Had Kuhn paid more attention to Brinton than to Rosenstock-Huessy, he might have anticipated one of the criticisms of *Structure* he would later encounter, namely, that political revolutions take different forms and sizes, and that there is never a sharp, clean line dividing everything old from what is new. Kuhn admitted some of this complexity, writing, for instance, that revolutions involve "*partial* relinquishment of one set of institutions in favor of another." But emphasizing complexity, variability, and vagueness would have been a problem for a book titled *The Structure of Scientific Revolutions*.

Soon after Conant left Harvard for a new diplomatic career in Germany, however, another example of political revolution likely presented itself to Kuhn: the transformation in postwar Germany. It was Conant's job to manage the ongoing replacement of National Socialism and its institutions with those of democracy and the free market (and it was McCarthy's job, as the nation's then most visible anticommunist, to charge that Conant's State Department was bungling that job by allowing pro-Communist books to circulate in Europe's libraries). "Nazism is completely dead, the legend of Hitler is gone," the *Crimson* announced proudly when Conant returned to visit the campus after three years of work in Bonn. When Conant returned to campus again in 1958 to deliver the annual Godkin Lectures, he reported that "the spirit of free Germany, today, is the spirit of a people who have turned their back on their Nazi past." Many Americans were understandably skeptical that Germany could make a sudden and irrevocable break. Given the demonstrated virulence of totalitarian ideology and the impressive economic recovery of the nation after the war, it seemed possible that German imperial and ideological ambitions might be stoked once again. But in lectures titled *Germany and Freedom: A Personal Appraisal*, Conant reassured them with confidence. "The myth of the Third Reich has been destroyed," he wrote. Modern Germany "condemns the Nazis, their methods and their goals."[26]

On this point, Conant himself seemed to make the point Rosenstock-Huessy had made:

> Some observers would report that in so doing many Germans attempt to banish the whole period from their minds; that for the present day German history begins in 1945.[27]

But in fact Conant saw German history differently. As far as he was concerned, it was crucial that Germans neither forget nor ignore their past. As he put it in a subsequent lecture, "[T]he public repudiation of Hitler is not the same thing as blotting out the twelve years of his rule from the history of the land." The nation's willingness to reject its past consciously and knowingly, and the willingness of its writers and historians to examine the Nazi period reassured Conant that the nation had not moved mechanically from one ideology to another. It had achieved a genuine liberalism that rose above ideologies, to see them for what they were and the effects they could have, and to never again succumb to blind, destructive ideological enthusiasm.[28]

Ideology, Consensus, and a Growing Manuscript

Whether Kuhn encountered Rosenstock-Huessy before or after he adopted "ideology" as the heart of his future monograph, his notes link the concept of ideology that he described to Morris in 1953 to the crucial idea that scientists, like the citizens of Orwell's Oceania, are misled by a false history of their tradition. Science textbooks, popularizations of science that reach out to the public, and philosophical treatments that define state of the art thinking about science in colleges and universities, he wrote in *Structure*, become an "authoritative source that systematically disguises—partly for important functional reasons—the existence and significance of scientific revolutions."[29] They help create and sustain something like Fleck's "harmony of illusions" about nature and—per Orwell and Rosenstock-Huessy—provide a false, rewritten history of how current paradigms were achieved.

This insight, however, was not sudden. It came to Kuhn after a great deal of doubt and second-guessing. When he first drafted chapters based on his initial outlines, only two things were clear to him: his monograph was growing, and it was taking much longer to write than he expected. The more detail he sketched, the more it seemed to grow beyond the size of most monographs for the encyclopedia—about seventy to one hundred pages. Instead of a pamphlet, Kuhn felt that his theory required a full-length academic book. Writing at the end of 1959, he told Morris that the four chapters he originally envisioned had become five. A new, final chapter would compare progress in science to progress in the arts. And there would be one more expansion, a new section:

> That section, if I write it, will return to the ground covered in section or chapter I and discuss in more detail the nature of the change in professional orientation that is produced by a scientific revolution. Here I should particularly like to deal with some of the subtler manifestations of the professional reorientation. It seems to me important to suggest that scientists "see" things quite differently after a revolution and there is a good deal of quite subtle historical evidence on this point.

"What then is the prognosis?" It was not good. It could well take five years to finish a first draft of this larger book, he told Morris; only then could he create a shorter, condensed version suitable for the encyclopedia. That is, Kuhn added, "if you still want it." Kuhn admitted that his monograph was "shamefully overdue" and he gave Morris every opportunity to gracefully drop him and find a replacement author. "I will entirely understand if you feel you cannot wait that long for so problematic a manuscript."[30]

They had been here before. In the summer of 1956, Kuhn had promised to deliver his monograph in two years and acknowledged that Morris might have to give up on him, either of necessity or pure frustration. But Morris would not let Kuhn go. "We would hate to give up a history of science monograph," he wrote back, "and I think you are the one to write this for us." He did not let him go in 1959, either. Yes, he wrote, he "would hate to see the completion of the monograph series postponed five or more years," but he didn't believe so much time was necessary. Based on what Kuhn was saying and the drafts of other papers he had sent, Kuhn seemed well underway, as far as Morris could tell. He suggested that Kuhn write a short introduction to scientific revolutions for the encyclopedia and later expand it into a book. "If you could put considerable time on the monograph next summer, I feel sure you could complete it, say by the end of 1960."[31]

No, Kuhn replied, he could not write just an introduction. "Too much of the most essential and most persuasive material has been reserved for later chapters. I wonder whether any part of my point would really be made if I followed your suggestion." Half of a theory of revolutions would not be convincing and would not have the revolutionary effect Kuhn aimed for. But, he told Morris, there was some good news to report. "A minor morning brainstorm" had pointed the way to a new, shorter outline of a book that Kuhn hoped to finish writing that summer. On December 3, Kuhn did create a "new outline" for the monograph, one that shows at least one condensation or simplification of his plan: "I now think that [chapter] 5 belongs in Chap. 4." If this is the "new and reduced outline" in question, all hope that the monograph would slim down was soon lost. *Structure* would grow to include twelve chapters, and then thirteen in its final, published version.[32] At the same time, another shift was underway. After spending the 1958–59 academic year at the Stanford Center for Advanced Study in the Behavioral Sciences, Kuhn dropped "ideology" or "theory-as-ideology" as his central theoretical concept. He planned to finish the monograph during his residency, and indeed the year was "extraordinarily fruitful," he told Morris, "but not in quite the way I had anticipated." The concept of ideology, for one, was fading from Kuhn's manuscript.

During the 1950s, the national fear and persecution of suspected communist intellectuals continued unabated. When he left Harvard for Berkeley, Kuhn hardly left behind the kinds of controversy that had swirled around

Wendell Furry, Robert Gorham Davis, and others. J. Robert Oppenheimer, one of Conant's most important physicists in the Manhattan Project, lost his security clearances in 1954. He had been a professor at Berkeley when Hoover's agents first began investigating his contacts and activities in the '30s and '40s. Having uncovered Oppenheimer's friendships with known communists on and around campus, Hoover and his agents remained certain, well into the 1960s, that the Berkeley campus, where Kuhn now taught, was a beachhead of communism's ideological invasion. Another Manhattan Project scientist who was repeatedly hounded in the '50s and ultimately driven out of public service was Edward Condon, a Berkeley graduate who later became president of the American Academy for the Advancement of Science.

Neither Oppenheimer nor Condon was connected to Kuhn's research or his year-long visit to the Stanford Center. But they illustrate a climate in which it hard to imagine Kuhn dusting off his outlines and notes about scientific theory-as-ideology and not raising his colleagues' eyebrows as he described his new theory of scientific revolutions.[33] Whatever his reasons were, by 1959 Kuhn had demoted "ideology" and adopted two new terms in its stead: "normal" and "consensus." One of his outlines dated to that year contains five chapters:

Chapter I—What are Scientific Revolutions?

Chapter II—The Normal Practice of Science

Chapter III—The Crisis State

Chapter IV—The Confirmation Debate

Chapter V—Revolutions & Scientific Progress.

Chapter II, formerly "The Functions of Commitment" that had focused on ideology, was now structured in five sections around the terms *normal* and *consensus*:

1. Consensus and the Maturity of a Science

2. Nature and Extent of Consensus

3. Normal Research—Theories

4. Normal Research—Facts

5. Effects of Consensus, etc.

228 | The Politics of Paradigms

The goal was to explain the ideological circumstances in which a scientific community comes to accept its shared historical myopia as "normal." On a separate sheet addressing "issues" that the chapter raised, Kuhn asked himself:

1. How much consensus?
2. ?? Implication about discovery and confirmation vs. small role in usual practice ??
3. Consensus is about what?

He then answered,

1. concepts, Laws, Instruments
   facts worth collecting
2. Problems, standards of solution, standards of agreement.
3. Ontology

Kuhn underlined "Ontology" three times, once again joining Fleck, whose saw a collective "harmony of illusions" sustaining scientific agreement about ontological facts of the world—what it is made of, what is fundamentally real, and how it works. In his entries for chapter 1, "What Are Scientific Revolutions?" Kuhn's notes continued to echo Rosenstock-Huessy: Why is a scientific revolution "so difficult to see" in science, when they are comparatively easy to see in the history of art? Kuhn listed these considerations:

a) Respect in which sciences and art are different. Relation to past
b) because of our [indecipherable]
c) Systematically disguised by writing history backwards. Where scientist is and isn't historical. Limits on his historical imagination.
d) Writing history backwards[34]

Kuhn would address science and art at the very end of *Structure*. One of the central differences between them, he would explain, is that science makes progress by losing sight of its actual past and embracing a false history.

The arts and humanities do not cut themselves off from the past and their textbooks do not rewrite history. As a result, however, they don't exhibit anything like scientific progress.[35]

With "normal" and "consensus" in place, however, Kuhn had not abandoned "ideology" entirely. Among the entries for chapter 1, one reads, "New ways of looking at the world. new words, new problems, new explanations, new uses of old words." Below that Kuhn inserted, "this is ideology." Next to notes about "the parallel to change of artistic style," he wrote in the margin, "Ideology tradition etc. can go here" but he then changed his mind. No, he wrote, "Ideology tradition etc." really belong in the discussion of "consensus" in chapter 2. His terminology was changing, but Kuhn's theorizing remained inspired by ideology.

"Why Should This Work at All?"

Structure's account of "normal science" was taking shape. But problems loomed. It may have seemed plausible to Kuhn that young students could be trained to suppress or devalue historical and scientific imagination and accept a narrow, illusory picture of science's past. But what about mature, productive scientists who may live through a scientific revolution, or who might themselves help create and articulate a new scientific consensus that replaces an old one? Conant had always insisted that there was no problem here. Not just older scientists but even *dead* scientists would be impressed by the forward march of scientific knowledge. But Kuhn believed that the intellectual gap between the old and the new was far bigger than Conant understood, and that there was no neutral ground from which the superiority of the new could be objectively and conclusively proved. Somehow, he reasoned, mature scientists must rationalize their choice to take the leap and embrace an understanding of the world that is new and completely different. Why and how do they do this?

Kuhn placed these questions in chapter IV, "The Confirmation Debate." This debate is characterized by

a) Irrelevance of concrete evidence (Creative Theories)

b) The ad hoc hypotheses.

c) the need to reject as well as gain
   Thus impossibility of proof. Propaganda.

d) the immense closeness. Whence bitterness

As item c) shows, Kuhn compared the contest between different systems of consensus to a propaganda war and asked how one side could win over the other. What could lead scientists to admit—without "concrete evidence"—that one system was superior to its rival? Writing "youth" and "the Pattern of Revolution," he noted that it's young scientists who usually accept change, while holdouts are older. The younger scientists will see "The new assimilating the old" as the new consensus becomes enshrined as textbook orthodoxy and supported by a new, rewritten historical narrative. Under "The Transition to Text," Kuhn wrote,

   a) purifying the new theory

   b) Thus real new world view.

   c) Rewriting history.

   d) The shifting significance of old experiments, etc.[36]

There was no proof that the new consensus was better than the old, but there was something for all scientists to gain—a "real new world view" purified of lingering disputes or criticisms from old-guard holdouts and a corresponding historical narrative that congratulates the community for embracing the new consensus and making scientific progress.

Still, Kuhn was unsure of his account and whether it was convincing. The last section of his outline, "Revolutions & Scientific Progress" would set out to explain, as he put it, "Why should this work at all?" It could work, he surmised, if the new scientific consensus and rewritten history does not destroy the past *completely*, if "destruction doesn't entirely extend to level of data." Some elements ("numbers" and "applicable stuff," he wrote as examples) scientists will find familiar in the new regime. "Thus there's some sense of cumulativeness across revolutions," Kuhn wrote.

Here the outline ends on a note of disappointment and evident frustration. The last two entries read:

   4. Thus partial truth of positivism. Important that scientists believe it, etc.

   5. But couldn't be right.

The entries are cryptic, but the flow of Kuhn's reasoning and other documents in his files suggest that Kuhn had painted himself into a corner: in order to theorize how scientists move *through* a revolutionary change of consensus, he found himself conceding "the partial truth" of the picture of science that he intended to refute. As a subversive, Trojan horse attack on the dominant view, the monograph Kuhn had just outlined would hardly amount to a revolution—just a qualification, a philosophical *maybe, sort of* that hardly matched the revolutionary implications of what Kuhn took from his Aristotle experience.

His dilemma is vividly charted on a separate, undated page of notes that point to these issues. It reads:

> Pattern of evolution—don't throw away the quantitative[.] Sacrifice the homogeneity of the world and its all[-]embracing quality for increasing professional specialization.
>
> Conclusion is that cumulativeness is real but partial. Profession won't let the core of what they've done before go, though they'll let much go. . . . Certainly significant quantitative predictions are held, etc. There's clear sense that's probably not whiggish that the significant accomplishment[s] of old science have not been lost.

Kuhn noted again that numbers, equations, and "quantitative" parts shared by the old and the new consensus will create a sense of progress and accumulation of knowledge. This was real continuity, he emphasized—not the subjective, "whiggish" celebration of scientific progress that he disliked in textbooks and popular histories of science.

For a moment, at least, Kuhn seemed to have solved the problem. But he soon returned to his notes, now writing in hand on the same sheet of paper:

> Not everything disappears—techniques of numerical computation stay, in particular. Here's the fullest locus of cumulativeness. But they're seen in new perspective and with new significance. Permanent points of accretion.

At some point, still unsatisfied, Kuhn loaded the same sheet of notes into a different typewriter and confessed that his theorizing was not working:

> I am not really happy with the preceding. I think it is not wrong and that it solves my problem, or resolves it. Yet there's still too

large a residue of cumulativeness directly in it. As a result, I feel that the novel portions of my approach have too indirect a role in explaining why science works. I sense that there should be more contributed at that point, and that this somehow retreats back to the objectionable position.

Kuhn had come full circle, back to "the partial truth" of Conant's view of progress, "the objectionable position,"[37] and that was not where he wanted to be.

With two inches left at the bottom of this page, Kuhn switched to single spacing for one last go-round to specify exactly what was bothering him. "That still isn't the real difficulty," he began. Yes, he confessed, "I still seem to be leaving the criterion of cumulativeness which is antithetical to the heart of my position." But hope was not lost: "Discussion with Mel suggests that this isn't so bad."

The Allure of Rewritten History

"Mel" may have been Melvin Kranzberg, who received his PhD from Harvard in 1942 and later led the Society for the History of Technology. Kuhn and Kranzberg had known each other at least since 1958 when Kuhn spoke at an early meeting of the society in Berkeley. Years later, Kranzberg wrote admiringly to Kuhn about *Structure* ("I am not in the habit of writing fan letters, but I couldn't resist telling you how helpful that book has been to me and how highly I regard it. It is a major work—a classic.")[38] But long before he congratulated Kuhn on his achievement, Kranzberg may have helped Kuhn puzzle through this quandary over whether and how revolutionary scientists would jump from an old consensus to a new one. "Discussion with Mel," that is, led Kuhn to realize that he had picked up the wrong end of the stick: the new, rewritten history does not help revolutionary scientists rationalize their scientific decision to join the new, revolutionary consensus. The new, rewritten history *itself* and the illusion it provides of a continuous, progressive history is what they stand to gain. "That is," Kuhn typed,

> profession as a whole can statistically strive for cumulativeness and buy only if it seems to be becoming available. . . . That is, cumulativeness (or the nearest approach to logical proof) turns

out to be institutionalized as the most powerful of all arguments for rejecting one world view and learning to use another.[39]

The appearance of cumulativeness from within a new, revolutionary perspective was not really a concession to "the objectionable position," after all. Kuhn had lost track of the question he had been asking: it was not a matter of what revolutionary scientists gain *after* they opt for a new worldview and a new history; it was a question of what that history itself provided and how much it was valued. History was immensely valuable and persuasive, Kuhn now understood—nothing less than "the most powerful of all arguments" leading scientists across revolutionary divides and driving science forward.

∽

This insight anticipates the "inversion of our normal view" of progress that Kuhn would offer in *The Structure of Scientific Revolutions*. We must "learn to recognize as causes what have ordinarily been taken to be effects," he wrote. We normally suppose that "a field makes progress because it is a science," but we should instead realize that a field becomes what we call "a science because it makes progress." Only science, Kuhn explained, destroys its past and creates the impression that as long as problems are solved and discoveries made, progress is achieved. Inevitably, he explained, we are led to think that "successful creative work" in science counts as progress—"How could it possibly be anything else?"[40]

On this, Kuhn was sure and confident. Science's history as it is understood by scientists, popularizers, and whiggish historians plays a crucial role in scientific revolutions. But Kuhn was still far from delivering his monograph to Charles Morris. He had only outlined it and it was still to be written. And when he began to draft the chapters he envisioned, he encountered a problem far larger than these puzzles about the power of history and ideology to move science forward from consensus to consensus. After replacing "ideology" by "consensus," Kuhn realized that in scientific communities there is usually no such thing.

12

From "Ideology" and "Consensus" to Paradigmania

A draft of *Structure*'s first chapter, titled "Discoveries as Revolutionary," begins with the claim that "every invention or discovery worth the name is itself a revolution or an episode in one." That should come as a surprise, Kuhn noted, because "we are all so accustomed to viewing scientific progress as continuous and cumulative."[1] In the book's final version, Kuhn would turn his plan around and raise the issue of progress only at the end. At this point, however, he wanted to put the issue of progress on the table at the very beginning. If he could convince readers that their ideas about scientific progress were mere prejudice, perhaps illusion, they would be more open to his new image of science as punctuated by discontinuous revolutions, incommensurable theories, and incompatible worldviews.

Kuhn framed the issue as Conant had, by contrasting the sciences and the arts. "Only in the sciences," Kuhn wrote, does it appear that new knowledge is added to old. No account of science's history or methodology, he added, should be acceptable "if it fails to take account of the phenomena that cumulativeness was designed to describe." But revolutionary subversion of this traditional view was now immanent, for this cumulativeness, Kuhn clearly implied, was only a phenomenon, an appearance that was indeed deceiving. Behind our habitual understanding of the sciences as cumulative, he wrote, there was a different, deeper story in which the sciences, unlike the arts, "embrace their history, render it functionless, and thus destroy it."

The language echoed Kuhn's notes on Eugen Rosenstock-Huessy ("each revolutionary movement starts calendar anew and destroys the past") as well as Fleck's "harmony of illusions" resting atop unseen, inaccessible realities. One of those unseen realities was revolutions in science. Kuhn explained,

> Even in metaphor, the term revolution implies the conflict of irreconcilable views, standards, or authorities, and the ultimate replacement of the old by the new. These are characteristics derived from political revolutions and anticipated in the arts and other non-cumulative disciplines where one standard of taste or style undoubtedly does conflict with and replace one another. But to the extent we are led by the concept of cumulativeness, we anticipate no similar characteristic in the sciences.

But we are wrong. This image of new discoveries as "bricks and mortar added to an ever-growing edifice" masks revolutionary dynamics within the sciences.[2]

The Elusive Consensus

This sure-footed confidence gave way as Kuhn drafted chapter 2, "The Nature of Scientific Consensus." The more he looked for cogent examples of historical consensus in the sciences, the less consensus he found. So far, he wrote, he had "spoken of scientific consensus as though it were a state that either characterized a professional group or did not. But clearly this is not the case." Consensus was a matter of degree, and there too could be different kinds of consensus in a community.

After a page of type contrasting consensus in physics with consensus in psychology, Kuhn grew frustrated and decided he was more or less making this all up. Next to his comment that "clearly this is not the case," he penciled in the margin "My problem starts right here." On the next page he filled the margin and confessed that what he had typed earlier "takes me right off the track." That is,

> I don't really have the data (nor am I sure it would be forthcoming) for remarks on degrees of consensus, etc. in the well-developed scientific fields. Probably physics, chem., biol., geo, do differ, but I can't handle that. All I can be sure of and handle is the gross consensus—no consensus distinction.

If Kuhn could not show that a consensus typically existed within each scientific community, then he could not support his idea that one function of that consensus is to write history backward and offer that history to scientists in return for joining the consensus in question. Even worse, the elusiveness of consensus challenged the starting point of his research: that science was possible only when communities were somehow held together through psychological, ideological, and sociological mechanisms and bonds. If there was no visible consensus, then what exactly were those mechanisms attaching to? What were they maintaining, if not some kind of agreement and consensus within the community? "I can't yet unravel this," he wrote at the bottom of a page of notes.[3]

He put a fresh sheet of paper in his machine and tried a second time. He changed the title to "The Nature of Consensus in Science" and set out to distinguish consensus in natural science from that in social science. Hoping that this rough distinction would suffice, he wrote that students in social science "may emerge with quite different conceptions of the field they have studied" because different textbooks used by different students will convey "a different conception of the field." Yet in physics, this was not so. Textbooks "cover much the same ground" and they "leave the student with much the same impression of the range of problems that physics does solve as well as of what the physicist means by a solution."

This crucial difference could be explained historically, Kuhn wrote. The natural sciences achieved their first "professional consensus" long ago, while the social sciences—psychology, anthropology, sociology, and so on—remain a disunified array of schools and approaches. The difference pointed to a general pattern: before any science could undergo its first revolution, there had to be a consensus that could be replaced. In each science, the first consensus formed slowly, he explained, and it is therefore sometimes difficult to date precisely. Still, the pattern was real and visible. "Few historians," he added, would be likely to disagree with "the following description" of consensus formation.[4] Kuhn again seemed nervous and unsure, as if other historians were looking over his shoulder. He retreated a bit, crossed out "description," and penciled in "impressionistic sketch."

His sketch was indeed impressionistic and vague. Different areas of science and different specialties within them, he explained, achieved consensus at different times, in different historical settings, and in different ways. There was Hellenic astronomy in the fourth century BC, physical optics at the beginning of the seventeenth century, and there was geology and parts of biology that reached maturity in "the first third of the nineteenth century." Within physics itself, Kuhn made further discriminations. Electricity formed a consensus in 1750, and then again in 1830, while theories of heat did so around 1780. More qualifications followed: these were all matters of degree, for there were different "levels of consensus." And when these periods of consensus occurred, Kuhn wrote, that did not always mean that "old debates and disagreements" went away completely.[5]

This was a problem. Kuhn wanted to argue that consensus formed around the "core meaning" of concepts (as he had described them in his Lowell Lectures, for instance) so that the community could turn to the all-important "fringe meanings" and undertake research and experiments to refine those meanings. As he put it here, a robust consensus allowed scientists to "take the conceptual, experimental, and instrumental bases of their field for granted and direct their attention to the peripheral areas . . . about which disagreement had endured." But these historical qualifications put this in doubt. It can't be true that scientists take their shared concepts for granted if they also continue to engage in "old debates and disagreements." Perhaps, he figured, he could instead say, "Uncertainties and debates continued in the later period" (he was comparing nineteenth-century theories of heat to those of the seventeenth century) "but in no area <u>then admitted to be part of the science</u> can one discover the range and multitude of opinion that had characterizes so much of physics two centuries before." Perhaps, that is, by defining science in a certain way the argument could be made to hold up.[6]

After ten pages, the draft ends in mid-thought. Whether or not it continued into additional pages Kuhn discarded or now missing from Kuhn's archival papers, those that remain document his frustration. He scolded himself in his marginal notes. To handle this chapter, he wrote, "I must know more than I now know about the transition from non-consensus stage in development of science." And, "I think I'm getting additional confusion by failing to separate maturity of science as an enterprise from that of any particular science"—an observation that seems to question whether any one-size-fits-all model of consensus and revolutionary change is possible.

"What can I say?" he asked himself. If there really was no identifiable pattern or structure of scientific revolutions to speak of, he could not say much. Kuhn took a breath and regrouped, his notes suggest: "Take physics pre-Newton as exemplar," he wrote. Here, in the science that Kuhn knew best, the pattern seemed clearer: at first, there are "many problems about which no agreement" exists. They were presented by everyday life and could be observed by anyone, including "tides, ores in veins, weather and winds, lightening [sic] and thunder, speed of light, colors." After Newton's great achievement, however, the problems that intrigued physicists and captured their attention became more specialized; they became "internal to the science and about which [the] layman has no idea. Therefore field can be closed." Mature, professional science begins at this point, Kuhn reasoned, when research is no longer guided by commonsense observations and analysis. It is governed instead by the internal professional ideology that he described earlier to Morris, one that isolates the community from the layman and the sociological pressures of public life, as he had described earlier in his letter to Frank.

Now, could this pattern be generalized to fit other sciences or science in general? The next day Kuhn returned to his notes and took stock: "I now suspect yesterday's wrestling did the trick," he typed. Yes, there was a pattern, but it could only have two steps: from immaturity—in which a science is not yet professionalized, not yet separated from "everyday life," and in which there remain "lots of competing solutions" to known problems—to a mature phase in which a science moves from one consensus to another via revolutions, independently of the demands of nonscientific life. Kuhn's working idea of consensus had shifted and expanded. The turning point he was seeking, the point at which scientific revolutions become possible, was no longer simply when "old debates and disagreements" ceased; it was also when the community's debates and disagreements could no longer be understood by laymen. Kuhn found a way through his confusion: there is clear-cut consensus in each science, but only when it is mature and isolated from "layety and craftsmen."[7]

From Paradigms to *the* Paradigm

As he refined his account of how this transition from immaturity to maturity occurred, Kuhn again worried that he'd merely replaced one impossible problem with another: "It seems likely that the way the transition is made may vary so much from science to science that it defies treatment in this monograph," he typed. But his breakthrough was at hand. For even if individual sciences did not manifest a shared pattern as they matured, there might still be some event or marker to signify that the transition had been made. That marker would turn out to be paradigms: the achievements that attract a group of scientist practitioners "away from competing modes of scientific activity" that mark this early, immature, and more public kind of research.[8]

Paradigms solved Kuhn's problems about the consensus he was seeking as he drafted his second chapter. In *Structure* he wrote,

> Men whose research is based on shared paradigms are committed to the same rules and standards for scientific practice. That commitment and the apparent consensus it produces are prerequisites

for normal science, i.e., for the genesis and continuation of a particular research tradition.⁹

Beneath the "apparent consensus" that foiled Kuhn's attempt to analyze revolutions as the replacement of one consensus by another there lay a different kind of consensus and shared commitment. What scientists in a community really agree on is how to solve certain basic and fruitful scientific problems. They learn these techniques together through coursework and especially through the drills and end-of-chapter exercises in textbooks.

Kuhn had used the word *paradigm* before. In his Lowell Lectures, for example, he offered his audience a thought-provoking puzzle about covering a chessboard with dominoes—a task the solution of which served as a "paradigm," he suggested, for understanding how a scientist's psychological "orientation" guides their reasoning. By 1959, however, Kuhn spoke of paradigms as parts of science itself. Attending a conference that year in Utah about science education and creativity, he used the word to describe this, so far, ineffable transition from an immature science that lacks consensus to a professional science that has it. Looking briefly at the history of optics and theories of light, Kuhn wrote, "From remote antiquity until the end of the seventeenth century there was no single set of paradigms for the study of physical optics." In a mature science, by contrast, there would be a single set that guide the scientist's education and subsequent career. A scientist, that is, "continues to work in the regions for which the paradigms derived from his education and from the research of his contemporaries seems adequate," as if adding detail to a map of nature that is already known in rough outline. To an audience of educators convened to discuss the cultivation of creative, original, and "divergent" thinking, Kuhn pointed out that big discoveries in science are not likely to come about by leaving behind one's educational training and charting new and unknown territory; it was far more likely that the scientist would come to notice a problem or incongruity *within* the map (what he would later call a recalcitrant "puzzle" or "anomaly") that was so far unnoticed. The scientist would locate, that is, "a fundamental weakness in the paradigm itself."¹⁰

Kuhn's paradigms for teaching science thus became the single paradigm that keeps a branch of science intact as a historical community. The functions he had earlier assigned to "meaning systems," "predispositions," "professional ideologies," and "consensus" within science he now assigned to a paradigm. A similar shift from plural to singular occurred in his draft of chapter 1, in which "paradigm" appeared as an adjective describing those several problems that sustain a tradition's shared sense of unity. He wrote,

> For reasons that are both obvious and highly functional, science textbooks (and too many of the older histories of science) refer only to that part of the work of past scientists that <u>can be made to seem</u> contributions to the paradigm solutions of the text's paradigm problems.[11]

Once these pluralities—a mature community's solutions to its shared problems—became a single thing, a paradigm, the difficulties stemming from Kuhn's elusive "consensus" began to dissolve.

In order to describe scientific revolutions, Kuhn no longer had to specify when the minds of individual scientists convened around some consensus or body of scientific belief; he needed only to specify when a paradigm was recognized and embraced by the community as a whole. That was relatively easy: it occurred when a textbook or textbooks teaching the same problem-solving techniques had been adopted by a community. The stubborn, lingering "old debates and disagreements" now had a new and important role to play. For if the scientists in a mature community never disagreed with each other, if they understood their shared paradigm in exactly the same theoretical ways, then that community could never advance toward a revolution. When it encountered a "fundamental weakness in the paradigm itself," all the members of that community would be puzzled and stymied for the same reasons. There would be no strength in numbers, no reservoir of different, even unformed hunches about what could be wrong with prevailing beliefs.

The point had come up in Kuhn's "conversation with Mel." There, Kuhn realized that not all the members of a community would react the same way to the new, cumulative professional history on offer during revolutionary times. He wrote,

> That is, profession as a whole can statistically strive for cumulativeness and buy only if it seems to be becoming available. Must insist, however, that many individuals don't wait and if they did science wouldn't work.[12]

"Science wouldn't work" because revolutions would be impossible and science would effectively cease changing. One or a few individuals must ignite the revolutionary debate by leaping ahead of the rest to buy into a new theory, its new view of the world, and the new, rewritten history it promises.

Kuhn's theory of paradigms offered a middle way through the extremes that framed his earlier theorizing. If all scientists thought differently and independently, science could never get off the ground and there would be no revolutions. Only with some shared predisposition to carve up experience in similar ways, and to understand it in similar ways, could modern science and revolutionary change become possible. On the other hand, if consensus were total, rigid, and controlling, something like modern science might exist—but it would never change. Its first paradigm would be its last.

Kuhn's transition from the multiple paradigms of education to *the paradigm* that defines a branch of science was therefore crucial. It reconciled the powerful grip that theories (as ideologies) had on the scientific mind with the lesson of his Aristotle experience: the mind can move from one kind of scientific understanding to another, but usually suddenly and abruptly during scientific revolutions. And it reconciled the fact that scientists in any community do agree on many points—that the current paradigm is superior to those of the past, that certain key experiments or breakthroughs in the past brought science to its current state, that there is no scientific need to examine science philosophically—with the stubborn fact Kuhn confronted when drafting chapter 2: there is very much on which individual scientists in a community do not agree.[13]

"Post-Partum Collywabbles"

Kuhn had last promised Morris that he would write his monograph during the summer of 1960. At the end of the summer, he wrote from Berkeley to report the good news:

> Freshman Orientation in which I was heavily involved started here early on the morning of Monday, September 12th. At two o'clock in the morning of Sunday, September 11th I put the last word on a first draft and breathed a long sigh of relief.

The bad news? The essay had ballooned to nearly fifty thousand words, well above the forty thousand that Morris had set as an outer limit. Kuhn's suspicion that it would become a book-length project had proved correct, and there was no fluff or filler to be cut.

In a way, Kuhn insisted to Morris, the matter was out of his hands—as if Clio herself had taken a particular interest in scientific revolutions and told Kuhn how the subject had to be treated. She arrived a little late, however: "After I dragged myself through the first forty pages," he wrote, "the rest just suddenly began to come. Though I know much of it is still rough, I retain the feeling that I have at last come very close to saying what has been on my mind for some time." Indeed "for the last three-quarters of the draft, I attained a writing pace that I have never been able to sustain before, and there are even a few parts that I have not yet reread myself." Kuhn was not quite finished because he wanted some feedback from colleagues before sending it in. Still, "despite a mild case of post-partum collywabbles, I retain a good deal of confidence in the manuscript," he wrote. "I hope that, when you see it, you will share some of my pleasure."[14]

As he earlier promised, this larger monograph examined the way "that scientists 'see' things quite differently after a revolution."[15] Backed up by experiments in the psychology of perception that Kuhn described, this allowed him to argue against the possibility of a neutral, paradigm-independent observation language for science (one pillar of the dominant logical empiricism approach he aimed to demolish). This addition also supported his now-celebrated claim that postrevolutionary scientists not only understand the world differently; they perceive it differently, as if they had entered a different world or traveled to a different planet. With additional chapters dedicated to writing history backward (i.e. the "invisibility" of revolutions) and the question of progress, the monograph was now twelve chapters long.

Kuhn had not shown these philosophical cards to his readers earlier when he wrote *The Copernican Revolution*, and that was by choice. That was a historical, more orthodox monograph. But *Structure* was the culmination of all of Kuhn's work in the new, interdisciplinary area of research he had first presented in his Lowell Lectures. It was a book "closest to my ultimate purpose as a scholar," he had told the Guggenheim Foundation. And now a draft was finally complete. True, it was "extremely condensed and schematic," and he still envisioned writing a "full-scale book" that would present additional "historical evidence."[16] But the debut of his revolutionary ideas was at hand.

Paradigms versus Rules and Propositions

Kuhn had good reason to be nervous, for the late addition of paradigms changed his ideas about scientific revolutions in at least three ways. One concerns the sociological "glue" that holds scientific communities together. In his earliest sketches of "meaning systems" and "professional ideologies," he insisted that these made modern science possible. In his early draft of chapter 1, he reiterated that agreement among scientists was paramount:

> [T]hey remain in close agreement about the validity of existing research theories, the problems remaining to be solved with these theories, and the nature of acceptable solutions. They share, that is, what we may now recognize as a single professional ideology.

The function of maintaining "this period of ideological stability"[17] was now passed to paradigms. But paradigms are not simply ideologies; they are not systems of interlocking propositions, rules, and definitions that mark a space of possible thoughts and beliefs. They are in part practical achievements—things that scientists *do*—and around which shared beliefs adhere.

Kuhn described the difference in the draft he finished. It remained tempting even then, he confessed, to believe that "periods of normal science were periods of consensus, during which the entire scientific community agreed about the rules of the game. And scientific revolutions were then the episodes through which the rules of the game were changed." But that is wrong, for if you

> ask any random group of physicists, chemists, astronomers, or geologists what sorts of problems are worth undertaking, what makes a problem scientific, or when an explanation or solution is incomplete. Probably you will get almost as many different answers as there are scientists in your sample.

Practicing, "normal" scientists simply do not agree with their colleagues about these things. But they do share a commitment to a paradigm that brings shared beliefs in its wake, such as what nature is like, what things are made of, and how events come to occur and be observed.[18]

Kuhn was able to complete his monograph, that is, by connecting scientific practices to beliefs and ideas within each community's paradigm. This hinge gave him flexibility to lay out his theory of revolutions in a way that made sense; but it laid a foundation for criticism he would later hear frequently: his theory of paradigms seemed too vague and flexible.

Paradigms as Unconscious

The shift from consensus to paradigms also meant that some important scientific beliefs remain unconscious and unarticulated. Scientists cannot give independent reasons for why they adhere to a paradigm; they adhere because they have been trained to. In a passage that did not make it into the final version, Kuhn described how scientists' "professional work is no sharp and sudden departure" from the exercises they repeated over and over as a student. True, the problems he or she pursues "are invented by the scientist for himself" and do not come from a teacher or a textbook.

> But they continue to be closely modeled on the existing applications of a paradigm, a paradigm which is thus both more and less than a set of rules for the conduct of scientific life. It is because they learn in this way that scientists can so regularly agree in their evaluations of particular problems and particular solutions without manifesting any similar agreement about the full set of rules that appear to underlie their judgments. One can model work upon a paradigm or recognize work modeled upon one without being entirely able to say what it is that gives the model its status.[19]

Paradigms inherited this power to guide individual behavior from the unconscious "predispositions" Kuhn had earlier described to Professor Owen, and the "group unconscious" shared by any scientific community that he theorized in his letter to Lawrence Kubie. On the surface, any group of scientists tends to be argumentative about fundamental questions, as Kuhn stated. But despite these differences they can understand each other, collaborate, and agree on scientific matters because of their shared education and the shared paradigm at its core. Not unlike Freud's psychoanalytic complexes, these early life experiences become controlling in later, professional life. "Paradigms," Kuhn would write in the final version, "are not corrigible by normal science at all."[20]

From Debate to Conversion

The way paradigms control scientific thought and behavior reflects one of the greatest fears of cold war liberalism—the image of the human mind transformed and captivated by an external agent or authority. Though he had long described them in his notes and drafts as "transformations" and replacements of one body of thought by another—such as meaning systems, predispositions, or theories-as-ideologies—the monograph finally came together when Kuhn chose to describe these events as "conversion experiences."

The change was not merely terminological. For Kuhn was no longer struggling to understand how revolutionary scientists might rationalize their way across the logical, epistemological, and ontological gaps between radically different orientations or worldviews. With paradigms understood to be working partly unconsciously in the scientific mind, these transitions were no longer necessarily deliberate and calculated. They became conversions that simply take a scientist from one paradigmatic worldview to another. They recalled William James's reports and case studies of the mystical, religious experiences in his *The Varieties of Religious Experience* that Kuhn read in 1943. And they reach back to Kuhn's own shocking experience reading Aristotle. Kuhn was as emphatic as James that these experiences are more than intellectual conclusions.

First, they cannot be understood as reasoned, step-by-step changes of belief or reinterpretations of data or observations. Scientists and the worlds they know seem to convert together:

> [N]ormal science ultimately leads only to the recognition of anomalies and to crises. And these are terminated, not by deliberation and interpretation, but by a relatively sudden and unstructured event like the gestalt switch. Scientists then often speak of the "scales falling from the eyes" or of the "lightning flash" that "inundates" a previously obscure puzzle. . . . On other occasions the relevant illumination comes in sleep. No ordinary sense of the term interpretation fits these flashes of intuition through which a new paradigm is born.[21]

Precisely because the conversion is not an experience that can be decomposed, like a logical argument or a geometric proof, communication between those who have converted and those who have not is often difficult and partial:

> [B]efore they can hope to communicate fully, one or the other must experience the conversion that we have been calling a paradigm shift. Just because it is a transition between incommensurables, the transition between competing paradigms cannot be made a step at a time, forced by logic. Like the gestalt switch, it must occur all at once (though not necessarily in an instant) or not at all.[22]

"Theory as ideology" lay in Kuhn's terminological past; but with paradigms and conversion experiences now playing important roles in explaining scientific revolutions, Kuhn's revolutionary scientists seemed to reflect and recapitulate the well-known dynamics of brainwashing and political conversion of the early cold war.

Kuhn was not alone in this. European and American cold war intellectuals, especially those who had been socialist or communist and wanted to confirm their liberal credentials after the war frequently described their personal conversions in similar terms. In *The God that Failed: A Confession*, published in 1950, Arthur Koestler, Richard Wright, Andre Gide, and other former leftists described their growing doubts, their crises, and then their decisive, revolutionary breaks with communism. Koestler's story is suggestive of Kuhn's, for he too was a scientist who became a historian of science. His book *The Sleepwalkers* of 1959 remains a classic account of Copernicanism alongside Kuhn's *The Copernican Revolution* of 1957. In *The God that Failed*, Koestler recalled his original conversion to Marxism, one that occurred—like Kuhn's Aristotle experience—when he was reading:

> Tired of electrons and wave mechanics, I began for the first time to read Marx, Engels, and Lenin in earnest. By the time I had finished with [Marx's *Theses on*] *Feuerbach* and [Lenin's] *State and Revolution*, something had clicked in my brain which shook me like a mental explosion. To say that one had "seen the light" is a poor description of the mental rapture which only the convert knows (regardless of the faith he has been converted to). The new light seems to pour from all directions across the skull; the whole universe falls into a pattern like the stray pieces of a jigsaw puzzle assembled by magic at one stroke.[23]

Kuhn did not *become* a dogmatic Aristotelian in 1947 in the same way that Koestler became for some years a convinced Marxist. But his Aristotle experience showed him that this kind of experience, familiar to contemporary intellectuals who had converted either to or from communism, could help explain the history of science.

Kuhn never suggested that these ideological dynamics in science had anything to do with ongoing debates over academic freedom and communist professors, but for Koestler the connection was obvious. It was precisely this exquisite "rapture" of mental captivity, he wrote, that explained how "with eyes to see and brains with which to think" there could remain unreconstructed communist intellectuals who had not defected from Stalinism by the end of the 1940s.[24] Intellectuals like these still existed because political ideas were powerful and seductive, as Koestler and other anticommunists like Sidney Hook and James Conant often observed. But so too are scientific ideas, Kuhn now added, as he described scientific revolutions as conversion experiences. And in science, as in politics, there will be those who convert and those who do not. From this point of view, Koestler's unreconstructed communist intellectuals circa 1949 were not unlike Kuhn's unreconstructed Aristotelians, phlogistonists, or biological creationists. They too failed to convert long after their former colleagues had left them behind.[25]

"A Special Circle of Hell"

Paradigms were a late but momentous addition to the manuscript. They revolutionized the monograph, taking the image of science it contained beyond what Kuhn had learned from Conant and his conceptual schemes. "A paradigm," Kuhn wrote in his first complete draft, "is at least in part a theory or, more loosely, a conceptual scheme. But, and this will later prove critically important, a paradigm must also be something more."[26] That something more would come to haunt Kuhn—first as he fended off objections from Conant and other colleagues, and later as he struggled to manage the sprawling layers of meaning, the magical things paradigms could do, at least according to Kuhn's readers.

In October 1960, however, this something more was gratifying and exciting. Kuhn promised Morris he would have a draft for him by the end of the year, but the length of the monograph was a problem. Because it had come together so well, with each piece of the puzzle playing its part, he saw no way to condense it. Fearing that Morris would require drastic cuts, Kuhn decided to ask the University of Chicago Press to publish a separate, stand-alone version of the book, one that included all the sections, footnotes, and scholarly apparatus that Kuhn figured he would remove for the encyclopedia.

Kuhn pitched the idea to Morris first, since he did not want to insult or annoy his editor. But Kuhn was blunt in his strategic reasoning. So far as he could tell, he told Morris, the encyclopedia in 1960 was not what it was "twenty years ago." It had fewer readers, he believed, and was therefore no longer an ideal venue for his purposes. On the other hand, Kuhn felt obligated to Morris. His prodding and encouragement over the last half of the 1950s helped him to finish the monograph. So Kuhn was unwilling to turn his back on the encyclopedia or the University of Chicago Press. Instead, he offered more than they had asked for—not one book, but two.[27]

Morris did not know what the press would say, but he discouraged Kuhn from pursuing it. "Your monograph in the Encyclopedia might well have a wider circulation than you realize. Some of the monographs have gone well over 5,000 copies," he countered. He urged Kuhn to shorten it to about forty thousand words. As for a bigger book, of course "you can explore that later if you wish."[28]

Kuhn didn't make his end-of-the-year deadline, but he was close. "How is your monograph coming along?" Morris wrote from his new position at the University of Florida the next April. In fact, Kuhn was just about to send it and remained firm in his plan to offer the press two books. He had read the manuscript again and reported "no chance of getting the text down to 40,000 words without omitting whole sections." And Kuhn would not allow it to be published without additional, "copious annotation." The integrity of the manuscript forced his hand:

> Under the circumstances I think I must ask the Chicago Press about the possibility of some special publication arrangement. If they'll bring out the full text with notes as a separate book or with some of my other articles, then I'd gladly supply a fairly drastic abridgment for the Encyclopedia. I might, for example, omit all footnotes, drop sections IV, IX, and X, and introduce the further alterations that these would require. Or I could try something else of a similar sort.

If Morris had not felt slighted by Kuhn's earlier comments, here was another opportunity. Kuhn insisted that only his full-length monograph would be an important scholarly contribution, yet he was now offering to Morris an abridged version from which entire sections—"Normal Science as Rule-Determined," "Revolutions as Changes of World View," and "The Invisibility of Revolutions"—had been removed. And his plan was to move ahead on this regardless of what Morris thought. "I shall very likely be writing to them," he told Morris. "Please forgive me for it."[29]

A month later Kuhn wrote to Carroll Bowen, the head of the university's press. Knowing that Bowen must have "a special circle of Hell reserved for authors who write to request special treatment," Kuhn asked for an indulgence: "Let me try to persuade you to be slow in assigning me to it." He explained the difficult situation concerning the length of his monograph, and offered evidence that it deserved special treatment. For besides Morris and Carnap, Kuhn had sent a mimeographed copy of the manuscript to about a dozen friends and colleagues. The feedback was almost uniformly, even hyperbolically, positive. The physicist Pierre Noyes, a colleague of Kuhn's at the University of California, described it as "an extraordinarily exciting intellectual experience" and wrote, "My intellectual world will never be the same." Ernest Nagel at Columbia, despite some prescient worries about some of Kuhn's characterizations of paradigms and revolutions, "found it engrossing and instructive reading."[30]

Kuhn passed on some of this praise to Bowen. He quoted the sociologist Bernard Barber ("It is genuinely exciting, new, and I find I come out with what is fundamentally a different conception of science than the one I started with") and this music to any publisher's ears from Nagel: "I am confident that the book will have a wide audience of readers, not restricted to the professional historians or philosophers." Kuhn felt obligated to give the monograph to the University of Chicago Press, but praise like this—along with his worries about the reputation of the encyclopedia—made it tacitly clear that Kuhn could take it to publishers elsewhere. "I have had a couple of inquiries from Great Britain," he noted.[31]

So, Kuhn asked, "would the Press consider simultaneous publication in another format?" He suggested the Press's Phoenix series of paperbacks, or perhaps a larger collection of Kuhn's own writings that included the monograph along with other essays. If Bowen would agree to either of those possibilities, Kuhn wrote, then he would agree to condense the monograph for Morris and Carnap's *Encyclopedia*.[32]

Paradigmania

For an evaluation of the monograph, Kuhn suggested that Bowen contact any of those he had quoted. Or, he could speak to Robert Merton, James Conant, or Kuhn's colleague at Berkeley, Paul Feyerabend. This could have turned out to be a momentous mistake, for just two weeks before, Conant had sent Kuhn a long and upsetting letter questioning the wisdom of his theory of paradigms. Feyerabend would not have applauded the manuscript either, for when he read it he almost viscerally detested the very idea of "normal science" as Kuhn presented it.[33]

As it happened, Bowen telephoned Indiana University professor Norwood Russell Hanson, the historian of physics whose book *Patterns of Discovery* anticipated some of Kuhn's views. At Bowen's request, Hanson wrote a report on the manuscript that raised serious questions about the logic of Kuhn's central claim. "For Kuhn, it appears that the claim 'scientific revolutions result from overthrown paradigms' is either circular or tautological." Were one to point to a historical revolution that was *not* the result of a paradigm shift, Hanson feared, Kuhn might reply that this was—by definition—not a revolution and therefore not a challenge to his claim. This led Hanson to ask, "Is Kuhn's thesis falsifiable?" He sensed that it was not, that "Kuhn has overplayed his hand somewhat," and that readers will sense this and be justifiably disappointed. After all, Hanson emphasized, a thesis that is unfalsifiable, "a thesis that fits everything thinkable, says nothing."[34]

The criticism could have come from Karl Popper himself, the Austrian philosopher who championed falsifiability—the criterion that a valid theory must be refutable by some possible observation—as the essential difference between genuine knowledge and its imposters.[35] Hanson's criticism was trenchant because Kuhn's monograph was headed into the heart of Popperian orthodoxy by claiming that scientists did not "normally" aim to falsify their theories, as Popper claimed they should. Popper would later dismiss *The Structure of Scientific Revolutions* as dangerously mistaken. But Hanson was on Kuhn's side. As serious a problem as unfalsifiability was, he told the press that Kuhn could easily take care of it. He should simply add a few paragraphs to address this complaint and specify "what, in his opinion, would or could falsify his main thesis." Until he does that, Hanson remarked, Kuhn "has not shown us in what sense his book—besides being interesting, refreshing, ingenious and imaginative—is also true."

Even with this potentially fatal flaw, Hanson plainly adored Kuhn's book. It was a masterpiece, "a work of surpassing merit, scholarship, and creative ingenuity" that simply must be published, even in its present, suspiciously unfalsifiable form.[36] That was one reason why Hanson urged the book's publication. Another was Kuhn himself, whom Hanson knew and obviously liked very much: "The Style is the man. Kuhn is himself an intricate embroidery of learning, interest, skills and passions." "His book," he continued in the report, "is no less intricate and no less an embroidery . . . suited both to the man and his objectives."[37]

Shortly after Bowen called him, Hanson telegraphed Kuhn at Berkeley:

Bowen of Univ of Chicago press called me here in Colorado asking me to referee your typescript. I told him I was favorably prejudiced—sight unseen, which pleased him. I will read it while in England wiring him reaction on July 15 will be at Oxford we could work out together the best setup for you.

Like Kuhn, Hanson was headed to Oxford, England, for Alistair Crombie's symposium on the history of science—an event where Kuhn would fire a shot across Popper's bow by reading his essay "The Function of Dogma in Scientific Research."[38] Hanson's suggestion that he and Kuhn "work out together" what his report would say suggests a different kind of suspicious circularity, one that Bowen might liked to have known about. But in the end it did not matter. Hanson's report came to the press late, well past July 15, and Bowen had already decided to accept Kuhn's manuscript on even better terms than those Kuhn proposed. Almost immediately after receiving Kuhn's letter from the "special circle of hell" in which he pleaded for redemption from the press, Bowen called Kuhn to accept the manuscript without any reservations, cuts, or condensations at all. The press would publish it both as an independent book and as an installment in Morris and Carnap's encyclopedia. "Wow, what an experience to have with a publisher!" Kuhn exclaimed some thirty years later.[39]

Paradigmania was at hand. In the summer of 1961, Kuhn polished his manuscript and worked out the last remaining details. "Does a volume of this sort want an index?" he asked Bowen. "I wouldn't be surprised," he continued, "but I'm damned if I know how one would devise one." In fact, Kuhn had a pretty good idea. It would be short but comprehensive. Something like, "Paradigm, pp. 11–179, *passim*."[40]

Part IV

The New World of Paradigms

13

"If Mr. Kuhn Is Right . . ."

Paradigms and Dogmas in Cold War Science Education

One of the most dramatic moments of the cold war occurred in October 1957 when the Soviets launched their first "Sputnik" satellite. The day after the launch, *New York Times* headlines burst with confirming details: "Soviet fires earth satellite into space; it is circling the globe at 18,000 m.p.h.; sphere tracked in 4 crossings over U.S." A week later, reporter John Finney explained that the one hundred and eighty-four pound weight of the first Sputnik, in particular, was unequivocal "evidence of Soviet superiority in rocketry."[1]

Anyone reading Finney's article, titled "U.S. Missile Experts Shaken By Sputnik," knew that ordinary Americans were shaken too. Since the earliest days of the cold war, most of Moscow's successes remained distant in the Far East—in China, North Korea, and Eastern Europe. The closest Moscow had come to invading the West were spies, some real and some imagined. Now, however, a mechanical agent of the enemy was circling the globe overhead. Anticommunist warnings that Moscow aimed to control the entire world now had a sinister and alarming confirmation.

For Conant and other liberal anticommunists, however, the Sputnik crisis had a silver lining. It led the nation to finally agree that public education, properly reformed and financially supported, could be a powerful weapon in the cold war's ideological struggle.[2] Since the end of the war, Conant had been arguing that public education would bring the nation closer to its ideals of freedom, equality of opportunity, economic growth, and cultural superiority while countering Soviet claims that America was on the verge of collapse. By cultivating scientific talent from the nation's enormous and varied population, reforms might also help close the science gap suddenly revealed by Sputnik. For ten years, Conant's general education program had used Harvard's resources and prestige to demonstrate that citizens who never set foot inside a scientific laboratory could still learn about science's tactics and strategy, how science strives always to move ahead. After Sputnik, congress moved public education ahead by passing the National Defense Education Act. It funded the Biological Sciences Curriculum Study (BSCS) and Physical Sciences Study Committee (PSSC) to reform and strengthen high school biology and physics. The reforms emphasized scientific creativity and exploration and downplayed drills and exercises designed to help students memorize facts. They looked at science as a creative intellectual adventure, one that should thrive in the free culture of the West.

Yet Kuhn and his emerging theory of science were on a collision course with this consensus. By the mid-1950s, Kuhn had come to doubt whether Conant's general education program really could give students a "feel" for the strategy and tactics of science. And in the cold war depths of 1957, when Sputnik plunged the nation into a crisis of educational and scientific soul searching, Kuhn seemed not too surprised at the Soviet achievement. For the stereotypical communist mind—closed, obedient, and beholden to the authority of accepted "theory as ideology"—played an important and necessary role in science as he understood it.

The Problem with General Education

Early in their collaboration, Kuhn had been something of a roving ambassador for Conant. While Conant wrote articles and lectured about the importance of general education, Kuhn carried the torch to educators elsewhere who hoped to emulate Harvard and create their own programs. In 1949, for example, he traveled to St. Louis to attend a conference on physical science in general education.[3] Kuhn delivered a lecture, "The Sciences in the Harvard General Education Program," that survives in his papers along with his invitation to the conference. He praised Conant's new program for showing students how scientists cultivate and choose central concepts, for teaching them some small area of science in impressive depth, and for providing future citizens and their leaders "an initial basis for intelligent evaluation of situations in which science or scientists are involved."[4] Despite the dry, scholarly style that Kuhn adopted during these years, he seemed to endorse the civic importance of Conant's educational mission.

Two years later, attending the Faculty Conference on General Education at the State University of New York, Kuhn assumed a higher profile as an expert in his own right. He addressed "The Sciences in General Education" in the first time slot on Monday morning and, along with two other expert "consultants" from the University of Chicago, attended study groups during the week. Conant had recently left Kuhn and Leonard Nash in charge of teaching *Natural Sciences 4*. Judging from Kuhn's detailed lecture notes, he now spoke not only as a representative of Conant but as a faculty member more relaxed and confident than before.[5] He also loved teaching this course. "I believe it and in it," he said, as he explained how the course aimed "to recreate [the] process of discovery for the student."

Unlike his short presentation from 1949, however, Kuhn now discussed the difficulties and limitations the course faced. There were large differences among students' aptitudes, and they sometimes lacked interest in scientific ideas and the case histories. Speaking to confusions within the Harvard faculty about *Natural Sciences 4* and other science-related courses, Kuhn emphasized that these courses were not courses in science proper, nor did they teach methods of clear or critical thinking (scientists themselves don't have such methods, Kuhn joked, and challenged his audience to correct him: "I'm ready to be shown, but I'm thoroughly skeptical").[6]

What, then, did *Natural Sciences 4* do well? It introduced students to "creative" as opposed to "textbook" science and in so doing provided "a picture of the scientist at work which I can guarantee leaves a lasting impression." This is important, his notes read, because "we meet science too much as text book [science]. This leaves lasting misapprehension." A historical impression of science-in-the-making can "teach our students to be intelligent about the issues which science raises in their daily lives without being participants." "This sounds hopeless and it may well be," Kuhn admitted. But there was a precedent for this kind of sophistication in scientists themselves. A chemist, for example, may have no formal training in physics or biology but will still be able to ask the right questions and form sound judgments about what the future holds for these areas of knowledge. If scientists can do this, then perhaps future lawyers and bankers can, too. "This is a distant goal," Kuhn admitted, "but one we should set our sights on."[7]

By 1955, however, Kuhn seemed skeptical about these ambitions. The problem was not simply that the goal was distant. It was that scientists had crucial knowledge and experience that nonscientists lacked; Kuhn's view that scientific theories function as "ideology" meant that something like an ideological, linguistic divide separated scientists from the larger public (from "layety and craftsmen," he would later write in an early draft of *Structure*[8]). At a general education conference in Bridgewater, Massachusetts, Kuhn explored the question "Can the Layman Know Science?" His opening remark hinted at the less-than-optimistic answer he would give:

> My topic today is "Can the Layman know Science?," and this question presents, I think, on one of the most vital and perplexing issues of our time.[9]

Vital, of course. But "perplexing"? Kuhn was about to suggest that general education in science may never meet the social and cultural goals Conant had set for it.

With the birth of professional science, he explained, science had become "more incisive in its techniques, more widespread in its consequences, more revolutionary in its impact than any intellectual enterprise has ever been before." He pointed to the new atomic weapons as well as radios, appliances, plastics and—presciently—artificial satellites that might someday orbit the earth. These developments posed an important question: How will traditional institutions such as law, the family, schools, and colleges, all of which are "notoriously slow to change," keep up with science and remain "in some sort of touch with the realities of everyday life"?[10]

Kuhn looked into this abyss and backed away. "I do not propose to answer this question, or even to discuss it," his notes read. But he admitted that the situation was urgent: there had to be some degree of mutual understanding and comprehension between the scientist and the layman. "In a democratic society," after all, it is "the citizen, the voter, the men and women who ultimately approve or disapprove policies." Yet because of this emerging cleavage between science and social and civic institutions, there was growing risk that citizens and voters would become even less informed about modern science than they are now.[11]

It was not always so. A century or two before, Kuhn explained, intelligent laymen could understand science as well as those whose names are now immortalized in textbooks. But today the gap has "gotten so large that sometime around the first world war both scientists and the layman stopped trying to bridge it." At the same time, the public lost interest (or perhaps hope) in keeping up with the advances: "People don't think they can understand, and they are not going to try," Kuhn explained. Without quite knowing it, they have settled for misleading stereotypes, such as "the advertising-scientist"—

The man who puts on a white lab coat and lays down his prejudices at the door and comes out AFTER SOME LABORIOUS but essentially mechanic process known as LOOKING AT THE FACTS with either Sulphadiazine or the statement that Cigarette X is milder, much milder.

Or, there is the scientist as "the ancient medicine man," Kuhn added. These images of scientists in the popular media illustrated how serious the problem had become: the public demanded the fruits of modern science, but failed to understand its complexities and difficulties. Could *Natural Sciences 4* overcome this divide? Not if that means teaching the layman to understand science in all its technical and mathematical detail, for "science is genuinely too tough and esoteric for that." If that is the goal, Kuhn's notes read, "We are licked before we start."[12]

Of course, that was not the goal of *Natural Sciences 4*. It would acquaint students with science by exploring developments centuries before, not the esoteric theories of today. But Kuhn seemed to have changed his mind about how far this kind of science education could go. In his talk four years before, he had said that if scientists could overcome disciplinary divisions to understand research in other fields than their own, then perhaps there was some hope for the educated public. But now he doubted his premise: a new PhD in physics, he explained, "is lucky if he can read half" of the articles in physics journals. So this problem is "much more acute" when it concerns different fields. Many physicists, Kuhn said, know nothing about organic chemistry or even elementary biology.

There remained hope for expert science administrators (Kuhn probably had Conant in mind) who could work effectively across the barriers dividing chemistry, physics, engineering and other areas:

> Particularly during the last war it was shown repeatedly that though [a] man in one scientific field may not know much about substantive knowledge of another, he was often able to ask very shrewd questions about it—sorts of questions on which intelligent administrative decisions could be based.

But this did not help the layman, for this "sort of knowledge or feel" that transferred from one area of science to another came only from "experience in doing research and watching ideas grow and interact with facts." What Kuhn described as a distant but important goal for general education in 1951 now seemed a shot in the dark: "I doubt that there is any real substitute for research experience," his notes read, so there's not much chance that laypeople outside of science can develop these skills. "But at least one can try. That is what we are after."[13]

Kuhn still championed *Natural Sciences 4*. He enjoyed it and took pride in teaching students something about science, in dispelling popular stereotypes, and drawing attention to scientific revolutions and "creative" (as opposed to "textbook" science). This showed students that science is "a far more human endeavor" than they knew—one that takes "crooked paths" into its future, and in which momentous discoveries can hinge on subtle matters of context and perspective. They learn that scientific research is often not "a process of discovering something which has never been seen before," but rather of "seeing something in a way in which it's not been seen before."[14]

The inquisitive and surprised students Kuhn described, however, were no longer Conant's future leaders of America. They were future readers of *The Structure of Scientific Revolutions* and this was not the kind of understanding Conant hoped his general education program would cultivate, as would become clear to Kuhn in the future when Conant returned his comments on his first complete draft. Understanding the strategy and tactics of science for Conant meant studying scientists in action, wrestling with problems, and finding success as conceptual schemes were improved or replaced. The sociological and ideological aspects of science that fascinated Kuhn, on the other hand, were impediments to scientific cooperation, effective communication across disciplines, and collective progress. And they emphasized the world of difference between scientists who work within a paradigmatic tradition or a "theory-as-ideology" and those on the outside who were never so indoctrinated. If these were the main features of modern science, then Kuhn was surely right to doubt whether the layman can know science. He did not, however, fully confess these doubts, for at the end of his talk he dodged the question. Can the Layman Know Science? "I'm quite aware that I've not answered the question," he confessed. "I don't know the answer," he said, "at least not yet."[15]

Science and "the Unknown"

Though he described some if its implications, Kuhn did not unveil his ideological theory of modern science in his talk at Bridgewater. He also did not unveil it in *The Copernican Revolution*, in which he understandably sculpted his words so as not to overtly disagree with Conant, the man whose educational ideals and intellectual prestige supported his budding career. Decades later, however, Kuhn distanced himself from *The Copernican Revolution*. He wrote it, he said, because his career required it: "I had been giving it as lectures. I needed a book. I had this material, I could do a book, and I didn't think it was a stupid book to do. I mean, it was not what I mainly wanted to be doing, but it was something worth getting done."[16]

Kuhn spoke of Conant's conceptual schemes and their cyclical process in science's history, one that may seem familiar to readers of *Structure*:

> A conceptual scheme, believed because it is economical, fruitful and cosmologically satisfying, finally leads to results that are incompatible with observation; belief must then be surrendered and a new theory adopted; after this the process starts again.

Conant had defined progress in just this way. It fit his definition of good scientific work as fruitful and always moving ahead. Kuhn, however, believing that the scientific mind is normally immersed in one and only one "theory as ideology," could not have endorsed Conant's image of scientists freely utilizing different conceptual schemes at the same time. So he crafted a kind of compromise that honored both Conant's liberalism and the popular image of the pioneering, scientific explorer. Yes, Kuhn wrote, conceptual schemes are tools for "exploring the unknown." But they are also tools for predicting what the unknown will turn out to be; how it will become known. That is, conceptual schemes and the theories within them tell a scientist "where to look and what he may expect to find." This is "the single most important function of conceptual schemes in science," he wrote.

In Kuhn's presentation, Knowledge moves forward not because of the roving, curious light of the free, liberal mind as Conant portrayed it; it moves forward because conceptual schemes—functioning as Kuhn's "unconscious predispositions" or "theory as ideology"—effectively create a coherent world of experience for scientists to explore and understand. That world may be "unknown" to them beforehand, but it is in a way already charted and defined by the conceptual scheme in question.[17] Kuhn did not use the word *dogma* as he would in just a few years; but he quietly implied that most scientists are unadventurous and intellectually afraid of the dark. If they are said to explore unknown areas of nature or experience, they do so only with the light and guidance provided by a conceptual scheme that they trust and do not question.

"Sputnik and the American Public Mind"

Reading *The Copernican Revolution* without the hindsight provided by *The Structure of Scientific Revolutions* and "The Function of Dogma in Scientific Research," few would have discerned the deep, subversive critique of liberalism forming in its pages. If Conant had reservations, they did not stop him from applauding the book. In the foreword he provided, he presented it as just the kind of general education–inspired scholarship that was needed to elevate science's prestige among liberal educators in the West.

Contemporary events conspired to make this critique difficult to see, as well. The book appeared in 1957, the same year that the Sputnik satellites put special urgency on Conant's educational initiatives. Then living and teaching in Berkeley, Kuhn addressed the crisis in a lecture titled "Sputnik & The American Public Mind." His notes suggest that it was a public, "evening" lecture attended by an audience eager to hear what a historian of science had to say about the sensational Soviet achievement and what it meant for America.

Kuhn did not sugarcoat the bad news. Sputnik had indeed punctured a myth of American scientific and technological superiority over the Soviets and led the country to an "agonizing reappraisal" of its scientific prestige. Still, it was healthy for the nation to face historical facts: "at least until about one generation ago," he explained, America had been a laggard in pure or basic science when compared to England and Europe. This was because American culture itself—the "American public mind," as Kuhn put it—did not sufficiently value the kinds of formal, abstract thinking that encourages scientific progress. If America wanted to best the Soviets in science (as opposed to practical inventions, such as Whitney's cotton gin or Edison's electric light bulb), it would have to cultivate a stronger appreciation for "things of the mind."[18]

The diagnosis complemented Kuhn's growing skepticism that laypeople could learn to understand modern science. But it was an awkward diagnosis of the scientific geopolitics at hand: the new man-made satellites, after all, were practical inventions. They extended current technologies of rockets and radio communication but did not revise basic scientific theory as relativity theory or quantum mechanics had. So it is difficult to see how an increased fondness for abstraction in American culture or a respect for pure instead of applied science might help close the satellite gap. The politics of the satellite gap was awkward for Kuhn, too. The National Defense Education

Act would soon pass through Congress and the nation would confirm the liberal consensus that scientific progress thrived on intellectual freedom and open-mindedness—precisely what Kuhn's emerging theory of science exposed as merely a popular myth.

Kuhn seemed to approve this consensus in his talk. But in light of his discovery that dogmatism and narrow-mindedness, not curiosity and creativity, were crucial ingredients of modern science, he was skeptical and it showed. His lecture notes read:

> Though there is little evidence on the point and that inconclusive, there is some reason to hope that the freedom from intellectual restraint implicit in at least the theory of democratic society may provide more fertile climate for basic science than autocracy.

This was not a confident claim, just a "reason to hope" that American freedom would help close the gap with the Russians.[19] The evidence at hand was slim: if it were true that America could close the satellite gap by squeezing better science out of its political and intellectual freedoms (freedoms that existed in theory, Kuhn noted), then how had Soviet scientists, widely believed to have little intellectual freedom under the dictates of dialectical materialism, managed to get ahead in space science? The evidence was also inconclusive: Soviet science under dialectical materialist orthodoxy had a mixed record. The Sputnik satellites were a success, but Lysenko's biology was a failure. While Kuhn could join his audience in the patriotic hope that America's freedoms would help, he could not take a definite, confident position that these freedoms would eventually vindicate the superiority of Western science and technology. As he saw it, science's success and its revolutionary progress depended upon the relatively authoritarian and rigid sociological and educational structures of "normal science" (as he would soon come to call it). But this was not what intellectuals, education reformers, or the public wanted to hear when Western liberalism seemed to fight for its very existence and the Kremlin's avatars zoomed overhead.

"The Function of Dogma in Scientific Research"

Three years after Sputnik, Kuhn finally unveiled his new theory of science and unmistakably pushed back against the liberal consensus. He did so at Alistair Crombie's conference on the history of science in Oxford, England, just weeks after the University of Chicago Press had agreed to publish *Structure* both as a monograph in the *International Encyclopedia of Unified Science* and as a separate book. He could therefore proudly tell his largely British audience that his lecture, titled "The Function of Dogma in Scientific Research," previewed his forthcoming monograph, *The Structure of Scientific Revolutions.*

While *Structure* takes aim at "the image of science by which we are possessed," "The Function of Dogma" aims at a popular image of scientists. Not the advertising scientist, and not the ancient medicine man, Kuhn's target was the popular "image of the scientist as the uncommitted searcher after truth." This is

> the explorer of nature—the man who rejects prejudice at the threshold of his laboratory, who collects and examines the bare and objective facts, and whose allegiance is to such facts and them alone. . . . To be scientific is, among other things, to be objective and open-minded.[20]

The truth, Kuhn explained, is precisely the opposite:

> Though the scientific enterprise may be open-minded, the individual scientist is very often not. Whether his work is predominantly theoretical or experimental, he usually seems to know, before his research project is well under way, all but the most intimate details of the result which that project will achieve. If the result is quickly forthcoming, well and good. If not, he will struggle with his apparatus and with his equations until, if at all possible, they will yield results which conform to the sort of pattern which he has foreseen from the start.[21]

Conant's conceptual schemes may be tools for *predicting* the unknown, but the paradigms that Kuhn unveiled here went farther: they *constitute* the unknown by making nature itself seem to conform to preexisting scientific beliefs.

Paradigms, Kuhn explained, are the "concrete problem-solutions" that science students learn and repeat through drills and exercises. They expose this mythology of science pursuing and slaying the unknown:

> In all these problems, as in most other that scientists undertake, the challenge is not to uncover the unknown but to obtain the known.
>
> Given that paradigm and the requisite confidence in it, the scientist largely ceases to be an explorer at all, or at least to be an explorer of the unknown.[22]

Paradigms cultivate confidence and dogmatism for three reasons. One, Kuhn explained, they are mutually exclusive and therefore have no competitors. A scientific community, "if it has a paradigm at all, can have only one." Two, they are taken to ground true and unchanging representations of nature. "In receiving a paradigm," that is, "the scientific community commits itself, consciously or not, to the view that the fundamental problems there resolved have, in fact, been solved once and for all." Third, a paradigm is the practical basis of most scientists' careers. So they naturally "strive with all their might and skill to bring it into closer and closer agreement with nature." This helps explain why many become defensive and anxious during times of crisis and revolution. For they are defending nothing less than "the basis of their professional way of life."[23]

It is difficult to read "The Function of Dogma" without sharing Kuhn's enthusiasm for the wonderful irony that finally emerged from his Aristotle experience: the human mind could achieve stunning breakthroughs such as heliocentrism, biological evolution, or relativity theory only because its creativity and imagination were inhibited and restrained by inertia and dogmatism. Aspects of science that *seem* opposed and contradictory—progress and stubbornness, intellectual creativity and dogmatism, "moving ahead" and "staying put"—are in fact complementary parts of a larger mechanism, reaching from sociology to psychology and language, that produces revolutions over the course of a long and recurring historical cycle. Dogma and dogmatism are not obstacles to science or progress; they are, rather, "instrumental in making the sciences the most consistently revolutionary of all human activities."[24] They help ensure that scientists stay focused, that subversive "novelties" tend to be overlooked, and that puzzles eventually solved by determined scientists are not confused with revolution-inducing anomalies that require nothing less than a historic paradigm shift. Yes, this dogmatism means that scientific revolutions are rare and that they occur only when communities have no other options. But it also ensures that revolutions can and will continue occur.

H. Bentley Glass and the Fragile Liberal Ideal

Kuhn's audience was not particularly delighted by this irony. The British historian A. Rupert Hall worried that Kuhn's talk of dogmatism was, in effect, "an apology for weakness" on the part of insufficiently creative, free-thinking scientists.[25] The philosopher Stephen Toulmin said Kuhn failed to distinguish the necessary use of conventional scientific wisdom to formulate scientific questions from the requirement that nature be left "to answer questions for herself, without prompting." On this understanding, Toulmin insisted, scientific research is not really dogmatic at all.[26]

Kuhn stood his ground against these critics. But one caught him off guard—the American biologist Hiram Bentley Glass. Though thirteen years younger than Conant, Glass was in many ways like Kuhn's mentor. He was a respected scientist (a biologist) who became a university administrator, a public intellectual, historian of science, and influential education reformer. From the late 1930s to the early 1960s, Glass's writings in the *Bulletin of the Atomic Scientists* and his regular column in the *Baltimore Evening Sun* (when he taught at Goucher College and then Johns Hopkins) made him nearly as visible to the public.

Glass also grappled with the controversy over communist faculty and the specter of mind-controlled professors who indoctrinate their students with dogmas of Marxism and dialectical materialism. But Glass and Conant engaged these debates from different sides. As president of Harvard, Conant reassured his overseers, the public, and zealous anticommunists in Washington that his campus was free of subversive faculty. Glass, on the other hand, was an officer in the American Association of University Professors, who represented many faculty who resented the inquisitorial climate that Sidney Hook, Conant, and other anticommunists helped to create. Yet the liberal consensus about the necessity of scientific freedom bridged these differences. Glass fully agreed with Conant that intellectual freedom was essential to science and to a healthy democracy. As editor of the *Quarterly Review of Biology* from 1949 to 1965, he condemned Lysenkoism as a prime example of ideological interference in science.[27]

In his presentation at the conference, Glass examined the history of genetics to show how different biological theories function to criticize each other to move biology forward into its unknown, unpredictable future:

> The modern view of the relation of heredity to development is not Bonnet's, but neither is it Maupertuis's. It has something of

both, and something of neither. The two views, once held to be irreconcilable, have merged in a higher synthesis. As for our current views of heredity and of species, much the same may be said.[28]

As it was for Conant, "truth" played no part in Glass's conception of science's history. "It is in the dedication to a conceptual model which may seem to hold true," he concluded, "but cannot in fact describe nature in its fullness, that we find both the highest stimulus to current scientific investigation and the greatest barrier to ultimate knowledge."[29] On this point, Kuhn agreed as well: no theory can capture "nature in its fullness" because our raw, original experience is so complex. Science becomes possible only when it is simplified and reduced by paradigms.

But Glass could not abide the educational implications of describing scientists as "dogmatic" and their beliefs as "dogmas." In the wake of Sputnik, Glass had become a national leader in science education reform, a director of the Biological Sciences Curriculum Study based at the University of Colorado at Boulder. By producing new textbooks and teaching materials, he and his colleagues aimed to make biology classes more progressive, less beholden to the authority of the textbook, and less dogmatic.

As he listened to Kuhn's presentation, Glass could only have bridled. Kuhn described scientific education as "far more likely to induce professional rigidity than education in other fields." He said it "inculcates . . . a deep commitment to a particular way of viewing the world." Students routinely undergo "a relatively dogmatic initiation into a pre-established problem-solving tradition that the student is neither invited nor equipped to evaluate" and which creates a "mindset" and "professional rigidity."[30] Words and images like these compared scientists to Sidney Hook's Communists and blinkered fellow-travelers, or to victims of mind-control or brainwashing described in the popular media. By leaning as he did on the words *initiate, induce,* and *inculcate,* Kuhn seemed to walk around the word *indoctrination* that had surrounded America's perceptions of totalitarianism (and, for its critics, progressive education) since the 1930s. But the thrust of his remarks would have remained the same had he used that loaded word.

In response, Glass insisted that what Kuhn was saying was simply not true. He knew many scientists, he said, who were not at all dogmatic and closed-minded, who "often discuss the validity of their basic assumptions—I think more often today than when I was a younger scientist," he explained. Nor did Kuhn's thesis fit the increasing rate of scientific progress that Glass observed in his lifetime. Within his own career as a geneticist, he remarked,

I have already seen two very fundamental overturns of prevailing conceptual models, or to use Mr. Kuhn's term, paradigms. The young scientists of today, therefore, must be trained to expect relatively frequent overturns of his basic ideas within his own field.[31]

Kuhn believed precisely the opposite. Modern science was not like poetry or literature, whose students expect revolutions to occur sooner or later. Should science teachers dispel the dogmatism he had described, Kuhn believed, it would most likely pull the plug on future revolutions.

Speaking "as chairman of one of the science curriculum studies into which the National Science Foundation of the United States has poured some ten millions of dollars," Glass spoke "for my fellow biologists of this generation in America" and denounced Kuhn's analysis:

I have found complete unanimity among them in the belief that science must be taught—I do not say *has* been taught—as a variety of methods of investigation and inquiry rather than as a body of authoritative facts and principles. They also agree emphatically that students must be taught that scientific laws and principles are approximations derived from the data of experience and that they remain forever subject to alteration and correction or replacement in light of new evidence. I am appalled to think that, if Mr. Kuhn is right, we should go back to teaching paradigms and dogmas, not merely as temporary expedients to aid us more clearly to visualize the nature of our scientific problems, but rather as part of the regular, approved method of scientific advance.[32]

If Mr. Kuhn was right, the liberal consensus enshrined in these national educational reforms could be wrong and counterproductive. If Sputnik were in fact the fruit of "autocracy," the word Kuhn had used in his lecture on Sputnik, then Glass's reforms might even *widen* the science gap with the Russians.

The Invisibility of "Dogma"

Kuhn reassured Glass that that he was sympathetic to his educational reforms—"The system they aim to change," he agreed, "is often no more than a parody of what scientific education should be."[33] And he announced that he would stop using the terms *dogma* and *dogmatism*. Glass's difficulty in understanding precisely what a "paradigm" is led Kuhn to his decision. In his remarks, Glass said that

> the paradigm looks backwards while moving forwards, whereas the dogma, a related creature with which I am more familiar, also looks backwards but stands its ground.[34]

Paradigms and dogmas were not the same thing, in other words, and paradigms—rightly understood—are similar to Conant's conceptual schemes, always moving ahead and never resting content in dogmatism. Kuhn agreed that the distinction was important and agreed to speak instead "of something like 'commitment to a paradigm.'"[35] Offered as an essential, functional component of normal science, the words *dogma* and *dogmatism* never appeared again in Kuhn's writings. Along with "dogma," Kuhn's essay, "The Function of Dogma in Scientific Research" was evidently a casualty of this encounter. Though it was published in the conference proceedings, Kuhn forbade it from being reprinted later in other venues, including collections of his own essays.[36]

In the long run, however, Kuhn had surrendered nothing but a word. The substantive claim about the nature of scientific research that he intended to make remained. Nor did Kuhn's retraction affect *The Structure of Scientific Revolutions*. The manuscript he circulated to his colleagues months before did not rely on the word in the same way and neither does the final, published version. "Dogma" appears briefly in reference to theology and ancient astronomy, but it carries none of the functional burden Kuhn gave it in this controversial paper at Oxford. Instead, that burden is carried by an array of synonyms and substitutions, such as *binding, commitment, rigid, accepts without question, relatively inflexible [theoretical] box, take for granted, assurance,* and *confidence in their paradigms*.[37] At Oxford, Kuhn said that "dogma" makes the scientific community "an immensely sensitive detector of the trouble spots" from which revolutions may be sparked. In *Structure*, it's the "rigidity" of individual scientists that "provides the community with a sensitive indicator that something has gone wrong" with the dominant paradigm.[38]

Only Kuhn knew precisely why he used this provocative word exclusively for his presentation at Oxford. Had it occurred to him to use "dogma" shortly after he had finished drafting *Structure*, his encounter with Glass may have dissuaded him from introducing it into the final draft he would soon deliver to the University of Chicago Press. Were "dogma" his first terminological choice for articulating his new theory of science—as the confidence and swagger of "The Function of Dogma in Scientific Research" seems to suggest—Kuhn may have reasoned that a largely British audience—removed from the American preoccupation with dogmatism, authoritarianism, and brainwashing—would be less distracted by the word's unsavory political connotations and better able to understand Kuhn's point. Conversely, he may have reasoned, American readers might be better served and less addled by the euphemisms for dogma and dogmatism that abound in *Structure*. If so, Kuhn was exactly right. For it was Glass, a fellow American, who quickly saw the geopolitical implications in play. In the ongoing struggle for men's minds with the Russians, a theory of science that exalted dogmatism would be no help at all.

14

The Magic of Paradigms

In June 1961, Thomas Kuhn was especially busy. Some things were going well—his monograph had finally come together, and he was preparing to move his family to Denmark for research on the recent history of modern physics. In just weeks he would present "The Function of Dogma in Scientific Research" in England, where his new theory of science would challenge the reigning philosophy of Karl Popper and meet H. Bentley Glass's wrath about dogmatism.

Finally taking his historical insights into philosophy's court, Kuhn could have used all the support he could find from philosophers—but that search was not going well. Paul Feyerabend, the philosopher of physics (and Popper's former student) intensely disliked *Structure*'s first draft. And weeks before, in May, the good news of Kuhn's promotion at Berkeley to full professor in the history department had come with a stinging philosophical insult: the philosophy department no longer wanted him.[1]

Some philosophers, such as Ernest Nagel, remained extremely supportive about *Structure*. But the support and feedback Kuhn wanted most was Conant's. Ten years after their collaboration teaching *Natural Sciences 4* had ended, Conant remained something of an intellectual father to Kuhn. Their letters still began with asymmetric greetings ("Dear Mr. Conant" and "Dear Tom") and Kuhn wanted Conant's support and approval not only for personal, emotional reasons. For Conant was powerful, and if he liked the manuscript he could quickly solve the problems Kuhn then faced regarding its excessive length and his worry that the *International Encyclopedia* was a dying venue that might bury his new book. A letter from Conant to the University of Chicago Press—or any publisher, for that matter—would quickly persuade editors to publish it outside the encyclopedia as an independent book.[2]

Kuhn asked Conant for this favor when he mailed him the manuscript and asked for his honest and critical assessment. Nagel and others had gotten back to Kuhn with their praise and encouragement, but after several weeks he had heard nothing from Conant and he was getting worried. Finally, he received a very short letter revealing how busy Conant was. He had been in Japan. He was busy wrapping up work for the Educational Testing Service and presided over a conference in Washington dedicated to problems of unemployed youth. All this was good news to Kuhn, for the nonstop itinerary explained Conant's silence. But this reassurance probably faded when Kuhn read this: "This is only a note to tell you that I am in the midst of it and before very long you should get a very long letter from me."[3]

A *very long* letter? Obviously Conant was finding problems too involved to mention casually. In a few days, Kuhn received that letter—four single-spaced pages of typed comments that were as bad as he could have imagined. Conant did not like Kuhn's sweeping, oversimplifying generalities about scientific revolutions. He did not understand why a revolution should necessarily involve a new worldview, nor why Kuhn had given scientific ideas and theories the starring role in his account, while experiments, instrument makers, and craftsmen only had occasional walk-on parts. And he did not like the puzzling, incessant, and frankly *annoying* reference to "paradigms."

Conant did not dislike everything. He liked the book as a whole and agreed that the issues it raised were important. But Kuhn knew Conant well enough to know that these compliments were formalities. Of course Conant would say he was "sympathetic to your unorthodox interpretation of science" for he had taught Kuhn to eschew the reigning caricature of science as a compendium of facts, to focus on scientific change ("science moves ahead") instead of orthodoxy, to grasp the driving power of ideas (conceptual schemes), and to distrust any history of science written in a science textbook. But these polite remarks did not extend to Kuhn's efforts to innovate and improve Conant's theory of science and they could hardly have softened Conant's sharp, dismissive remarks such as, "To my mind, the page on which you sum up your point of view without recourse to the word 'paradigm' is the clearest page in the whole document." As for the experiments in perception "which have so deeply impressed you,"—clearly they had not Conant—"I think this whole section of your document complicates your fundamental argument."

Conant agreed to write to the University of Chicago Press as Kuhn had asked. But he hoped Kuhn would be submitting a rather different manuscript. Kuhn pointed out that the last two chapters needed more work, but Conant nudged him toward a thorough rewrite. "I should be much happier with it if in the revision you could clarify your presentation and, if you think any of my points are well taken, if you modified certain sections to eliminate my objections." In other words, eliminate the paradigms. Kuhn should spare the reader who after being told about this vague, so-hard-to-define word *paradigm* will be "ready to cry out with pain when he encounters it six or eight times in a single page!" And he should spare himself the professional risk that scholars may "brush you aside, I fear, as the man who grabbed onto the world 'paradigm' and used it as a magic verbal word to explain everything!"[4]

Foo!

Kuhn and Conant had argued before, but this was new territory. He read and reread the letter and scribbled defensive, contrite, and sometimes just despairing notes in the margins. "I *say* this! p. 16," he replied to Conant's point that the practical arts play enormous roles in science's early stages. "Foo!" he exclaimed over Conant's remark that all this created "needless trouble about progress." Next to rapid-fire of questions from Conant about "paradigms"—Conant pointed to page 24, then 28, then 30, then 34 . . .—Kuhn figured that Conant was simply *making* trouble for him. "I think these are not problems except in so far as they encourage these reactions,"[5] he wrote. Kuhn was defensive and evidently unsure of what to make of Conant's misgivings. Were his innovations really so confused and needless? Did Conant resent paradigms because they stole the spotlight from his own writings about science's conceptual schemes? Was Conant perhaps just as impressed as Kuhn's other, effusive readers, but hiding it behind the crusty demeanor of a military scientist—or a stern father-figure—for whom no achievement is quite good enough?

Or, did Conant simply *fail to see* the new image of science, the new *gestalt*, Kuhn had painted? *Structure* was, after all, part of an ongoing revolution in the history of science itself. Kuhn might well have seen Conant as an Aristotle to Kuhn's Newton—a brilliant man who nevertheless could not find his way beyond his own historiographical paradigm. Conant himself said in his letter that "a choice between two different sets of theories . . . is made slowly by the younger men in the scientific community." The parallel was obvious: Conant was older and perhaps simply clinging to his own, unexamined dogmas about what science is and how it works.[6] So used to explaining how science moves ahead confidently and heroically, he could not see the point of the paradigms and dogmatic worldviews that *prevent* science from moving ahead until genuinely revolutionary advances become inevitable.

Kuhn stewed for about three weeks before replying. In a letter twice as long as Conant's, he confessed that he was surprised, upset, and saddened by the criticisms. "I have had reactions to the dittoed ms from about ten people beside yourself, and none of them had at all prepared me for your letter," he wrote. One thing was clear. Unlike Conant's helpful feedback about *The Copernican Revolution* four years before, Kuhn would be on his own with this book. "The points you bring up now are not mainly about style and presentation." Kuhn explained:

> A number of them reflect fundamental disagreements; others reflect misunderstandings which have not arisen with other readers and whose source I cannot locate; a few of them I simply cannot understand. That does not mean that I do not find your letter helpful. . . .

A handful of Conant's points were good, Kuhn said, and he would adjust the draft accordingly. But Kuhn could not and would not rework the whole book. Conant's fundamental objections meant they may never see eye to eye about paradigms, worldviews, and progress in science. "That will make me quite sad," Kuhn admitted, but he did not let on how sad he was. For Kuhn wanted to dedicate the book to Conant, not only as a gesture of his admiration but of his hope that Conant would like the book and perhaps even join the historiographic revolution now afoot. That was now impossible and Kuhn didn't even bring it up. Why would Conant want his name on a book that would make readers cry out in pain or harm his student's scholarly reputation?[7]

Progress and the Citadel of Learning

The disagreement at hand was fundamental. In the years since they collaborated directly in Cambridge, Kuhn continued to move away from the passionate politics of his youth into a world of scholarship that the struggle for men's minds had helped to professionalize and depoliticize. His ever-changing notes and outlines illustrated the trend, as "theory as ideology" became "scientific consensus," which finally became "paradigms." Hand in hand with his doubts that laypeople could have a genuine understanding of modern science, the public mattered less and less as Kuhn's scholarly interests and career evolved.

Conant had moved the other way. He left academia for cold war diplomacy and then returned to his longstanding interests in public education and its roles in the nation's civic welfare. When he read Kuhn's manuscript, Conant was studying schools in the nation's growing cities, and—as he had done in Germany—fighting communism by promoting liberal education. His post-Harvard career helped to prepare his visceral, negative reaction to Kuhn's theory of paradigms.

While working for the State Department in Germany, Conant had frequently returned to the United States and spoke publicly about the civic and political landscape in Germany that he aimed to rebuild into a modern, vibrant democracy. The difference between his vision for West Germany and the realities of East Berlin repeatedly amazed him. He told the story of how he walked through the Brandenburg Gate into East Berlin and into the bookstore of Humboldt University, which he had visited before the war when it was part of Berlin University. He perused the philosophy books on the shelves and witnessed the aftermath of a disappointing transformation:

> Here one hunts in vain for copies of the famous German philosophers, except for a slim volume of selections from Hegel's logic. But the shelves are full of volumes of the Marx-Lenin Library published in uniform bindings, primarily selections from the writings of Marx, Engels, Lenin and Stalin.

Copies of Lenin's influential *Materialism and Empiriocriticism* abounded, but this was hardly a book "worthy of attention by students of philosophy in the free nations of the world." The problem continued into an article Conant found in the magazine *Einheit*. It was about "the philosophical significance of the new discoveries of the Soviet astronomers" and it began, oddly, not with discussions about telescopes, planets, or galaxies but instead with the epic battle between philosophical materialism and idealism championed by Lenin and Soviet philosophers in his wake. Wherever one looked behind the Iron Curtain, Conant explained, the lines of reason and cultural knowledge circled back to Lenin, Stalin, and the dogmas of dialectical materialism, the official philosophy of the Soviet Union.[8]

Conant offered these reflections at Yale, where he had first delivered the lectures that became *On Understanding Science*. Under the title *The Citadel of Learning* he denounced the ongoing siege against intellectual freedom in Eastern Europe and the Soviet Union. In Stalin's antiliberal citadel, he explained, "one and only one philosophical point of view is tolerated." All students must pass examinations based on Marxism and Leninism. It was important for Americans to understand this intellectual myopia and constriction, he insisted, for "by realizing what happens when the citadel of learning has been captured, we may be better able to understand what is the essence of the activities within this citadel when it remains free."[9]

That liberal "essence" was the pluralism of competing perspectives, clashing theories and conceptual schemes in science, and the pluralism of debates in politics, society, and economics. It required the mind to stand apart from any one theoretical authority, to independently evaluate different, competing points of view, and never to succumb to the false comforts of dogmatism or conformism. Conant had offered the message repeatedly to Harvard's undergrads and to the public. It was the crux of his defense of the liberal arts years before when he fended off criticism that his militarization of Harvard's campus might consign them to oblivion. It supported his condemnation of Communist faculty beholden an ideology. It grounded his vision of a national foundation devoted to pure, esoteric research undertaken by experts (admittedly, prone to "autointoxication") but funded according to decisions made by "laymen who understand science and scientists."[10] And it lay at the core of his definition of science, where progress occurs when some scientists, less dogmatic than others, see unnoticed connections among different conceptual schemes that lead to new ideas, yet unthought, and new experiments, yet unperformed.

One version of Conant's imaginary proof of historical scientific progress suggests why he would later recoil at Kuhn's paradigms. Writing in the *Journal of Higher Education*, Conant wrote that important and influential scientists of the past, if brought back to life to see the science of today, "would exclaim with delight and wonder at the progress which has been made."[11] These resurrected scientists may not immediately understand the conceptual schemes that had come to replace their own, but in looking at contemporary knowledge with "delight and wonder" they were plainly eager, able, and ready to learn. Far from feeling disoriented or transported to a new, strange planet, Conant's scientists were pluralists with interests and perspective more expansive than either the Kremlin's dialectical materialists or Kuhn's normal scientists.

As a public intellectual, to be sure, Conant sometimes moderated his liberalism. It came to the fore in times of peace, but in times of war, or of great threat or danger, he instead championed unanimity and consensus in thought and behavior. When he first read *Structure*, the nation's pendulum had swung toward liberalism and he urged Americans to eschew any form of intellectual and cultural regimentation; to embrace pluralism and tolerance for opposing beliefs and approaches. As he put it in *The Citadel of Learning*,

> Freedom and tolerance go hand in hand in matters of the spirit. The premises of an argument cannot be truly examined unless alternative premises are believed and defended by one or more persons of integrity and scholarly competence. It is the absence of dissenters from the official dogma that signalizes the capture of the citadel of learning. Conversely, it is the presence of defenders of different sets of premises that has in the past ensured the vigorous development of the Western tradition that stems from the ancient Greeks.

Pluralism's ever-fruitful clash of ideas was the engine of progress—and not only in science. Every lawyer and doctor must understand the "the significance of dissent," Conant wrote, by studying—if not in fact experiencing—a "battle royal of ideas" in their chosen fields.[12]

When "The Citadel of Learning" was first published, Kuhn read it and noted Conant's observations of "Russian suppression of free inquiry in Eastern Zone of Berlin," and his argument for "the essential role of controversy in research of all sorts." As for Conant's thesis that Western learning was unified, that there existed (in Kuhn's words) "a fundamental parallelism of method and nature between all forms of human endeavor," Kuhn did not buy it. The essay was "good but confused," he wrote. In Kuhn's eyes, Conant and other cold warriors may be right about politics, but they were overlooking something important about science. At least normally, apart from times of revolution, scientific inquiry thrives *not* in a free, open, and tolerant citadel of learning.

Kuhn criticized Conant's thinking only privately, however. As far as Conant understood it, he and Kuhn still worked together to advance the understanding of science among the public as well as among science-averse intellectuals. When Conant wrote a foreword to *The Copernican Revolution*, he began by situating Kuhn's new book squarely in a geopolitical setting: "In Europe west of the Iron Curtain," he wrote, "the literary tradition in education still prevails." That was no accident, Conant acknowledged, for he wrote it when he lived and breathed cold war tensions as ambassador to West Germany. His observations of how Stalinist philosophy had harmed science (through Lysenkoism, for example) and other areas of intellectual and political life underscored why it was important that "every educated person of our times"—not just Harvard undergrads in his general education program—learn something science's history. Kuhn's new, highly readable account of the Copernican revolution drew on the case study taught at Harvard and fit the bill perfectly.[13]

That was 1956, and *The Copernican Revolution* remained committed to Conant's "conceptual schemes" and, at least overtly, his preferred image of the open, nontotalitarian scientific mind. As for Kuhn's new manuscript that Conant read in 1961, even though "ideology" had been replaced by "paradigm" and there was less emphasis on the functions of dogma than in the lecture Kuhn would deliver at Oxford, Conant could hardly have missed the fact that it headed in a different political direction. Though Kuhn's manuscript started out on the same note with which Conant began his foreword (it began, "The study of history has not been a usual source for the West's conception of science, and it might usefully become one")[14], it stepped closer and closer, with each invocation of "paradigm," to the official image of research that Conant had spent the 1950s denouncing on the far side of the Iron Curtain.

Kuhn Replies to Conant's Criticisms

Despite the depth of their disagreements, Kuhn typed on for five more pages in an effort to persuade Conant to see things his way, to see the importance of paradigms and their hold on the scientific mind. "Some of your criticisms are wide of the mark," he announced as he began his detailed reply, though he had to suspect that no matter how effective his rebuttals would be, he might be talking past his former mentor.

A Single Point of View?

Conant said Kuhn made too much of scientific revolutions. After all, they were not always alike. "Minor scientific revolutions" were simply not the same as "the major ones," he wrote. And he doubted that worldviews always changed with revolutions, even with the momentous transition from Newton's world of gravity and material bodies to Einstein's world of curved spaces and energy fields:

A "new world view" is implied by your treatment of all scientific revolutions but I query if this is not far too grandiose a characterization of most of the revolutions you cite as examples. I doubt if the world view of science was "new" after the adoption of the atomic theory or even after the acceptance by physicists of Einstein's relativity. You tend to treat the scientific community far too much as a community with a single point of view. In short, I believe you dodge some of the difficulties of the detailed analysis of the application of your doctrines by taking refuge in the word "paradigm."

"A single point of view"? Which community did Conant have in mind—some physicists, all physicists, or all scientists in all fields? Perhaps because Conant referred to "the world view of science"—as if chemists, biologists, and other kinds of scientists saw the world the same way (which Kuhn did not mean to say)—it seemed that Conant had simply misunderstood. "Must make local world view clearer," Kuhn wrote in the margin.

But Kuhn probably misunderstood Conant's objection. His point was that Kuhn's picture of individual scientific communities was too uniform, too much dominated by the "single point of view" manifest in its single paradigm. Conant had not seen Kuhn's essay "The Function of Dogma in Scientific Research" that Kuhn was then preparing (as far as their correspondence goes, there is no evidence that Conant ever read it). But he objected as H. Bentley Glass would in a matter of weeks to this image of dogmatic scientists who think alike and never question their assumptions.

Progress and the Practical Arts?

This like-mindedness is why *Structure* raised "needless trouble about progress." For no good reason that Conant could see, Kuhn had left out the tinkerers, experimenters, artisans, and even businessmen whose practical, nontheoretical contributions pushed science forward alongside the abstract, mathematical symbols and formulas conjured by theoreticians. This was a hallmark of Conant's (and, before him, L. J. Henderson's) view of conceptual schemes as essential guides to the practical, working knowledge of research scientists.[15] Scientific theories, especially those within the last three hundred years, Conant understood, may *seem* to have an abstract intellectual life of their own, but they are usually framed and rooted in practical problems.

There was "a symbiotic nature of the relation of science to industry" he wrote in *On Understanding Science*; for biology, there was "a two-way street" connecting theoretical advances with those in agriculture, for example. Yet Conant saw very little of this complexity in the manuscript:

> By leaving out any reference to technology and advances in the practical arts (including the practical art of experiment and observation) you distort the picture of science and get yourself into needless trouble about progress.

Conant had called it "false snobbery" to suppose that theories and formulas were somehow more important in science's history than the practical arts. But now that Kuhn had written a book so tilted to one side, Conant decided to make his point once again and then leave it alone. "Since you are familiar with my own preoccupations and my emphasis on the close relation between progress in the practical arts (and this is progress) with advances in theoretical science I shall push this point no further."[16]

Does Perception Really Matter?

Because progress emerged from contrasts and competition between clashing ideas and proposals, Kuhn's discussion of scientific perception seemed only to compound *Structure*'s error. In Conant's eyes, it falsely implied that groups of scientists were so like-minded they even saw nature and the world in the same ways. In astronomy, Kuhn had written that "shifts in scientific perception" occur when paradigms change—at which point Conant rejected this discussion as a distraction: "I don't think it is what people see that matters. What matters is the guide to action which they accept."

> Even in astronomy is it seeing that really counts? Isn't it the design and placing of the instrument and a comparison of what one sees at one point of time and space with what one sees at another? In other words action is necessary. It is even in the psychological experiments which have so deeply impressed you.

What scientists do, in other words, is more important than what they see, Conant insisted. But Kuhn seemed to have turned this around. "I think this whole section of your document complicates your fundamental argument," he complained. Kuhn, however, was baffled: "But this is The fundamental argument," he wrote in the margin.

Conant ended his letter on a positive, gracious note by admiring Kuhn's attempt to delineate a regular pattern to historical revolutions:

> If I read you at all correctly, you emphasize the fact that in each of the fields of the biological and physical sciences . . . there has been an "immature" or "pre-natal" or "adolescent" stage. The emergence of an area of concern from this stage is the wide acceptance among investigators of (a) a theoretical framework (b) certain unexpressed premises (c) a number of unformulated rules of correct experimentation.

But stuffing all of these items—theories, tacit premises, and rules of experimentation—into *one* box, into paradigms, seemed pointless and historically misleading. Conant therefore urged Kuhn to treat these elements separately, "to distinguish each of the three components and the way they change in a revolution."[17] Thus, Conant's criticism circled back to its central, thematic point: not all scientific revolutions are the same, and there is no reason to expect that they would be.

A Matter of Priority

Kuhn replied that he *had* to introduce paradigms because they are the very basis of scientists' educations. Theories, models, techniques, and presuppositions come later and are not as fundamental as the problem solutions, the paradigms, in which these separate aspects of science are "often inseparably mixed together." Paradigms, he might have told Conant the chemist, are compounds, not mixtures. And the individual elements he urged Kuhn to disentangle cannot be disentangled when paradigms are doing their functional work. Kuhn explained, "The problem-solution, the achievement, comes first; the analytical separations come later and are always incomplete."[18] Paradigms don't obscure science's history, therefore; they help reveal the underlying sociological and ideological dynamics that animate scientific communities.

Did Kuhn neglect the practical arts? Yes, he admitted, his manuscript did shortchange them. But he felt hemmed in by the need to keep his monograph short enough for the encyclopedia. "On the other hand," he countered, "I cannot believe the distortions are as bad as you take them to be," or that they really create "needless trouble about progress." On this, Kuhn felt sure that Conant badly misunderstood the issues surrounding the idea of progress. A history dedicated to the practical arts will invite the same (alleged) trouble, he explained, because "technological progress is only progress to the man who accepts more specialized and effective machinery as a good in itself. Others can say it's merely a feature of social evolution and a retrograde one at that." Judgments about progress, that is, are inevitably subjective.

Kuhn had been alert to these issues about progress ever since the start of his collaboration with Conant, when his Aristotle experience—now hovering between the lines Kuhn typed to Conant—guided him to the main insight that *Structure* aimed to explain and elaborate:

> This is the most fundamental issue of all and the one on which we are and have almost always been furthest apart. You opened On Understanding Science by discussing cumulativeness as the distinguishing feature of science. You then sent me off to look at pre-Newtonian dynamics. I returned from that assignment convinced that science was not cumulative in the most important sense. Newton was not trying to do Aristotle's job better; rather Aristotle had been trying to do a different job and one that Newton did not do so well.

This Aristotle, resurrected from the dead, would not agree that Newton was doing his job better; he'd conclude that Newton was doing something else, a different kind of job. "Progress is easy to define and evaluate only for a cumulative process," Kuhn explained:

> As long as the goals are fixed and agreed upon, one stands a quite good chance of deciding whether one machine or another (or one theory or another) meets them more closely. But if the goals change with the machine (or theory) as they do and if people nonetheless ultimately insist that progress has been made, then one does need to ask what progress is about.

If Kuhn worried that Conant was unlikely to understand his new perspective, this impasse about progress seemed a case in point. "I doubt that this will convince you," he admitted.[19]

Kuhn was not sanguine about their disagreement over perceptions, either. "This one again confuses me entirely," he wrote, "for I am not clear what you are disagreeing about." Conant had objected that "passive seeing proves nothing" and Kuhn agreed emphatically because one thrust of his book was that there is "no such thing" as passive seeing. As he put it in the manuscript, research in human perception "makes one suspect that something like a paradigm is prerequisite to perception itself. What a man sees depends both upon what he looks at and also upon what his previous visual-conceptual experience has taught him to see." This was "the fundamental argument," Kuhn told Conant, because it promised to spark the historiographic revolution at hand. As long as scholars believed that something like passive, paradigm-less seeing was possible, Kuhn explained, "we shall be stuck with traditional epistemology" of science. The idea that scientists register perceptions passively as light falls on their retinas "has been the philosopher's paradigm for objective perception, and in most circles it still is." So this discussion was no unnecessary complication; it was storming the Bastille at the heart of the reigning epistemological regime. "If that notion can be licked," Kuhn urged Conant to see, "the rest of the structure falls" and this new image of science stands as a revolutionary alternative.[20]

"Again, I doubt I shall convince you," Kuhn wrote. He had convinced the majority of his early readers, but some (such as Hanson) already had one foot in the revolutionary circle; others like Nagel and Merton seemed eager to jump in. Kuhn dearly wanted Conant to respect his revolutionary efforts, to see that "the section on perception is the fundamental one in the monograph." But it was now clear that Conant was just not impressed. Five pages of clarifications, explanations, defenses, and justifications left Kuhn sounding exhausted and uncertain. "I don't know quite where all this leaves us," he wrote. Could Conant really not have grasped the points that his other readers had no trouble with? Were his many objections, he hinted, "not really substantive rather than tactical"?

Kuhn could at least end his long letter with some good news. He no longer needed Conant to draft a letter recommending the manuscript, for by this time Carroll Bowen had called to say that the University of Chicago Press would be happy to publish the monograph exactly as Kuhn wished. With that, Kuhn signed off and wondered if his intellectual ties with Conant were forever frayed. Offering his "fondest greeting to both you and Mrs. Conant," he added, "I hope they bring us closer together than my substantive responses are likely to."[21]

James B. Conant: Reluctant Revolutionary?

Now sixty-eight years old, Conant may have sensed that his blistering criticisms of Kuhn's new manuscript were a first reluctant step in a passing of a torch. Despite his reservations about paradigms and worldviews, Conant seemed to have no doubts that he and Kuhn remained colleagues united by their "joint efforts," as Conant once called them, to educate the public about science.[22] Writing back about two weeks later, Conant enthused over Kuhn's good news from the University of Chicago and immediately softened his criticisms, saying "the most important was a matter of style and presentation, largely turning on what I regarded as your use and misuse of the word 'paradigm.'" Perhaps sensing that the scholarly winds were now poised to blow in Kuhn's direction and not his, Conant did what he usually did when he lost one of his battles. He capitulated and went along with the new consensus—at least as far as his long letter of criticism was concerned.

Kuhn must have been surprised, for Conant's reservations were fundamental and not matters of style. But there was no mistaking that Conant was backing down. On these disagreements over matters of perception and epistemology, he wrote, "I hoist the white flag or show the yellow feather,—however you want to judge my hasty retreat!" It seems likely, however, that Conant had not dropped his reservations. Rather, he did not want to debate Kuhn further. He would let readers and reviewers do that. "I shall be curious to see what philosophers of various kinds have to say about this part of your document that you consider so fundamental," he wrote almost presciently.[23]

Kuhn had just returned from the Oxford Conference where H. Bentley Glass had attacked his use of the word *dogma*, so he was surely relieved that Conant had relented in his attack on "paradigm" and the manuscript's fascination with scientific perception. By all means, Kuhn wrote back, let us agree to disagree and talk more about these issues when opportunity arises. The lowered tension meant that Kuhn could now ask about a dedication:

> Would you be willing to let me dedicate the book to you? For both personal and historical reasons I should like very much to do so. Historically, you are the one who, perhaps inadvertently, started me on the road of which the present manuscript describes the early stages. Even our disagreements, or some of them, have roots in older differences about the structure of case histories. And personally, you are the one who taught me that the turtle travels fastest when his neck is out.[24]

"I shall be delighted to have you do so," Conant replied, adding that he was amused to see his favorite adage about risk-taking turtles. Kuhn was certainly taking an intellectual risk and Conant knew it. For a second time, he asked to be kept apprised of the book's reviews. But Kuhn's daring claim was also ironic: He was taking an intellectual risk by claiming that scientists *don't* and *shouldn't* take intellectual risks that might upset the dogmatic workings of normal science. If Conant saw the irony, he didn't say it. Writing to Kuhn from his summer home in New Hampshire, and having spent much of the past two months with his grandchildren, he was starting to accept the book, its flaws, and perhaps its ironies as only a grandfather can. As Kuhn suggested with his dedication—To James B. Conant/who started it—Conant had given life to something he could no longer control.[25]

As the months went by, Conant seemed to like the book even more. Returning home to New York City from a trip to Europe in late 1962, he found the just-published book in his mailbox. He read it again and sent his congratulations:

> I have just finished reading it and congratulate you most fervently. Needless to say I am grateful to you for the dedication and more than proud to have my name associated in this way with what is a truly important book.
> You have not only presented a challenging and unorthodox interpretation of scientific history, but you have documented what you have written in a most impressive manner. . . . Quite apart from the impact of your novel ideas, the setting forth of the scientific revolutions as you have is going to help many readers to understand science better.

Like *The Copernican Revolution*, Conant saw *Structure* as an extension, even a progressive continuation, of his own work. He even began to rewrite the history of his own criticisms:

> [M]any of the points I found difficult a year ago now seem to me to present no difficulty. Perhaps you have changed the manuscript somewhat or perhaps my receptive powers have increased. In my experience every manuscript is easier reading once it is set forth in print.[26]

In fact, Kuhn had changed the manuscript. He had added the chapter "The Priority of Paradigms" and addressed it to historians who might question why paradigms were necessary for the understanding of science. But it was tacitly addressed to Conant, for it elaborated the reply Kuhn had given in his long letter: paradigms are necessary because they are "prior to, more binding, and more complete than any set of rules for research that could be unequivocally abstracted from them." Kuhn had also warned readers away from other confusions he saw lurking in Conant's long letter, such as the claim that revolutions in one science would spark revolutions in others or in all of science. And Kuhn had rewritten sections of the final chapter, "Progress Through Revolutions."[27]

These changes do not explain Conant's change of heart, however. Kuhn had neither revised nor softened the troubling implications about progress in this final chapter. He approached the issue exactly as Conant had, through a comparison of science and art. Conant said science makes cumulative progress while the arts spin in place and multiply perspectives, but Kuhn denied this. Citing art historian E. H. Gombrich, he pointed out that "for many centuries painting was regarded as *the* cumulative discipline" because advances in techniques such as perspective and chiaroscuro "made possible successively more perfect representations of nature." Quietly rejecting Conant's all-important distinction, Kuhn wrote that "scientific progress is not different in kind from progress in other fields." It merely *appears* that way, he explained, because of the ideological dynamics of normal science, including the rigid education of scientists and the lack of alternative paradigms. As for progress through revolutions, Kuhn turned to George Orwell and the power of history's winners to rewrite textbooks, to make scientific revolutions invisible and "to make certain that future members of their community will see past History in the same way"—that is, as progressive and cumulative.[28]

Conant said nothing about these new additions or their lingering differences. He drew attention instead to one momentous point Kuhn made about truth: that "we may have to relinquish the notion, explicit or implicit, that changes of paradigm carry scientists and those who learn from them closer and closer to the truth." Kuhn had made the point in the manuscript that Conant read earlier, but this sentence in the final, published version grabbed Conant and allowed him to stake a claim in the new image of science. Indeed, he wrote, "I have never thought of science as an example of accumulative knowledge in the sense of a process approaching 'the truth.'"[29] Perhaps like H. Bentley Glass, who said he could live with paradigms as long as they had nothing to do with "dogma," Conant chose to live with paradigms and worldviews as long as it was clear that science always moved forward and never rested satisfied with any dogma understood as "the truth."

If Conant had found his way inside Kuhn's revolutionary circle, however, he still had some doubts. "I must admit frankly your choosing the word 'paradigm' still troubles me a little," he wrote. And he seemed to flatly reject any suggestion that his own criterion of scientific progress had been pushed aside: "If you question whether Aristotle or Archimedes [brought back to life] was working in 'the same field' as a modern biologist or physicist," Conant egged him on, "I would say yes in spite of all the revolutionary changes in the paradigms which have occurred!"[30] Conant was willing to go along with his revolutionary student, even to use the new terminology of "paradigms" that he once disliked. But nothing was going to persuade Conant to abandon his ideal of scientific progress and its political implications.

Liberty, Learning, and Pluralism in a World Transformed

When he received the published version, Conant was writing a book titled *Two Modes of Thought: My Encounters with Science and Education*. He told Kuhn that he would find Conant's reflections familiar and that he would refer to *The Structure of Scientific Revolutions*. "If nothing else this little volume may stimulate people to read yours."[31]

The book really was little. It was an entry in the Credo series, published by Pocket Books, known for small, inexpensive versions of nonfiction classics. The series was created for leaders in different fields to reflect on their careers and present what they have finally come to believe. Other authors in this series included René Dubos, Erich Fromm, and William O. Douglas. Pocket Books had promised to print an impressive 150,000 copies, Conant reported, so quite a few readers may have learned about Kuhn's new book in this way. When it came up, Conant called it simply "brilliant" and his praise was only enhanced by the company it kept. Conant mentioned only a handful of other authors, including W. V. O. Quine, Karl Jaspers, and Charles Darwin.

Still, Conant invoked *Structure* to support, not replace, his long-standing view of science as restless, dynamic, and progressive—"an interconnected series of concepts and conceptual schemes that have developed as a result of experimentation and observation and are fruitful for further experimentation and observation."[32] To explain why his definition does not include the notion of truth (something many readers would find strange, he admitted) he turned to Kuhn's new book, which shows that "the history of science does not afford evidence for such a widely accepted view." That is, as Kuhn had put it, "we have to relinquish the notion, explicit or implicit, that changes of paradigm carry scientists and those who learn from them closer to the truth."[33]

Those who had already read *Structure* could have noticed the narrowness of Conant's praise. He applauded this particular conclusion and quoted his favorite sentence. But did he accept the assumptions, the reasoning, and the model of science that led Kuhn to this conclusion? There is no mention of normal science, perceptions influenced by ideas, incommensurability, or revolutions and new worlds of perception or understanding. Kuhn wondered whether Conant's first criticisms were more tactical than substantive, and the same could be asked of this praise. It is not difficult to read *Two Modes of Thought* as a quiet but sustained critique of *Structure* from a man divided by his loyalties to its author, on one hand, and to a vision of liberalism and pluralism that *The Structure of Scientific Revolutions* cast into doubt.

Consider the basic thesis *Two Modes of Thought* set out to defend:

> A free society requires today among its teachers, professors, and practitioners two types of individuals: the one prefers the empirical-inductive method of inquiry; the other prefers the theoretical-deductive outlook. Both modes of thought have their dangers; both have their advantages. In any given profession, in any single institution, in any particular country, the one mode may be underdeveloped; if so, the balance will need redressing. Above all, the continuation of intellectual freedom requires a tolerance of the activities of the proponents of the one mode by the other.

Kuhn had made a similar point in his essay "The Essential Tension: Tradition and Innovation in Scientific Research." He even used the phrase "two modes of thought" to describe how both "convergent" and "divergent" thinking are ingredients of scientific progress. But Kuhn had not recommended the reciprocal tolerance of methods and practices that Conant had in mind.[34] These empiricists were the tinkerers, artisans, and instrument makers that Conant found missing in *Structure*, and these theorists were scientists drawn to grand hypotheses that all too often seemed to captivate the mind.

Conant's argument drew on *The Citadel of Learning* as well as some of the points he raised in his long, critical letter about paradigms. Conant admitted, for example, that according to Kuhn "scientific revolutions involve such a drastic reorientation of a scientist's ways of thinking that the idea of a continuous process over long periods of time can hardly be maintained."

But once "the advances in the practical arts" were added to the picture, science's history remained progressive and Conant's imaginary test for scientific progress delivered the result he always expected:

> One can scarcely imagine . . . a metalmaker two centuries ago refusing to admit, if brought back to life today, that our methods of making metals were superior to his.

That would not be case, he insisted, if it were painters, sculptors, or poets brought back to life. Conant rescued his ideal of progress along with his view that the sciences and the arts were fundamentally different in this respect.

Progress in science could occur during revolutions, Conant wrote. But there was also routine progress as these plural sensibilities and methods combined, clashed, and sparked new ideas. Modern science, he explained, was rooted in a "merging of two traditions" and the creative tensions that arose. One tradition

> derives from Euclid and Archimedes and was carried forward by Stevin and Pascal. The other comes from the generations of unnamed artisans who advanced the practical arts by trial-and-error experimentation. In Galileo's work we see reflections of both traditions, as we do also in Newton's writings and those of the founders of the new chemistry in the late eighteenth century.

Even Darwin, whom Kuhn invoked at the end of *Structure* to support his new idea of progress, became an illustration of progress as Conant understood it—not the replacement of one paradigm by another, but the ongoing interaction of multiple conceptual schemes and methods. Darwin's great breakthrough, that is, came when years of patient empirical-inductive research spent classifying the specimens he collected were leavened by Malthus's abstract, theoretical-deductive view of population dynamics.[35]

As for the worldviews that Conant disliked in *Structure*, Conant addressed them in his final chapter, "Scholars, Scientists and Philosophers." Kuhn believed that normal science had no use for philosophical analysis, but Conant invited philosophy back into the normal, everyday process of science as a hedge against the dangers of monolithic worldviews or *Weltanschauungen*. Of course, Conant admitted, "few, if any, people can carry on their lives without accepting some theoretical framework into which common-sense narrow generalizations can be fitted," some "worldview" or other. But those among us "who demand an all-embracing answer to the deep problems of human life and the nature of the cosmos" court a special danger, Conant explained, as speculation and fantasy could captivate the mind in the guise of "high-sounding pseudo-philosophic ideas." Here, academic philosophy had an important and powerful role to play in modern education:

> Unless one has wrestled with problems of epistemology and ontology, it is easy to succumb to the temptation to accept at face value those high-sounding pseudo-philosophic ideas which are always current.[36]

During times of crisis and revolution, Kuhn said, philosophy helps to "weaken the grip of a tradition upon the mind." But traditions and particular worldviews should *never* have a tight grip on the human mind, as Conant saw it.

Conant's comparison of ideological worldviews to fantasy or magical thinking was not original. In the essay that made him one of the eras most compelling anticommunist writers, the Polish poet Czeslaw Milosz described the all-embracing theory and worldview of Stalinism as a "happiness pill" that magically transformed—and corrupted—the minds of those who swallowed it. Michael Polanyi, in his essay "The Magic of Marxism," similarly diagnosed the allure of dialectical materialism as a fantastic synthesis of scientific ambition, moral passion, and certitude packaged inside one, all-embracing worldview.[37] Conant praised "The Magic of Marxism" in *Two Modes of Thought* as another "brilliant essay"—one that perhaps balanced the excesses he found in *Structure*. "Paradigm," he had earlier exclaimed to Kuhn, seemed like "a magic verbal word to explain everything," a word that "you have fallen in love with!"

Conant chose not to openly criticize *The Structure of Scientific Revolutions* in his *Two Modes of Thought* for wrapping our understanding of science within all-encompassing worldviews. But he did explain what was at stake should the nation's understanding of science, learning, and progress accept Kuhn's grandiose claims about paradigms and worldviews. For Polanyi's essay, Conant remarked, deftly exposed the dangerous path leading from alluring, intoxicating worldviews to the living nightmare of totalitarianism. "I need not point out the practical consequences of living in a totalitarian state with an *all-embracing* official dogma," Conant wrote. But he did so anyway as he returned to the observations he offered in *The Citadel of Learning*: Just a glance over the wall that now divided Berlin will show anyone why so many Germans left the East for the West, he noted. And "Even a superficial study of the books and magazine articles that have been published in the Soviet Zone in East Berlin in the last twenty years highlights the dangers inherent in the complete devotion to the theoretical-deductive mode of thought."[38]

This is why Conant urged Kuhn to abandon his paradigms, to unpack them and treat their separate parts differently. Denying that they could be separated, Kuhn doubled down and composed a new chapter to explain why paradigms existed prior to their separate parts. Conant probably regretted that, but it gave him time to articulate more precisely—if indirectly—why he so urgently disliked Kuhn's talk of paradigms and worldviews when he first read them. And it gave him an opening in *Two Modes of Thought* to fairly promote his student's brilliant new book, to indicate precisely why he liked it, and—indirectly—why he didn't. While he seemed sure that philosophers would have things of their own to say about *Structure*, Conant took the opportunity to tell educators and fellow citizens that there was another way to think about modern science that was very different from Kuhn's, one that finds progress not in a succession of insular paradigms but rather in intellectual pluralism and the political freedom it depends on.

15

Spies, Prisons, Mobs, Bandwagons, and Beasts

Cold war America was suspicious. Even the British, who carried on as best they could when Hitler's bombs rained down, raised questions in some leaders' minds. General Leslie Groves, Conant's military counterpart in the development of the atomic bomb, blamed the British for the atomic spy Klaus Fuchs. Fuchs was German-born, but he became a British citizen and moved undetected through England to Los Alamos. "I have always felt," Groves wrote in his autobiography, "that the basic reason for this was the attitude prevalent in all British officialdom that for an Englishman treason was impossible." In Fuchs they saw a fellow Englishman who could never have been seduced and controlled by a malicious, conspiratorial ideology.[1]

Congress blamed the British, too. In its 1951 report "Soviet Atomic Espionage," the Joint Committee on Atomic Energy concluded that Fuchs had been a Communist in Germany and that he moved in British Communist circles, as well. But Fuchs was nonetheless released from an internment camp on the Isle of Man because administrators figured that this Communist was "quite immersed in his academic studies and his work as a research worker, and was taking no active role in politics."[2]

Later in the 1950s and '60s, several British intelligence officers were revealed to be Soviet moles. Groves thought this revelation may have awoken the British from their unsuspecting slumber. But in fact England never shared the anticommunist paranoia that gripped the United States. Naturally, especially after the Russian Revolution, they kept watch on communism in Russia and elsewhere. But even in the depths of Britain's own "red scare" of the 1950s, as one historian has described it, officers in Whitehall worried more about Spain, France, and Italy than they did about England's cities and countryside.[3] Groves had it right the first time: the British simply did not believe they were susceptible to the powers of ideology that so terrified the American public and its leaders.

293

When McCarthy's assistants Roy Cohn and David Schine arrived in Europe, British headlines ridiculed their daily ruckus. "McCarthy 'Kids' Make America Gasp," read the pro-labor *Daily Mirror*. The conservative *Daily Sketch* blared, "Cohn-Schine Shindy Rocks United States."[4] British leaders saw McCarthy's crusade as "a sort of psychological reign of terror" that had taken root not in facts about subversion but rather American "fear of being accused of disloyalty," the *New York Times* explained. By no means would they permit this paranoia into Britain. When Cohn and Schine arrived in England to interview officers at the BBC about the integrity of its overseas broadcasts, the House of Commons declared that they threatened England's sovereignty and would not be permitted to "blacken the reputations of British subjects who perhaps during the Spanish Civil war contributed to the Loyalists or in their salad days perhaps flirted with Marxism."[5] For the British, the fact that one may have flirted with Marxism, or even embraced it in one's past, did not mean that brainwashing or treason was afoot.

The attitude was entrenched in British culture. The imperial anthem declares, "Britons never, never, never shall be slaves." And British philosophy had long been a bastion of hard-nosed empiricism and skepticism. "A wise man," the great Scottish philosopher David Hume wrote, "proportions his belief to the evidence"—*only* the evidence, that is. One must not get caught up in *hypotheses,* Isaac Newton declared, as he liberated physics from Cartesian surrender to the powers of reason. With only a few exceptions, Kuhn's book would be received skeptically by British philosophers, most of whom upheld Karl Popper's falsificationist conception of science.[6]

The Structure of Scientific Revolutions in America

In the United States, Kuhn's picture of the captive, "normal" scientific mind and its geopolitical significance went largely unnoticed. Partly, this is because of the book's famous first sentence:

> History, if viewed as a repository for more than anecdote or chronology, could produce a decisive transformation in the image of science by which we are now possessed.

What Kuhn originally wrote in his preliminary draft provides more perspective:

> The study of history has not been a usual source for the West's conception of science, and it might usefully become one. Viewed

as a repository for more than anecdote or chronology, history could produce a decisive transformation in the image of science by which we are now possessed.[7]

Reference to the West nods tacitly to the East and situates Kuhn's new theory within the divided geopolitical world analyzed by Conant and other anticommunists. Remarkably, Kuhn addressed his manuscript to only some of the scholars in it—those in the West whom he described as being possessed by a mistaken, unhistorical picture of science. Did that mean Soviet philosophy of science required no such transformation; that it was in some ways more correct? *Structure* and Kuhn's essay "The Function of Dogma in Scientific Research" imply this may be so. But perhaps because of H. Bentley Glass's emphatic objections, this is not a point that Kuhn intended to press. After this first sentence, neither his original manuscript nor the published book returns to this geopolitical distinction.

What most critical reviewers found in *Structure* was not some implied sympathy with Soviet philosophy but instead the specter of relativism that portrayed science as merely one human enterprise among others, no closer to objective knowledge and truth than religion or superstition. Most dismissive was a short, anonymous posting in *Scientific American* that described *Structure* as an excursus on conceptual relativism. What the author presented as "startling new revelations about the nature of science," the reviewer wrote, were "at best wild exaggerations" about incommensurability and the malleability of scientific perception. In short order, the reviewer dismissed the book as a collection of "sound but familiar reflections on the nature of science; it is also much ado about very little."[8]

Dudley Shapere, a philosopher at the University of Chicago, was the kind of established, respected philosopher of science whom Kuhn hoped to engage (and whose reactions Conant pressed Kuhn to report). But Shapere was in some ways just as negative and dismissive as *Scientific American*—especially when it came to paradigms. He wrote that "paradigm" seemed too vague and flexible a concept to convey anything but the illusion of understanding. The word simply covers too much ground, he complained, with the result that "anything that allows science to accomplish anything" has some relationship or other to the reigning paradigm.[9]

Things only got worse in "The Priority of Paradigms," the chapter Kuhn added after digesting Conant's criticisms. Here, Kuhn explained that a paradigm is prior to other parts of the scientific enterprise. It is not a body of rules or beliefs that can be specified, but instead a practical "achievement" from which particular methods, presuppositions, and scientific laws are later abstracted. If so, Shapere reasoned, it becomes "difficult to see what is gained by appealing to the notion of a paradigm."[10] One might as well attribute the achievements of human knowledge to a distant and advanced race of mind-controlling space aliens, or to some supernatural deity, or any postulated agency that is both difficult to specify or define and, at the same time, is presumed to govern and guide the conduct of science.

For all his hope to refute logical empiricism's Spartan, antimetaphysical approach to understanding science, Kuhn seemed to have indulged in a vague, but all-controlling metaphysics of paradigms that eluded precise specification yet somehow controlled the history of science. Shapere had no way of knowing that paradigms had descended from Kuhn's interest in "unconscious predispositions" and "theories-as-ideologies." But he too chalked up this wrong turn to a kind of enthusiasm on Kuhn's part, as if he were "carried away by the logic of his notion of a paradigm." Shapere then got carried away in his review, writing,

> For his view is made to appear convincing only by inflating the definition of "paradigm" until that term becomes so vague and ambiguous that it cannot easily be withheld, so general that it cannot easily be applied, so mysterious that it cannot help explain, and so misleading that it is a positive hindrance to the understanding of some central aspects of science; and then, finally, these excesses must be counterbalanced by qualifications that simply contradict them.

Shapere granted that there were open, unsolved questions in philosophy and history of science, but that did not "compel us to adopt a *mystique* regarding a single paradigm" that seemed to explain everything.[11] Somehow, Shapere sensed, it had compelled Kuhn.

The Invisibility—and Ubiquity—of Mind Control

What unites these criticisms with those of Conant and Norwood Russell Hanson is a shared perception that Kuhn's new book was asserting too much and reaching too far. It was all "far too grandiose," as Conant put it—or "purple," as John Earman and Kuhn himself later agreed. But none of these critics appeared to suspect that this *too much* was inspired by the dynamics of ideology and political mind control. One possible reason for this is the way Kuhn toned down his central interest in ideology that he first described to Charles Morris in 1953. Besides the depoliticized first sentence of the monograph, the once-central term *ideology* was no longer prominent. Set in context, Kuhn's interest in *revolutions* might have pointed these critics to the ongoing cold war and the struggle for men's minds as an inspiration, but the relevant contexts were changing and multiplying as academic life became less political and more specialized during and after the McCarthy years. Despite Kuhn's comparison of them in *Structure*, political revolutions and scientific revolutions became two different kinds of things in postwar scholarship.[12]

At the same time that some markers were missing, others were perhaps too ordinary and ubiquitous to be noticed. Kuhn's suggestion that the mind of any modern scientist was deeply impressed by education and the sociological pressures of professional life—so much so that creativity and imagination were reined in and perceptions themselves were constrained by paradigmatic beliefs—was no departure from the conventional wisdom on which popular anticommunism rested. The imperative to purge schools of Communist faculty, for example, reflected a universal belief, shared by progressive educators and their reactionary critics, that students are naturally susceptible to ideological manipulation and indoctrination into specific ways of life. In this regard, academics and educators in the United States might well have been the last readers to notice the cold war inspiration within Kuhn's new theory of science.

Conant's case is instructive, for he recognized and objected to the geopolitical significance and possible effects of Kuhn's new theory of science. But he never doubted the powers of education to shape and influence the human mind. It was partly because Conant's future citizens of America—like Kuhn's future scientists of America—were sure to be mentally shaped and manipulated by their educations that he registered his initial objections to paradigms and worldviews, just as Glass had so objected to dogmas in science. If education is a kind of mind control, these two cold war educators believed, it must be harnessed to support liberalism and open minds, not restrictive worldviews and closed minds.

Conant was also familiar with then-popular theories of mind control and high-level research in brainwashing. His Committee on the Present Danger formed a subcommittee to look into the emerging science of "psychological warfare." Hopes that Communists in the East could be manipulated to rise up and overthrow their totalitarian overlords excited some cold warriors in Washington, but Conant, then stationed in Germany, expressed doubts—

not about the power of ideas to control the mind and behavior, but about the competence of the "P.W. Boys" in Eisenhower's State Department to execute their plans effectively. As he put it in one cable to Washington, psychological manipulations intended to incite uprisings behind the Iron Curtain could instead backfire and cause "the fetters of Communism [to be] tightened again and Commie propagandists to fasten responsibility upon USA."[13] Even within the Conant household, these concerns were taken for granted. Months before *Structure* was published, Conant's wife, Patty (to whom Kuhn unfailingly sent his regards in his correspondence) wrote an article for *Modern Age* magazine about "the cold war of the mind" raging behind the Iron Curtain and "the harsh ideological straight-jacket in which 17,000,000 Germans have to live."[14]

The View from England

A month after Shapere's review was published, the *British Journal for the Philosophy of Science* echoed some of his observations. "One's first impression" of the book, noted Harry Stopes-Roe, was the "enthusiasm and vitality" of Kuhn's prose. Like Conant, Stopes-Roe found that many passages were clearest and genuinely informative if "paradigm" were removed and substituted with "basic theory." This would remove not only the word's vagueness but also the dubious cognitive functions Kuhn assigned to his paradigms. Was it true that paradigms were necessary for "anomaly and crisis" to occur? And was it necessary to suppose that a paradigm will "suppress their recognition" until an alternative paradigm emerges to engage in revolutionary battle?[15]

All this is stimulating, Stopes-Roe admitted. But it seemed doubtful and, in a way, redundant to the wisdom of common sense:

> Who would change a theory unless there was something wrong with it? Can one recognise any result as odd unless one has some framework of expectations into which to fit it?

Reacting as it did to Conant's restless liberalism, in which science always moves ahead to new and different theories and experiments, Kuhn's paradigms perhaps reasonably applied some brakes to the scientific process. But Stopes-Roe found these brakes unnecessary:

> The facts are, the inertia of mind, inability to work except with a fairly restrictive framework, inability to see the inadequacy of

one's own well-established opinions, all these are general qualities of mankind. It is misleading to explain the source of these limitations by reference to shared paradigms. Kuhn, however, accepting this explanation, seeks to explain the regulative force of paradigms; and he finds this in the scientist's education.

Stopes-Roe looked carefully at the narrow scientific education Kuhn described, as well as his claims about the Orwellian rewriting of science's history. He then quietly concluded, "It seems to me that the author's good ideas get rather submerged."[16]

If Stopes-Roe did not suspect that Kuhn's otherwise good ideas were saturated by America's preoccupations with loyalty and ideology, other British philosophers did. Kuhn probably met Karl Popper in 1950, when Popper lectured at Harvard. In some ways, Popper's career was similar. *Structure* was inspired by Kuhn's encounter with Aristotle, but Popper's signature idea was sparked by an encounter with Einstein. Attending a lecture in Vienna, Popper was struck by the great physicist's remark that he'd consider his new space- and mind-bending theory of gravity false were the path of starlight passing by the sun not slightly bent by the sun's gravitational warping of space-time. (In 1919, the British physicist Arthur Eddington observed an eclipse from an island off the west coast of Africa, detected the "bending" effect, and helped to make Einstein a household name around the world.) Popper was dazzled by the daring new physics Einstein had created. But he later concluded that he had been most impressed by Einstein's methods, in particular by

> Einstein's own clear statement that he should regard his theory as untenable if it should fail in certain tests. . . . Here was an attitude utterly different from the dogmatic attitude of Marx, Freud, Adler and even more so that of their followers. Einstein was looking for crucial experiments whose agreement with his predictions would by no means establish his theory; while a disagreement, as he was the first to stress, would show his theory to be untenable.

In Popper's eyes, Einstein showed that real scientists were the very opposite of dogmatic, true believers like those who defended the politics of Marx or the psychological theories of Freud or Alfred Adler. Unlike dogmatists, Popper's scientists use all the resources at their disposal, from logic to the laboratory, not to defend and promote ideas but to criticize them, refine them, and—if necessary—abandon them.[17]

Popper and Kuhn did not disagree about everything. Alongside Conant, they knew that science was more than fact collecting and they were both suspicious of logical empiricism, albeit for different reasons. Popper argued that his fellow Viennese philosophers (in their Vienna Circle) failed to appreciate the logical asymmetry inside the so-called problem of induction: one can never conclude by fact collecting that "all swans are white" for there may be green swans somewhere no fact collectors have ever been. But one can be sure that finding one green swan anywhere will falsify the proposition. Such is the power of falsification to move science forward past ideas that are false and quite possibly lodged in the minds of uncritical believers.

Yet Popper remained a formalist for whom logic could illuminate the workings of science. Kuhn rejected that approach along with Popper's vision of philosophical criticism as a regular, "normal" feature of science and political life. On this view, philosophy and critical thinking are not to be confined within the ivory tower and surely not taken away from day-to-day activities of scientists, reserved only for times of crisis and revolution. Especially given the dangerous ideologies circulating in the modern world, Popper believed that philosophy of science should properly help to liberate humanity—to illuminate and dispel dark conspiracies, unseen metaphysical forces, and misunderstandings that drive zealotry, war, and suffering.

Writing in Christchurch, New Zealand, in the late 1930s, Popper published influential liberal critiques of Marxism, totalitarianism, and scientific planning. In his book *The Open Society and Its Enemies* and his long essay *The Poverty of Historicism,* he attacked Marxist "historicists" who proposed to manipulate the springs and levers of history to create a new, better world. These were the dangerous descendants of Plato, Popper argued, who pined for a world regimented and controlled by philosopher kings. An understanding of how real, scientific knowledge is acquired punctured this mythology, Popper said. Science thrives only in an unregimented, open society, in which beliefs and theories are always subject to criticism and falsifying tests. Plato and his utopian descendants were simply but tragically wrong:

> Our dream of heaven cannot be realized on earth. Once we begin to rely upon our reason, to use our powers of criticism, once we feel the call of personal responsibilities, and with it, the responsibility of helping to advance knowledge, we cannot return to a state of implicit submission to tribal magic. For those who have eaten of the tree of knowledge, paradise is lost.

Blinded by their dogmatism, historicists can only lead us only to political disaster, Popper argued. The more we dogmatically favor certain theories, doctrines, and worldviews, "the more surely we arrive at the Inquisition, at the Secret Police, and at romanticized gangsterism." With Hitler's National Socialism in mind, he argued, historicism could lead us only back to "the beasts."[18]

"Ideology Covered Up As History"

After the war, Popper moved to England, where he became famous as a defender of intellectual liberty, not unlike James B. Conant in the United States. There, Popper took on a brilliant and precocious student, Paul Feyerabend, who also came from Austria and whose interests lay in physics, philosophy of science, and politics. In the early 1950s, while Kuhn was articulating his understanding of "theory as ideology," Feyerabend studied with Popper across the Atlantic where he developed his own critique of logical empiricism and a visceral aversion to dogmatism and ideology.

Feyerabend arrived to teach at Berkeley in 1958, two years after Kuhn. They were colleagues in the philosophy department but did not have much contact. As soon as Feyerabend arrived, Kuhn left to spend a year in Palo Alto at the Center for Advanced Study in the Behavioral Sciences (where, he promised Charles Morris, he would finally write his monograph).[19] Along with those copies he sent to Nagel, Conant, Morris, and others, Kuhn sent the first draft of *Structure* to Feyerabend, who was likely the first Popperian to comment on it. He may have wished he hadn't, for Feyerabend was more negative, more blunt, and more aggressive in his criticisms than Conant. For the next several months, the surviving correspondence implies, Kuhn and Feyerabend talked in person and exchanged letters about their disagreements.[20]

Where Conant abhorred "paradigms," Feyerabend abhorred "normal science." Good scientists could not be dogmatic, uncritical, and unquestioning of their paradigms, he felt; and Kuhn could not be correct that normal science was a surefire path to scientific change and revolution. Feyerabend offered many examples of fruitful, important ideas in the history of physics emerging not after a paradigm broke down, but when scientists engaged in active and critical debate with each others' beliefs, when multiple and conflicting paradigms are on science's table at the same time. This was hidden, however, behind Kuhn's monolithic paradigms that obscured historical debate among advocates of different theories. Your paradigm of "classical physics," Feyerabend complained, is "in fact a bundle of alternatives."[21]

As much as he disliked the book, Feyerabend sensed that it was going to be popular. It would work as propaganda, he charged, to fool readers into accepting its claims: "What you are writing is not just history," Feyerabend told Kuhn flatly. "It is ideology covered up as history." He did not complain that Kuhn's book had its own point of view; a book that had none, Feyerabend acknowledged, "would be the most drab and uninteresting affair imaginable." He objected instead to Kuhn's claim that he had discovered an objective historical pattern—normal science, anomaly, crisis, revolution, and normal science—through his research. Instead, Feyerabend was sure, Kuhn had selected historical facts, and described them in the requisite ways, in order to present this pattern as the essence and mark of scientific progress. "You use a kind of double-talk where every assertion may be read in two ways, as the report of a historical fact" or, very differently, "as a methodological rule" for good, effective research.

Feyerabend was right that *Structure* is less a product of historical research than Kuhn's theorizing and wrestling with his drafts. Most of the historical examples it presents were first presented by others, such as Annaliese Meier, Marshall Clagett, Norwood Russell Hanson, Alexander Koyré, and Conant himself (with the help of I. Bernard Cohen) in *On Understanding Science*. Kuhn found his revolutionary image of science not by exploring dusty archives or undiscovered manuscripts but by looking at known historical episodes in light of his psychological, sociological, and ideological interests. Yet it was "history," Kuhn's book announced to its readers, and not its author's creative intellect, that had produced these insights. Kuhn himself criticized science textbooks for giving a misleading picture of science's history, so it may have seemed ironic or puzzling to Feyerabend that Kuhn's new manuscript was doing the same thing. "It is this bewitching way of representation to which I object the most," Feyerabend explained—"the fact that you take your readers in rather than trying to persuade them."[22]

Feyerabend read the draft closely, often commenting line by line to expose the ways Kuhn seemed to bewitch his readers. His treatment of Kuhn's central distinction between pre-paradigmatic science and the normal science that marks the start of any science's history illustrates the deluge of frustrated criticism Kuhn received from Feyerabend's long, detailed letters:

> Here it is very obvious that your "history" is ideologically infected double talk. What is the criterion according to which you distinguish between "history" and "prehistory"? . . . when differentiat-

ing between the "history" and the "prehistory" you must, at least implicitly, refer to some ideal of scientific procedure by which you measure progress (or regress). If I understand you correctly, the ideal is "normal science" or pattern guided science (science guided by a single pattern which everybody accepts with the sole exception of some people you would perhaps be inclined to call cranks). But you never state clearly that this is your ideal. Quite the contrary—you insinuate that this is what historical research teaches you.

A few pages later, he claimed "you are a mystic, an irrationalist."

And by this I mean that you not only hold certain beliefs (conservative character of normal science) but that you are not prepared to let these beliefs speak for themselves; you rather present them in a matter which suggests they are facts and thereby force people to swallow them without criticising them. What are you afraid of? Are you afraid that people will oppose at once when your beliefs are presented to them in their proper form, viz. as demands as to how science ought to be run? When discussing such demands you are very careful and give the appearance of a critical person ("probably" etc. etc.) But this is just a trick (of which you yourself may not be aware) For you present the very same demands a little later as facts ("what scientists never do" etc. etc.) and then with the assurance of the historian who knows. Again, it is this kind of double-talk to which I object most. You really are like a witch doctor.

Feyerabend sometimes apologized for tirades that may have seemed unscholarly or "a little violent" and he asked Kuhn to "please take this in the proper spirit." "My violence is really the result of an effort to say things as clearly as possible."[23]

As for paradigms, Feyerabend's remarks illustrate the very different ways that these two philosophers situated scientific methodology in this age of mind control and propaganda. Paradigms are necessary to understand science, Kuhn explained, because they are "the source of coherence for normal research traditions." Rules could not perform this function, Kuhn reasoned, so there had to exist some other source of positive and collective guidance—something to "define the puzzles" for scientists to solve, "tell them what the universe contains," and to "restrict scientific attention" so that the revolutionary process stays on track. This maneuver mystified Feyerabend, for he took rules to be essentially prohibitive and restrictive; they tell scientists only what they may *not* do. Of course, Feyerabend noted, no amount of study of the rules of chess will allow the course of actual games to be predicted or particular strategies or methods to be defined, for these depend on the choices made by the individual players. By this analogy, then, "it does not follow at all from this that science is not played according to certain rules," and it did not follow that understanding science required the addition of paradigms, over and above the usual rules of science and common sense.[24] Feyerabend saw professional scientists as free agents who had no need for the dictates of paradigms. Their job was to think creatively and critically, to produce new ideas, theories, and experiments, and to carefully sort the good from the bad.

These free scientists, or course, were also free citizens, which helps explain why Feyerabend became so agitated over *Structure* and why, as he put it, he could not "leave you in peace." Like Conant, Feyerabend believed that whether and how the public understood modern science could be historically decisive. "This conviction is the motor which makes me go on with the discussion," he explained, because "some of your ideas, when published, may have a disadvantageous influence." Feyerabend had Kuhn's students in his own classes and he saw the immense appeal of Kuhn's ideological norms cloaked behind a veil of factual, historical reporting. "You say you are a historian who describes what happens," Feyerabend wrote. "You will not describe it differently, for this would mean being untruthful." But this was "a very dangerous attitude," Feyerabend insisted,

> for you are not only a defender of truth; first of all you are a member of humanity and have obligations towards humanity; these obligations come first; these obligations do have the character of certainty; but whatever truths you may unearth—they have the character of hypotheses and therefore ought to be treated with great care when confronted with the certain obligations you possess.[25]

Much of what Kuhn was saying about the nature of science was "simply mistaken," Feyerabend insisted. But his objections—and those of other readers, he worried—were sure to be blunted by *Structure*'s beguiling mixture of facts and ideological values about science and its place in modern society.

With far less bluster and agitation, Shapere, Hanson, and Conant had offered similar criticisms—deep, potentially fatal criticisms—of Kuhn's manuscript. And like them Feyerabend joined those criticisms to genuine praise: despite my sometimes "violent" reactions, Feyerabend reassured Kuhn, "you ought to know that despite all this I find your essay very important, very stimulating."[26] This pattern may have encouraged Kuhn to stick to his paradigmatic guns, as if these reported problems and complaints belonged more to the idiosyncratic reactions of his colleagues than to the manuscript itself. If a revolution in the understanding of science was afoot, after all, one would of course expect dire, urgent complaints like those Feyerabend repeated in his long letters. Decades later, Kuhn confessed as much when he described his debates with Feyerabend as strange and disconcerting:

> I think he liked it in one sense, but he was terribly upset by this whole business of dogma, rigidity, which of course is exactly counter to what he believed himself. And I couldn't get him to talk about anything except that. And I tried, and I tried: we would have lunch together, or something—he'd always come back to it. I got more and more frustrated and I finally just stopped trying.

He stopped trying because the problems were Feyerabend's own: "The quasi-sociological elements of my approach were overwhelmed by his desires for society in the ideal," Kuhn explained, so they never really were able to understand each other; they "really never made contact."[27]

Another Trip to England

After *Structure* was published, Kuhn returned to England in 1965 for a conference organized by the Hungarian philosopher Imre Lakatos. Lakatos greatly admired Popper, his colleague at the London School of Economics, whom he invited in order to stage an encounter: the leading philosopher of science in Britain would confront a representative of the new philosophical relativism from America. Having aspired to be a politician in his youth (and having fled Hungary on foot during the short-lived Hungarian Uprising), Lakatos was keenly aware of politics and rejected the subterranean currents he saw in *The Structure of Scientific Revolutions*—in particular, the logic of social mobs united by shared beliefs. If science goes through revolutions as Kuhn described them, Lakatos said in his lecture, then science ceases to be a rational, intellectual enterprise. For when competing paradigms do revolutionary battle,

[t]here are no rational standards for their comparison. Each paradigm contains its own standards. The crisis sweeps away not only the old theories and rules but also the standards which made us respect them. The new paradigm brings us a totally new rationality. There are no super-paradigmatic standards. The change is a bandwagon effect. Thus *in Kuhn's view scientific revolution is irrational, a matter for mob psychology.*[28]

Feyerabend attended the conference and seconded Lakatos's proposal. With characteristic flair and drama (Feyerabend was an amateur opera singer) he suggested that Kuhn's picture of science evoked a different kind of mob: "Every statement which Kuhn makes about normal science remains true when we replace 'normal science' by 'organized crime.'" After all, he added, being successful as in organized crime certainly involves puzzle-solving.[29]

Nearly every speaker rejected Kuhn's theory of normal science and its role in preparing for scientific revolutions. But before they weighed in, Kuhn as the guest of honor presented his brief against Popper. His paper "Logic of Discovery or Psychology of Research?" pitted Popper's logical insights about science against Kuhn's own psychological and, in fact, sociological approach. There was much in Popper's philosophy of science to agree with, Kuhn said, but he did not shy away from charging Popper with making a basic, fundamental mistake: Popper ignored normal science and paid attention only to revolutionary science.

Of course, during times of crisis and revolutionary science, Kuhn admitted, scientists treat theories critically and subject them to tests. Their faith in one paradigm has been shaken, so they scrutinize it before looking for a new one. But Popper mistook this kind of behavior for "the entire scientific enterprise" in a way that took him off track. His celebrated criterion of falsifiability missed its target and suggested mistakenly that only revolutionary, extraordinary science was truly scientific. "If a demarcation criterion exists," Kuhn wrote, "it may lie just in that part of science which Sir Karl ignores." What makes a field scientific may not be the critical, skeptical attitude of the falsificationist, but rather the dogmatism of the normal scientist. Indeed, Kuhn did "turn Sir Karl's view on its head," as he put it.[30]

Popper's Unlocked Prisons of the Mind

When Popper took the podium, he quickly righted himself and turned Kuhn's charge back against him. It was Kuhn, he said, who had misplaced his attention and built a spurious theory of science. Kuhn was simply mistaken to believe that "what he calls 'normal' science is normal." He had picked the wrong kind of scientist to reveal the engines of science's progress—as if he had studied children finger painting to understand something about Rembrandt. Of course something like normal science exists, Popper admitted. But it is the science of the second- or third-rate scientist;

> [I]t is the activity of the non-revolutionary, or more precisely, the not-too-critical professional: of the science student who accepts the ruling dogma of the day; who does not wish to challenge it; and who accepts a new revolutionary theory only if almost everybody else is ready to accept it—if it becomes fashionable by a kind of bandwagon effect.

The specter of the mob, the partisan, uncritical crowd, and the power of indoctrination emerged once again at the conference. This kind of scientist has been "badly taught," Popper said. "He has been taught in a dogmatic spirit: he is a victim of indoctrination."[31]

For the author of *The Open Society and Its Enemies* these issues were crucial. Rightly understood, science showed how the human mind can rise above the emotional reflexes of politics, the big lies used to motivate mobs or nations, or the dangerous promises of historicists and planners to chart and control the future of science and society. But instead of using scientific methodology to help liberate modern life from dogmatism, Kuhn embraced dogmatism as the feature of science that makes scientific revolutions possible. Like his student Feyerabend, Popper saw this as a tragic mistake. Naturally, there had to be a logical framework, some "edifice, an organized structure of science which provides the scientist with a generally accepted problem-situation into which his own work can be fitted"—something like a paradigm, Popper conceded. But to suppose that paradigms captivate the scientific mind and discourage it from critical, reflective analysis of the framework itself seemed both dangerous and mistaken: "If we try, we can break out of our framework at any time. Admittedly we shall find ourselves again in a framework, but it will be a better and roomier one; and we can at any moment break out of it again." It is therefore always possible to analyze, compare, and criticize different theories and research programs, he insisted, and scientists routinely do just that. "It is just a dogma—a dangerous dogma—that the different frameworks are like mutually untranslatable languages" that belong to separate, disconnected worlds. This "myth of the framework," Popper called it, was "the central bulwark of irrationalism" in the postwar world. And he found it at the heart of Kuhn's view of science.[32]

Defending Popper

Popper's supporters were quick to pile on and poke more holes in *Structure*, in part by examining and attacking its political foundations. Kuhn had encountered Stephen Toulmin four years before when Kuhn read "The Function of Dogma in Scientific Research" in Oxford. On this occasion Toulmin attacked Kuhn's comparison of scientific and political revolutions as outmoded, stereotyped scholarship. "At one time," Toulmin explained,

> historians faced with political changes of a peculiarly drastic variety were quite ready to say, '. . . and then there was a revolution,' and leave it at that: the implication was that, in the case of such drastic changes, no explanation could be given of the rational kind we rightly demand in the case of normal political developments.

But political revolutions are more complicated, drawn out, and never exhibit an "absolute and outright breach of continuity." Toulmin pointed to the French, American, and Russian revolutions and he cited Crane Brinton's *Anatomy of Revolution* to say that the dynamics of these revolutions were different. Matters of degree—not kind—separated "normal" from "revolutionary" change. So it was for Kuhn's normal and revolutionary science, Toulmin argued. This central distinction of Kuhn's account did not hold up.[33]

Neither did Kuhn's claims about "conversion experiences," as Toulmin understood them. If these occurred, then why is there no historical record of physicists reporting them as Newton's physics was superseded by Einstein's?

> If the complete breakdown in scientific communication which Kuhn treats as the essential characteristic of a scientific revolution had in fact been manifest during this period, one should be able to document it from the experience of the men in question. What do we find? If the conceptual change involved in the transition was as deep as Kuhn claims, these physicists at any rate appeared curiously unaware of that fact. On the contrary, many . . . were able to say, after the event, why they had changed their own personal position from a classical to a relativistic one.

So far as Toulmin could tell, Kuhn's enthusiasm for his model of science along with a measure of "rhetorical exaggeration" had backfired to obscure—and not reveal—historical realities about science.[34]

John Watkins, who had also studied with Popper, said he had read the early draft of *Structure* and was immediately struck by its political overtones; by the contrast between

> his view of the scientific community as an essentially closed society, intermittently shaken by collective nervous breakdowns followed by restored mental unison, and Popper's view that this scientific community ought to be, and to a considerable degree actually is, an open society in which no theory, however dominant and successful, no "paradigm" to use Kuhn's term, is ever sacred.

Watkins twice noted how the paradigms Kuhn described seemed to have "a sway over men's minds." And he described Kuhn's theory as "an epidemiological account" of scientific change. It shows "how, after a new paradigm has infected a few carriers, the epidemic is liable to spread among the scientific community."[35]

Watkins did not dwell on the image, but he returned again and again to reject the view that scientists' minds work properly in the tight grip of scientific paradigms. He wrote, for example, that each normal scientist works "under some mysterious compulsion to preserve the current theories of science against awkward results." Perhaps, he suggested, the inspiration behind Kuhn's view was religion. (Kuhn had, after all, stopped just shy of comparing scientific education to that of "orthodox theology.") "My suggestion," Watkins offered "is that Kuhn sees the scientific community on the analogy of a religious community and sees science as the scientist's religion." Whether in terms of anticommunist hysteria, the nation's traditional piety, or both, Watkins believed that *The Structure of Scientific Revolutions* was in some ways a reflection of postwar America; that Kuhn had Americanized the history and philosophy of science.[36]

The Rosenbergs and the Mob

Watkins borrowed this "infection" metaphor from the rhetoric of anticommunism that Roy Cohn and David Schine had tried to bring to England and which anticommunists like J. Edgar Hoover and Sidney Hook made a fixture in the United States. As the FBI director once put it, it is "a way of life—an evil and malignant way of life. It reveals a condition akin to disease that spreads like an epidemic." George Kennan's boss Dean Acheson once urged American financial support to Greek anticommunists because "like apples in a barrel, infected by one rotten one, the corruption of Greece would infect Iran and all to the east. It would also carry infection to Africa through Asia Minor and Egypt and to Europe."[37] In 1948, Conant himself spoke of "the virus of Soviet Philosophy" one week after being awarded the rank of Honorary Commander of the Most Excellent Order of the British Empire. The ambassador who bestowed the honor spoke in Lowell Hall of communism as a "new and loathsome virus, bent upon the destruction of the freedom and dignity of man."[38]

Watkins and other Europeans were familiar with the metaphor. But they were nonetheless amazed at the measures Americans would take to prevent this supposed virus from spreading. Before the headline antics of McCarthy, and of Cohn and Schine's circus-like tour of European libraries, there had been the case of Julius and Ethel Rosenberg, which had also crucially involved Roy Cohn. In the mid-1940s, the Rosenbergs were unremarkable, American Jews struggling to make a living in Brooklyn. Their membership in the Communist Party and sympathies for Stalin's Soviet Union were no longer as acceptable as they had been in the New York of the 1930s, but they were not unique. Since about 1942, however, Julius had been in contact with the Russian embassy in New York. When Ethel's brother, David Greenglass, was assigned to work as a machinist at Los Alamos, Julius arranged for Greenglass to provide information and sketches made from memory about the atomic bomb being developed.

Greenglass was motivated by a sense of fairness, his belief that America's ally against Hitler deserved to know something about America's weapons research. But in 1950, when FBI investigators followed a thread from the atomic spy Klaus Fuchs to Greenglass (the two had contacts with the same Soviet agent), the Soviet Union was no longer an ally. It was the source of the conspiracy to subvert democracy from within, as Sidney Hook described it, and the source of the "present danger" that Conant's committee addressed. Consequently, Greenglass and the Rosenbergs were in deep trouble. To protect his wife and children, evidently, Greenglass told the FBI everything they wished to know about Julius and Ethel. Both were convicted of espionage and eventually executed.[39]

Between their conviction in 1951 and their execution two years later, the Rosenbergs stirred controversies that in some respects have never abated. To this day, the extent and nature of Ethel's involvement remains in dispute.

She was believed merely to have typed one or a few documents that her brother wrote by hand. But that was enough for Cohn the assistant prosecutor to pursue her as doggedly as he pursued Julius. Her revenge came decades later in the hands of playwright Tony Kushner. In his play *Angels in America*, Ethel's ghost visits Cohn when he is hospitalized and delirious from his late-stage HIV illness. "I came to forgive," she explains. But face to face with her nemesis on his deathbed she confesses that "all I can do is take pleasure in your misery." Ethel's—Kushner's—enduring anger drew on reports that Cohn and presiding judge Irving Kaufman had illegal *ex parte* contact outside of court and that Cohn persuaded Kaufman to see Ethel as "the mastermind of this conspiracy." Cohn's undue influence as well as the reigning metaphor of communism as a contagion encouraged Kaufman's harsh sentence. Prosecutor Irving Saypol (who first recruited Cohn for the case) said in his closing argument that the root of the conspiracy was both Julius and Ethel, who "infected Ruth and David Greenglass with the poison of communist ideology."[40]

As their execution neared, public declarations and demonstrations—both for and against the sentence—were held around the world. (It was by signing a petition for clemency that Rudolf Carnap came to Hoover's attention as a philosopher worth investigating for possible communist connections.) After court appeals to reduce the sentence were finally exhausted, and after Eisenhower refused to issue an executive order, the Rosenbergs were electrocuted at Sing Sing Prison in Ossining, New York, on June 19, 1953. The next day, the French philosopher Jean-Paul Sartre attacked the United States in the newspaper *Libération*. Nodding to the Jim Crow south, he denounced the treatment of the Rosenbergs as "a legal lynching which covers a whole people with blood." "From every side," Sartre exclaimed, "people cried out: 'Be careful! In judging them you judge yourselves; you are deciding if you are men or beasts.'"[41]

Most Americans could easily dismiss Sartre. He was a foreigner, openly sympathetic with communism, and a philosopher in the subversive, atheistic mold of Socrates. But his outrage marks the international reputation McCarthy's America was acquiring. "Your country is sick with fear," he wrote, "You're afraid of everything: the Russians, the Chinese, the Europeans. You're afraid of each other. You're afraid of the shadow of your own bomb." Fear had seemed to corrode American intelligence and compromise its morality. "Do you remember Nuremberg and your theory of collective responsibility?" he asked, nodding to American leadership in the prosecution of leading Nazis. "Well, today you are the ones it ought to be applied to."

The British historian David Caute agreed in the 1970s that the nation became carried away by fear. In his monograph *The Great Fear* he set out to measure the size and contours of communism as a political force in the United States. "What was this Leviathan, the American Communist Party," he asked,

> against whose revolutionary and subversive designs it was evidently necessary to array not only the law, the FBI and the Congressional inquisition, but also a massive propaganda campaign?

It was merely "a flea on the dog's back, no more," Caute argued. Even at its peak in the late 1940s, the Party and its roughly sixty to eighty thousand members amounted to a sliver, a mere five-thousandth of one percent of the population with negligible electoral power.[42] Caute's metaphor was ironic, however, for one flea can indeed infect an entire population and that was one basis for the struggle for men's minds. Communism and its cousin totalitarianism were seen as contagious ideologies that imprisoned the mind with its rigid framework of meanings and perceptions—no matter what British or French philosophers said to the contrary.

The Problem of "Atom Spies"

The Rosenbergs' sons, Robert and Michael, five and nine years old, became orphans in 1953. But as Sartre would have predicted, there was no national introspection or hints of regret over that outcome. Fear continued to reign and make it a priority that "atom spies" never again threaten the nation's security. How had so many found their way into the inner laboratories of American military science? Besides Fuchs and Greenglass, there was Alan May, another British physicist who had passed uranium samples to the Russians in the early 1940s, and the Italian physicist Bruno Pontecorvo, who worked in Canada. Was there something about nuclear physics, about scientific research itself, or these individuals' psychologies that explained these lapses in national security?

Legislators in the early 1950s wanted answers and Congress's Joint Committee on Atomic Energy seemed to find some. On the one hand, it found that each spy had been recruited or indoctrinated as a young adult:

> The spies—during their formative years—had been pulled into a Communist apparatus which systematically destroyed their sense

of moral values and substituted the facile capacity for rationalization found in the code of totalitarian dictatorship.

On the other hand, some demonstrated an inflated sense of moral value and responsibility to mankind. Because Fuchs and May "were bachelors with few friends and scant interests outside science and communism," the committee reasoned that they may have reveled in secretly passing crucial, history-making information. Whether brainwashed into obedience or driven by narcissism, the committee struggled to understand how individuals like these could have blended into their scientific communities and remained undetected.[43]

On this question the committee turned to James Conant. Its chairman Senator Brien McMahon had known Conant and his views on the international control of atomic energy since the close of the war. Citing a long passage from Conant's new book *Science and Common Sense*, the committee explained how the demands of modern science made it possible for a traitor like Fuchs or Greenglass to work unseen. "Even an emotionally unstable person," Conant had written, can be professional, "exact and impartial in the laboratory."

Whether or not spies tend to be emotionally unstable, Conant probably did not have spies in mind when the wrote the passage in question. The quotation appeared verbatim years before in *On Understanding Science*—a book that appeared before Conant's cold warrior conversion, when he urged international cultural and scientific cooperation for peace and progress. He appealed to the strength and discipline of scientific communities to explain how an unstable person can nonetheless make important contributions to science. It was because of

> [t]he traditions he inherits, the instruments, the high degree of specialization, the crowd of witnesses that surrounds him, so to speak (if he publishes his results)—these all exert pressures that make impartiality on matters of his science almost automatic. Let him deviate from the rigorous role of impartial experimenter or observer at his peril; he knows all too well what a fool So-and-So made of himself by blindly sticking to a set of observations or a theory now clearly recognized as in error.

"But once he leaves the laboratory," the committee's quotation continued, a scientist can let down his disciplinary guard and perhaps "indulge his fancy all he pleases." "One would not be surprised, therefore, if, as regards matters beyond their professional competence, laboratory workers were a little less impartial and self-restrained than other men."[44]

Groves blamed the British over Fuchs. Now Congress, with Conant's writings in hand, sought to blame the rigors of modern science. These helped explain how dangerous individuals—with their "warped mentalities" and "lack of moral standards, combined with an overweening and childlike arrogance"—worked alongside their patriotic colleagues without being detected. These congressmen naturally assumed that individuality and idiosyncrasies, masked by the professional norms and pressures of modern science, would otherwise have been received as signs of potential disloyalty or treason. When he wrote *On Understanding Science*, however, Conant was making the opposite point as he basked in the postwar glow of victory and called on the nation—and Harvard's students specifically—to muster the civic courage to reject conformism and to tolerate, if not applaud, individual differences. Odd or unusual persons were good for science exactly because they bucked conformity and helped to keep dogmatism and collective complacency at bay. Were they likely to be a special danger to the nation or the world? Conant did not think so. "My own observations," he wrote at the end of the committee's long quotation, "lead me to conclude that as human beings scientific investigators are statistically distributed over the whole spectrum of human folly and wisdom much as other men."[45]

Because the potential for human wisdom and folly took so many different shapes and sizes, Conant was sure that any attempt to cull would-be subversives or discriminate against them—by imposing loyalty oaths, background checks, or other kinds of invasive investigations—would harm science and limit the nation's ability to make the most of its intellectual resources. Three years before his committee turned to Conant's book, Conant personally warned McMahon to avoid these measures. Congress must avoid "creating an atmosphere of distrust and suspicion in the scientific world, as I feel certain the loss to the country"—in the form of scientists repelled by the repressive atmosphere—"will far outweigh the possible hazards involved."[46] Though the Fuchs case and other developments would later heighten Conant's sense of alarm and imminent danger, his underlying philosophy of science consistently joined Popper's ideal of an "open," unregimented society. Science requires divergent attitudes and points of view to fuel debate, criticism, and to move knowledge forward.

The Structure of American Revolutions

Congress's Joint Committee had read Conant in the same way that Kuhn had. Instead of seeing scientific communities as microcosms of a tolerant nation with porous, open borders, the committee saw them as independent, isolated societies with internal sociological dynamics of their own—behind which dangerous subversives could effectively hide. The "discipline" of which Conant spoke perhaps fairly suggested a reading like this; but this discipline functioned to keep science's doors open to the wide spectrum of human talent and to manage the natural, centrifugal tendencies of the curious, brilliant, and competitive talents drawn to science.

With Kuhn, this professional discipline became stronger. He invested paradigms with a positive power over the mind, its creativity, and its perceptual mechanisms. A powerful discipline was necessary, Kuhn believed, lest the scientific community lose its sociological integrity and its collective intellectual focus—in which case future scientific revolutions would become impossible. This set Kuhn on his collision course with Popperian philosophy of science. His theory that the progressive, world-changing qualities of revolutions were the ironic fruit of normal science, that progress and innovation could not be understood apart from the paradigms that held "sway over men's minds"—as Watkins had put it—must have seemed both bound to fail and, from a British point of view, all too reminiscent of the nation's revolutionary self-confidence and expansionist ambitions.

Popper and his students lived and worked surrounded by the geopolitical realities that Conant himself had helped to create and sustain. He helped to invent the atomic weapons now stockpiled to their West and to their East. His Committee on the Present Danger helped to militarize Europe with American forces and personnel, and in the 1950s he was the face of American diplomacy and military power in Germany, where East and West struggled intensely for cultural and political control. From the American point of view, these were defensive measures responding realistically to threats of Soviet aggression and expansion. But from across the Atlantic they could readily be seen as aggressive and imperialistic. While the Marshall Plan aimed to rebuild European economies for commerce with the goods and services of the West, for example, the Congress for Cultural Freedom aimed to win the hearts and minds of European intellectuals, to indoctrinate them with liberalism and turn them away from communist philosophy.

In his indictment, Sartre denounced just this exercise cultural power. It overflowed across the Atlantic a dozen years before when Roy Cohn turned his wrath from Julius and Ethel Rosenberg to the books in America's European libraries. "Do you think we want to defend McCarthy's culture? McCarthy's freedom? McCarthy's justice?" he sputtered in anger—"That we'll make Europe a battleground to let this bloodstained idiot burn all the books?" "Don't kid yourselves. We'll never let the Rosenbergs' assassin lead the West."[47]

Kuhn's philosophical critics in England were more polite and less agitated. But they too reacted to his new theory of science as a cultural invader. They did not accept his claim that there was another revolution afoot in humanity's understanding of science, that most of what British and Austrian philosophers believed about science and epistemology was grossly misleading. With its crises and revolutionary conversions, and especially the dogmatic, devotional sensibilities of normal science, the new American theory of science that Kuhn defended seemed instead to say more about the enormous and now powerful nation that created it, its exceptional self-regard and postwar confidence, and its sometimes irrational, unpredictable fears and enthusiasms.

16

The Thomas Kuhn Experience

Thomas Kuhn had evidence in his hands that his life was about to change, but he didn't recognize it at the time. It was a letter he received in August 1962, just months before his monograph was to be published, and it was from the Leo Burnett Agency, the powerhouse advertising firm in Chicago. Puzzled, Kuhn opened it to learn that an executive at the firm admired an essay Kuhn had published in the magazine *Science*, in which he briefly introduced *Structure*'s new theories. The executive wanted to reprint an excerpt in Burnett's *Journal of Marketing*. The excerpt described how new discoveries in science "react back upon what has previously been known." It read,

> In a sense I can now develop only in part, they also react back upon what has previously been known, providing a new view of some previously familiar objects and simultaneously changing the way in which even some traditional parts of science are practiced. Those in whose area of special competence the new phenomenon falls often see both the world and their words differently as they emerge from the extended struggle with anomaly which constitutes that phenomenon's discovery.

"I am somewhat puzzled to know what your readers will make out of the excerpt," Kuhn replied, but he was happy to let them reprint it.[1]

Kuhn could not imagine why advertisers cared about the psychological dynamics of scientific discovery. But times were changing and *revolution* was fast becoming the kind of change that fascinated Americans. Though the McCarthy years were dominated by widespread *fear* of revolution, taste makers as different as Leo Burnett and Bob Dylan, whose first album was released months before, cultivated a new, exciting, and fashionable ideal of positive and transformative revolutionary change.

For Leo Burnett, it was the "creative revolution." Here advertisers no longer presented mere information, but instead used images and words to evoke associations or feelings that change how consumers feel about a product—words and images that "react back upon what has previously been known, providing a new view." Burnett's famous Marlboro Man campaign, for example, used icons of rugged masculinity (and, by the early '60s, exclusively cowboys) to transform perceptions of filtered cigarettes that were originally marketed to women.[2]

Other revolutionaries aimed to rebuild democracy. Beginning in the early 1960s, the student movement, led mainly by Students for Democratic Society, aimed to create "participatory democracy," while SDS's emerging radical wing, the Weathermen (and, later, the Weather Underground) called explicitly for a communist revolution. For those who shunned politics, other kinds of revolutionary possibilities became available—in the arts, in communal living, in sexual or hallucinogenic experiences, or in the self-awareness promised by psychoanalysis, an interest that Kuhn happened to share.

Even with this letter from Leo Burnett on his desk—just a puzzle, not even an anomaly, he might have figured—Kuhn expected and hoped that his new book would find a large but strictly academic audience. His first book *The Copernican Revolution* was praised not only for its compelling account of astronomy's greatest revolution, but for its style, clarity, and punch. "Professor Kuhn writes as he talks—brilliantly," enthused a review in *Renaissance News*, adding, "He has a keen and perceptive mind. He argues persuasively." "Mr. Kuhn has the ability to make the events he describes come alive," seconded the *Journal of Higher Education*. "We experience the issues presented as if they were now facing us," with the result that the book speaks not only to the past but to contemporary physics, the reviewer enthused.[3] Yet even readers outside of academia were impressed. One Mort Friedlander wrote to Kuhn on letterhead from Executives' Service, Inc. in Mystic, Connecticut:

> Reading your Copernican Revolution leaves in its wake that sense of discovery, that sort of suspended excitation of the mind which may well be likened to pure happiness. This is indeed revelation, the experience of knowledge so few impart.

"How rich both the past and present become," Friedlander enthused, "for the rest of us through your guidance."[4]

Structure's first complete draft garnered similar praise from fellow academics. Ernest Nagel exclaimed that it was "a first-rate piece of work!" and praised its clarity, organization, and the way it "give[s] the reader a vivid sense of important issues being at stake." Once it was published, Kuhn's hopes were confirmed. Robert Merton, Nagel's colleague at Columbia and the nation's premier sociologist of science, called himself "an enthusiastic fan" who considered it "merely brilliant." Charles Gillispie, historian of science at Princeton (where Kuhn would teach after leaving Berkeley), praised the "the erudition, the scholarship, the fidelity, and the seriousness that the enterprise reflects on every page." Harvard psychologist Edwin Boring wrote Kuhn to say, "I have come away from it with great excitement," and that "Your book is brilliant and deserves to have an enormous effect." In professional literature, Boring praised *Structure* as "brilliant" and described Kuhn's paradigms as a breakthrough in the understanding of science.[5]

This vividness and brilliance was just what Kuhn was aiming for—to shake his colleagues' by their tweed lapels, wake them from their intellectual slumbers, and lead them to the conceptual revolution he himself had begun to experience when reading Aristotle. They too should discover that what they had long believed about the nature of science and the world it explores is simply wrong. Part of the book's power, Kuhn knew, was its compactness and lucidity. *Structure*'s "effect," he told art historian Ernst Gombrich, "depends on its maintaining pace" as the reader takes a brisk but mind-changing tour of history's great conceptual revolutions. For the same reason, Kuhn told his publisher, the book would most likely benefit from "word of mouth advertising" as readers who experienced its dazzling, disorienting, and illuminating effects introduced it to others.[6]

Though he did not expect to convert many readers outside academia, his mother Minette disagreed and urged Kuhn to demand advertisements and reviews in the highbrow magazines she read in her New York home. At her request, Kuhn relayed the message to the press, explaining that his mother had yet to see an advertisement for the book in any of these venues. "I have never supposed that the book could be sold to the 'general public,'" he confessed. But this kind of promotion couldn't hurt. It would make his mother happy and he himself would appreciate "the sense of confidence and effort that more mention in the major media would provide."[7]

Two Toms in a Revolutionary Decade

Kuhn's mother and Connecticut's Mort Friedlander were right. Kuhn would soon become widely seen as an expert on the fascinating dynamics of scientific revolutions. Though he would be blindsided by this popular interest, he surely remembered his own youthful, exuberant fascination with revolutions and the dramatic personal and moral challenges they pose. At the Hessian Hills School, he co-wrote and performed in a play, "United We Stand." It follows a band of early Americans who are forced into revolutionary violence by an English king who "listens to naught but what these English merchants tell him." These characters chose to fight the British only because they must. Their patriotism was twinned with Elizabeth Moos's pacifism that inspired Kuhn that same year to praise the American Student Union and to editorialize against American "dictatorial" foreign policy. "War! I had hoped we could avoid it," the play's script reads. "But if the King refuses to think of ought but his profits then it must be so."[8]

When he graduated from Harvard in the spring of 1943, his personal crisis over interventionism lay behind him. But he still wrestled with the generational crisis at hand as he and his fellow graduates headed off to war. Delivering the Phi Beta Kappa address at his convocation, Kuhn acknowledged an acute sense of conflict that he and others embodied. "We have been called a generation of cynics," he remarked. The charge was not unfair:

> We were born into a world not yet recovered from the disappointment and disillusionment of one war, but determined that there should not be another. . . . Now, before many of us have reached voting age, we find ourselves in a world again at war; and because we occasionally feel and proclaim that these things were not of our making, we have been called bitter and cynical.

Still, Kuhn urged his classmates to throw themselves into the war, lest they be left on the sidelines when the postwar world takes shape. "We are at the edge of nihilism," he explained, and must pull back. We must not "step out of history" like those citizens of Germany who allowed "a strong man shape the world for them." If only on the basis of faith, he explained, Kuhn's generation must actively engage and defend the tradition of learning that Harvard represents.[9]

About two decades later, Tom Hayden would make a similar plea for activism to displace nihilism and apathy among his collegiate peers. Hayden was a student leader like Kuhn, who shared his intellectual passions for philosophy and literature, his gifts as a writer (honed, like Kuhn's, writing for his college newspaper), and a passion to rouse his peers for political engagement. Hayden's politics were also rooted in the prewar socialist Left. The organization he led, Students for Democratic Society (SDS), was founded within the League for Industrial Democracy, the surviving descendant of the Intercollegiate Socialist Society founded in 1905 by Progressive-era novelist Upton Sinclair. Prior to the cold war, the society attracted intellectuals such as Sidney Hook (a supporter but not a member) and John Dewey (who served as its president in 1939).[10] The league's student wing, SLID, merged in 1935 with other groups to become the American Student Union that Kuhn described and applauded as a thirteen-year-old pacifist on the stage of his Hessian Hills School.

After the war, SLID became independent. But as its unfortunate acronym suggests, it became moribund during the McCarthy years and the 1950s when dissent and political activism became rare on college campuses. In 1960, however, with students energized by the growing civil rights movement, SLID leaders at the University of Michigan in Ann Arbor took note of Hayden's powerful, eloquent editorials in the student newspaper. They recruited him and rechristened themselves Students for Democratic Society. Hayden quickly became the group's leader and wrote the first draft of its manifesto for "participatory democracy"—"The Port Huron Statement," named after the Michigan town where SDS met to finalize the document.

The two careers of Kuhn and Hayden chart the political and cultural landscape of the 1960s in which revolutionary change became a mantra and many readers would embrace *The Structure of Scientific Revolutions* as a roadmap for their aspirations and convictions. The first mimeographed copies of Port Huron circulated in New York City in the summer of 1962,[11] months before *Structure* was published in the fall, and they shared an alluring, slightly enigmatic revolutionary style. "Participatory Democracy" stands at the center of Hayden's manifesto, much like "paradigms" in Kuhn's—ringing with a sense of novelty, discovery, and promise. Each has a flexible penumbra of meaning (a "meaning fringe," Kuhn might once have called it) that invites devotees to articulate and refine their beliefs and values.

Politically, both catchwords embraced transformative change while keeping the subversive specter of Marxism and communism at arm's length. Kuhn's paradigms played the roles he had earlier assigned to professional ideologies. But he knew well enough (as he had told Charles Morris) that "ideology" was best left out of his monograph's title, just as "dogma" (as H. Bentley Glass had insisted) was best replaced by other words. Hayden was equally careful to disassociate SDS from communist overtones that would surely make the public nervous, invite investigation from Hoover's FBI, and complicate SDS's relationship with its parent organization, the anti-Stalinist League for Industrial Democracy (from which it would sever its ties in

1965). At the same time, however, one intended meaning of participatory democracy was the socialist ideal of workers assuming some control over the political institutions and mechanisms of contemporary life. Bob Ross, recruited into SDS alongside Hayden, remarked that it allowed SDS to "talk about socialism with an American accent."[12]

Philosophically, Hayden's participatory democracy and Kuhn's paradigms shared a deeper root in philosophical pragmatism insofar as they are not doctrines or propositions but instead practical *achievements* around which theories, doctrines, and methods grow and become grounded. Participatory Democracy was a practical goal to be articulated through activities such as community organizing, voter registrations, "teach-ins," protests of the Vietnam War, and experimental, communal living arrangements in various American cities. Participatory democracy joined "not one, but two distinct political visions," as James Miller put it. "The first is of a face-to-face community of friends sharing interests in common; the second is of an experimental collective, embarked on a high-risk effort to test the limits of democracy in modern life." The first was likely inspired by Hayden's reading of the sociologist C. Wright Mills and University of Michigan philosophy professor Arnold Kaufman, both of whom urged intellectuals to engage American culture *as intellectuals* and to advance reason in democratic culture. In Hayden's words, Port Huron was a manifesto for rational, pragmatic action, a "document of our convictions and analysis"—

> an effort in understanding and changing the conditions of humanity in the late twentieth century, an effort rooted in the ancient, still unfulfilled conception of man attaining determining influence over his circumstances of life.

Mills and Kaufman were Hayden's Conant. He respected their ideal of reasoned, deliberate engagement with politics, but found it unsuited to a modern world in which reason seemed exhausted and powerless against reigning absurdities and paradoxes. In this, Hayden also found inspiration in existentialism and Albert Camus's picture (in *The Rebel*, for instance) of "moral resolve as a bulwark against nihilism."[13] As Kuhn had put it in his commencement speech, Hayden believed that alienation and nihilism were best overcome by action, not just debate and contemplation.

Both registers of the Port Huron statement—its political reformism and its existential impulse to surpass conventional boundaries of political and personal experience—found cultural connections to *Structure*'s new image of science. If Kuhn was right that paradigm changes in scientific communities can effectively lead to different worlds of science and understanding, to remake the very metaphysics of nature and reality, then surely the nation's students could help dismantle Jim Crow, curtail the business-as-usual warmongering of the military industrial complex and Mills's "power elite," and revitalize American democracy. This is not to suggest that Hayden looked to Kuhn for inspiration (there is no evidence that he or other SDS'ers read *Structure* as a handbook). It is to say, instead, that both projects—Kuhn's to revolutionize the understanding of science and Hayden's to revitalize American democracy—were led by comparably ambitious and intelligent young men who drew on common political, cultural, and philosophical resources at hand in postwar America.

To be sure, Kuhn's career was in some ways the opposite of Hayden's. He became a scholar, not an activist. Not long after he sermonized about the power of faith to overcome contemporary nihilism among the youth, Kuhn's interests in democracy, citizenship, and international relations seemed to wane. His drafts, notebooks, letters, doodles and speeches no longer broached the evils of war and capitalist profiteering or the "the dignity of all peoples" that he noted in his graduation speech. His passion had become his scholarship and teaching, the new field of research he aimed to pioneer, and his insights about scientific change and their psychological, epistemological, and possibly metaphysical implications.

Kuhn himself, and his career, was no longer overtly political. Yet as the letters he began to receive from avid readers of *Structure* seem to suggest, the political interests of his youth and his fascination with revolutionary, transformative change were latent within his theory of knowledge and scientific progress. However scholarly and abstract it may seem on the surface, *The Structure of Scientific Revolutions* argues that the sociological and intellectual dynamics of momentous scientific change are similar to those driving political change. A scientific revolution begins the same way that political revolutions begin—when "a segment" of the relevant community believes "that existing institutions"—be they scientific traditions or political regimes—"have ceased adequately to meet the problems posed by an environment that they have in part created." Both scientists and citizens know "the sense of malfunction that can lead to crisis" because, in both settings, the institutions in question are conservative and resist reform. "Political revolutions," he wrote, "aim to change political institutions in ways that those institutions themselves prohibit." The only way forward, therefore, is to replace those institutions with others that, because they solve these problems, must be very different.

Scientific paradigms, too, resist change because their most devoted supporters, Kuhn's normal scientists, are dogmatic and loath to abandon them. A new paradigm will therefore attract younger scientists, typically, while their dogmatic elders drift into professional irrelevance.[14]

Hayden and his SDS followed this revolutionary logic. The nation required new forms of democracy precisely because it was failing to live up to its ideals and because established interests increasingly opposed reforms—often violently. Students from Kuhn's University of California at Berkeley were among those protesting in May 1960, when the House Un-American Activities Committee held hearings at San Francisco's City Hall. Police attacked protestors using high pressure water hoses and some who passively resisted arrest were dragged by their arms or legs down the hall's enormous stone staircase (see Figure 16.1).

Faculty who were sympathetic to the students and the embryonic Free Speech Movement fared no better against the university administration and J. Edgar Hoover's FBI. Since the 1940s, when rumors swirled around Berkeley physicist J. Robert Oppenheimer, Hoover and his agents surveilled not only suspicious students and faculty, but Berkeley chancellor Clark Kerr. Despite Kerr's record of anticommunism and his opposition to student activism, Hoover orchestrated public sentiment against him to help facilitate the rise of Ronald Reagan (whom Hoover had come to know as the likable, loquacious head of the Screen Actors Guild). In 1967, after winning his race for governor—partly by promising to "clean up the mess at Berkeley"—Reagan fired Kerr after their first meeting. Hoover and his agents pursued other Berkeley activists with gusto (such as the philosophy major Mario Savio, the spokesman for the Free Speech Movement) by collecting personal information that could be used against them in court. Or—one of Hoover's trademark tactics—that information would be sent to their employers who would typically fire them upon learning there was a potential subversive on the payroll.[15]

SDS and other student groups were willing to stand up to the anticommunist consensus, however, because they saw it beginning to weaken during the paranoid 1950s. McCarthy was censured by Congress in 1954, and in 1956 Stalin's successor Nikita Khrushchev denounced the brutality of Stalin's regime in a speech that was soon leaked to the West. In a replay of the Hitler-Stalin pact some fifteen years before, this news caused many stalwart Communist holdouts in America to finally leave the Party. With the radical Left virtually nonexistent in America, and Russia's new leader agreeing that life under Stalin had been a nightmare, it became once again possible to regard East and West not as metaphysical enemies but as different

Figure 16.1. HUAC protestors dragged down stairs by police in San Francisco City Hall. (Image courtesy of San Francisco History Center, San Francisco Public Library)

cultures sharing the same planet. By 1963, John Kennedy could publicly call himself a *Berliner* and justify his call for bilateral reductions in nuclear testing on the grounds that Americans and Russians "breathe the same air. We all cherish our children's future. And we are all mortal."[16]

Hayden and his peers had long perceived anticommunist fears as overblown and politically debilitating. They were tired of "Growing Up Absurd," as writer and social critic Paul Goodman had put it in his surprise best-seller of 1960. Like Kuhn's revolutionary scientists, no longer willing to abide these anomalies, they sought to move past them. "Every generation inherits from the past a set of problems—personal and social," Hayden wrote in his draft of what became the Port Huron Statement,

> and a dominant set of insights and perspectives by which the problems are to be understood and, hopefully, managed. The critical feature of this generation's inheritance is that the problems are so serious as to actually threaten civilization, while the conventional perspectives are of dubious worth. Horrors are regarded as commonplace; we take universal strife in stride; we treat newness with a normalcy that suggests a deliberate flight from reality.[17]

At Port Huron a year later, Hayden's peers found this opening too scholastic and stiff. They chose to begin the manifesto with what Hayden had written a page later, an introduction of themselves and their concerns. From then on, the arc and logic of the manifesto matches Kuhn's description of political revolutions, the problems they emerge to address, and the sense shared by a segment of society that something had gone very wrong. "We are people of this generation," Port Huron began, "bred in at least modest comfort, housed now in universities, looking uncomfortably to the world we inherit." Two enormous problems stood out: racial bigotry and the cold war's threat of nuclear annihilation:

> We might deliberately ignore, or avoid, or fail to feel all other human problems, but not these two, for these were too immediate and crushing in their impact, too challenging in the demand that we as individuals take the responsibility for encounter and resolution.

These problems were surrounded by "paradoxes" that seemed to disable rational discourse and problem solving: the rhetoric of freedom coupled to the reality of Jim Crow, the promise of nuclear energy coupled to the threat of nuclear destruction, and the nation's already immense wealth growing while "two-thirds of mankind suffers undernourishment."[18]

"In increasing numbers individuals become increasingly estranged from political life and behave more and more eccentrically within it," Kuhn wrote.[19] Port Huron's authors dwelled on the apathy they observed around them. This was "the most outstanding paradox," they explained: "we ourselves are imbued with urgency, yet the message of our society is that there is no viable alternative to the present." Even organized labor, once strong and progressive, had become a powerless band of "indifferent unionists, uninterested in meetings, alienated from the complexities of labor-management negotiating apparatus, lulled to comfort by the accessibility of luxury and the opportunity of long-term contracts." The American voter was beat up, as well:

> buffeted from all directions by pseudo-problems, by the structurally-initiated sense that nothing political is subject to human mastery. Worried by his mundane problems which never get solved, but constrained by the common belief that politics is an agonizingly slow accommodation of views, he quits all pretense of bothering.

Things looked bleak, but Port Huron was not humorless. Many students, it noted, "don't even give a damn about the apathy."

As "crisis deepens," Kuhn wrote, the disaffected will "commit themselves to some concrete proposal for the reconstruction of society."[20] Port Huron proposed that reconstruction begin with American colleges and universities. Just as postrevolutionary scientists educate for a new kind of science, Hayden and his colleagues argued that campuses must educate for a new, reformed and reinvigorated democracy. They will become venues in which students, labor, and civil rights organizations interact to create and maintain the new institutions of participatory democracy.

Revolutionary Tensions and Explosions

Hayden's talent as a writer, especially as Port Huron came to a close, was one reason to cast one's lot with the student movement:

> To turn these possibilities into realities will involve national efforts at university reform by an alliance of students and faculty. They must wrest control of the educational process from the administrative bureaucracy. They must make fraternal and functional

contact with allies in labor, civil rights, and other liberal forces outside the campus. They must import major public issues into the curriculum—research and teaching on problems of war and peace is an outstanding example. They must make debate and controversy, not dull pedantic cant, the common style for educational life. They must consciously build a base for their assault upon the loci of power.

As students, for a democratic society, we are committed to stimulating this kind of social movement, this kind of vision and program in campus and community across the country. If we appear to seek the unattainable, it has been said, then let it be known that we do so to avoid the unimaginable.

Port Huron's aspirations were not revolutionary in all respects. As this emphasis on reform and collective efforts suggest, participatory democracy would be a return the socialistic promise of the 1930s when this "fraternal and functional contact" between the nation's intellectuals and its ordinary citizens was possible, even normal—when university professors such as Sidney Hook and James Burnham would rub elbows with union workers and labor leaders and argue for socialism in popular magazines; when John Dewey and George Counts inspired teachers like Elizabeth Moos to build a new social order by educating students like Tom Kuhn.

Yet some of Hayden's generation disagreed with reformism and insisted on a total, revolutionary break with the past. Similar divides had long surrounded communism's history. In Russia it divided Menshevik reformists from Bolshevik revolutionaries. In the United States, communists seeking progressive reforms were derided by revolutionaries as "impossibilists." And it divided the Black Panthers from supporters of Martin Luther King Jr. and his faith in nonviolence. In *Structure*, Kuhn acknowledged both kinds of approach to change, but he focused on revolutionary, discontinuous change. When there exists "no supra-institutional framework for the adjudication of revolutionary difference," he wrote, "the parties to a revolutionary conflict must finally resort to the techniques of mass persuasion, often including force."[21]

As for those revolutionaries within SDS, they eventually broke away to form the Weathermen. Angered especially by the Vietnam War and the Johnson administration's manipulation of public opinion, the Weathermen denied that reforms or reasoned appeals for change could be effective. Their manifesto, *Prairie Fire: The Politics of Revolutionary Anti-Imperialism*, argued that SDS was itself ideologically blinkered and duped by its reformist faith. "It is an illusion that imperialism will decay peacefully," it explained, for "change is violently opposed every step of the way." Reform and persuasion must give way to force and, finally, an antithetical alternative that fully replaces and obliterates the status quo: "Our final goal is the destruction of imperialism, the seizure of power, and the creation of socialism . . . the total opposite of capitalism/imperialism."[22]

Most Americans would never understand this, however, for their thoughts and aspirations were controlled by the reigning imperialist ideology. It has "complete control over people's lives. It is economic power and far more. It involves and implicates people in a system over which they have little control." One mechanism of this control was history: "The real history of the US is almost totally unknown to the US people," *Prairie Fire* explained. "The most important parts have been buried, falsified, hidden from our view." The Weathermen therefore included history lessons in their manifesto, covering the early settlers to the Great Depression. The goal was to illuminate a political methodology not unlike the scientific methodology Kuhn unveiled in his Lowell Lectures—an awareness that the past had not unfolded inevitably; that history, and therefore the present, could have been different. Any responsible historiography, *Prairie Fire* explained, must show "the possibilities for liberation at any given time, how far these were carried, what held us back, what basis was laid for future struggles, including our own."[23]

In the end, however, the Weathermen did not choose history as their primary tool for liberating modern society. Bidding to shock Americans out of their normal, ideological prisons, they bombed a San Francisco police headquarters in February 1970 and the U.S. Pentagon in May 1972. Taking their manifesto's title from Chairman Mao—"A single spark can start a prairie fire"—they believed that violence could spark "leaps in confidence and consciousness" that would lead to revolutionary change.[24]

The Fall and Rise of Revolutionary Consciousness

Despite predictable attempts by reactionaries and Hoover's FBI to discredit SDS as puppets of subversive Communists, the student movement's size and credibility grew during the late 1960s. "The sense that something epic was happening was pretty hard to miss," Hayden once said. James Miller wrote that 1968 was "the only year in post-war American history when many intelligent people sincerely believed that a revolution was about to occur."[25] The movement's heady mixture of creative destruction seemed to have international momentum—from Columbia University in April and May 1968, where protestors occupied an administration building, to Paris and Prague where students fought police in the streets, to the cooperative, student-led construction of Berkeley's People's Park in 1969. Alongside the assassinations of Martin Luther King Jr. and Bobby Kennedy, and riots in Los Angeles, Newark, and Detroit, the nation seemed plainly in crisis.

Other kinds of revolution were brewing, too—spiritual, psychological, sexual, feminist, intellectual, and existential. "For those protesting," Miller wrote,

> the turmoil was intoxicating. Thrilled by the prospect of change, young people plunged across the frontiers of experience, boldly exploring altered states of consciousness, new types of bodily pleasure, nonhierarchical forms of community. Like Marx and Nietzsche before them, they dreamed of creating new men and new women, undivided, without shame, each one in tune with a unique constellation of animal instincts and creative ideals.[26]

These ideals would in the end survive much longer than the political goals of SDS and the Weathermen. Protesting the war and the business-as-usual sensibilities of the Democratic Party at their 1968 national convention in Chicago, SDS and other groups were denied protest permits and then attacked by city police wielding tear gas and billy clubs. An incredulous crowd chanted "The whole world is watching." But Chicago's reactionary political culture did not seem to care who was watching or what they might see. The next evening Mayor Richard J. Daley boiled over when Connecticut senator Abraham Ribicoff took the podium to recommend George McGovern as the party's presidential candidate. With McGovern as president, Ribicoff ventured to say, there would be never be "Gestapo Tactics on the streets of Chicago." Television cameras then showed the Mayor shaking his fist and sputtering in rage at Ribicoff from the convention floor, "Fuck you, you Jew son of a bitch you lousy motherfucker go home."[27]

Chicago was not far from Ann Arbor or Port Huron. But ideologically, Daley's city of broad shoulders and sharp elbows disliked change. Labor unions and civil rights activists had made progress in the city decades and centuries before, but it was always incremental and met with ferocious resistance and reaction. At Haymarket Square in 1886, where demonstrators rallied for an eight-hour work day, a bomb lobbed at police led to a city-wide crackdown on labor, anarchist, socialist, and immigrant groups. Four suspects were tried and executed, even though it was never convincingly determined who threw the bomb.[28] Two years before the Democratic Convention, Martin Luther King Jr. and followers marched peacefully through South Side neighborhoods and parks in support of fair housing laws. He and his entourage suffered bricks and bottles thrown by "thousands of whites" who also set cars on fire behind King's procession. Speaking to reporters after being hit in the head with a rock, King noted that he was familiar with social resistance, but, "I have never seen anything so hostile and so hateful as I've seen here today."[29]

The whole world *was* watching Chicago in 1968. But only idealists like King were surprised at the seemingly limitless and open resistance to change within America's biggest Midwest city. When CBS's Dan Rather, reporting on the convention floor, tried to interview a delegate from Georgia whom he observed being forcibly removed by convention security, local security guards surrounded and assaulted him. Anchorman Walter Cronkite and CBS's cameras watched from above as Rather tried to continue his on-air conversation with Cronkite. "I'm sorry to be out of breath but someone belted me in the stomach during that," Rather gasped after the scuffle. "I think we've got a bunch of thugs here, if I may be permitted to say so," Cronkite added.[30]

To a majority of Americans, the problem was not Daley's thugs but student protesters who had excited the nation's deep fear of political change. The violence in Chicago, the victory of Republican Richard Nixon, the nation's tolerance for the escalating Vietnam War, and the shooting of students by National Guardsmen at Kent State University in 1970 signaled that the nation would ably resist political reformers and revolutionaries. SDS itself began to fall apart after the Chicago convention. In December, Hayden was called to testify before the House Un-American Activities Committee, and within months he and six other organizers of the protests were on trial in Chicago for inciting riots. Hayden now had more fame and attention than he ever expected, but far from the intellectual and political reformer he set out to be, he was perceived by many as a dangerous, violent agitator. The Weathermen's aggressive and sensational tactics soon eclipsed SDS-style teach-ins and community organizing. The "Days of Rage" they scheduled for Chicago the next year were handily suppressed by Daley's police and the National Guard.[31] To avoid the fate of Hayden and the rest of the Chicago

Seven, the Weathermen's leaders went into hiding. In March 1970, three were killed by an accidental explosion in a New York City townhouse where they were living and building bombs.³²

You'd Better Free Your Mind Instead

Even popular culture rejected the militant turn that these political revolutionaries had taken. As early as 1968, John Lennon had scolded the counterculture over where things were heading:

> We all want to change the world
> But when you talk about destruction
> Don't you know that you can count me out?

In 1971, Hayden changed his name and moved to Venice, California. After fleeing the exploded Greenwich Village townhouse, Kathy Boudin and other Weatherman holdouts survived by theft and robbery until 1981 when they were captured after holding up an armored truck in Nanuet, New York.

If you want a revolution, Lennon advised, "you'd better free your mind instead." Taking his advice, revolutionary politics gave way to revolutionary culture, or "counterculture," which abandoned politics for intellectual, psychological, or spiritual change inspired by drugs and experiments in literature, film, and music. For feminist writer Vivian Gornick, the insight came in 1968 with "the realization that social change had more to do with altered consciousness than with legislated law." She wrote, "It was as though the kaleidoscope of experience had been shaken and when the pieces settled into place an entirely new design had been formed." For musicians, it may have been a year later at Woodstock when Jimi Hendrix played an incandescent version of the national anthem on his electric guitar, or upon first hearing The Beatles' recording "I Am the Walrus." In each case, like the paradigm shifts that Kuhn had described in science, these experiences would "gather up large portions of that experience and transform them to the rather different bundle of experience that will thereafter be linked piecemeal to the new paradigm but not to the old." The Beatles and their producer George Martin re-bundled familiar auditory experiences—the string quartet, the carnival barker's bullhorn, the beat poet's free verse ("expert texpert choking smokers")—to create music at once unfamiliar and riveting (at least for those who could free their minds from the conventions of "normal" music).[33] Hendrix's band debuted as "The Jimi Hendrix Experience," and the title of its first album, *Are You Experienced?* knowingly asked listeners to prepare for something new and very different.

With his expertise about paradigm shifts in this revolutionary age, Kuhn might be compared to the psychologist Timothy Leary, whose experimentation and promotion with LSD in the 1950s led him to worldwide fame as an explorer of altered consciousness. Leary's academic career was also rooted at Harvard and Berkeley, but unlike Leary, Kuhn was averse to interviewers and television cameras and never believed that his expertise in history and philosophy of science had a legitimate place in popular culture. Popular culture disagreed and freely appropriated crises, revolutions, and paradigm shifts for myriad purposes. "The more aware we are of our basic paradigms, maps, or assumptions" Stephen Covey wrote in his best-selling business book *The Seven Habits of Highly Effective People*, "the more we can take responsibility for these paradigms, examine them, test them against reality."[34] Others who adapted paradigms for success in business, fiction, cultural history, and the arts could do so not only because "paradigm" and "paradigm shift" were flexible concepts, but because Kuhn himself effectively composed his book to elicit a transformational intellectual experience.[35] He did not tell them to "tune in, turn on, and drop out" to revel in whatever one's psychedelic mind conjures up, as Leary had. But he did promise to liberate readers from their preconceptions about science and join him in fascination at the almost world-changing powers of paradigms and scientific revolutions.

"My most fundamental objective is to urge a change in the perception and evaluation of familiar data," he wrote in the preface. For reasons the book itself would make clear, however, Kuhn could not lead readers to it step by step (as Covey presumed to do). But he could instead lay the groundwork by introducing certain ingredients of his new image of science, including paradigms, simultaneous discoveries, and sudden shifts of scientific vision. He made these ingredients seem natural and increasingly familiar by visiting and revisiting crucial moments in the history of astronomy, chemistry, and physics that illustrated them. At the same time, Kuhn minimized terms that return readers to their familiar preconceptions about sciences, such as *fact, progress, truth,* and *reality*. He offered instead a novel vocabulary of *normal science, paradigm, puzzle* and *puzzle-solving, anomaly, crisis,* and *incommensurable*. Unencumbered by prerevolutionary associations and meanings, *Structure* became a canvas on which the reader's new, revolutionary understanding of science and knowledge could take shape.[36]

Traditionally, scholars present a thesis and then provide arguments and evidence for why it should be believed. Kuhn's early drafts adopted that approach. But after arriving at his theories of paradigms and conversion experiences, Kuhn turned the plan of his book around so that the reading experience itself recapitulates the kind of conversion experiences the book describes. In the earlier chapters, the reader learning about paradigms and the formation of normal science begins to experience a crisis, an awareness of incongruity between Kuhn's claims about how science works and what their science teachers or popular science writers had taught them to believe. He structured the chapters as Elizabeth Moos might structure a field trip, a woodworking project, or the production of a day-long pageant. He knew what lay ahead. But if the conversion was to be authentic, readers would have to encounter the crisis, feel its force, and struggle toward a solution actively on their own. Only in the later chapters does Kuhn address the reader's conversion and help them tie up remaining loose ends. He warns them away from the almost reflexive belief in some "neutral" language of observations or retinal sensations, away from the confounding notion of "truth," and toward a new and different conception of scientific progress that makes better sense in this new world of understanding.

The Trouble with Brilliance

Kuhn was naturally pleased that so many would describe *Structure* as "brilliant." But any physicist knows that brilliant white light contains a cacophony of colors and shades, many of which Kuhn never intended to convey.

"Dear Thomas Kuhn," wrote one Barbara Decker-Ritchey, a student at Ohio State University in 1968:

> I am just reading your book on Scientific Revolutions. I feel as if a great weight has been lifted from my shoulders. I am so happy to hear, to read, rather, that the various problems I have long encountered, in combining, or perhaps jumping back and forth re behavioral and biological sciences, have real substance.

Many others wrote to say that *The Structure of Scientific Revolutions* overflowed with meaning and personal significance. "It's a great book, confirming a great many thoughts of my own which, however, I did not find clearly expressed—and documented—anywhere up to now," wrote a sociology professor at Rutgers. An attorney in Chicago could "no longer refrain from acknowledging my great debt to you and to two of your writings"—*Structure* as well as *The Copernican Revolution* of 1957, both of which he encountered "at a time when they were sorely needed in my life."[37] A biophysicist at the University of Pittsburgh said that the book hit him "with particular force" and inspired him to write an essay (enclosed for Kuhn's perusal) for his colleagues on campus. A devotee of communication theorist Marshall McLuhan at the University of Missouri said he experienced a "gestalt flip" reading McLuhan and now, after reading Kuhn, finally understood why he had trouble effectively communicating with his colleagues. "Now I know," he said.[38]

Kuhn had told the University of Chicago Press that *Structure* would benefit from word-of-mouth publicity and several pieces of fan mail he received in the early and mid-1960s told how readers discovered the book through friends or colleagues. Many wanted Kuhn himself to become a friend or colleague. Yet those who seemed most effusive and enthusiastic sometimes understood Kuhn and his intellectual goals the least, especially those who read *Structure* as a manual or guide for bringing about revolutionary change. In 1963, for example, one Edward Dewey wrote on letterhead from his "Foundation for the Study of Cycles" to say that *Structure* "has been a great help to me because we are attempting to create a revolution in several sciences simultaneously." One Roy Woodmansee thanked Kuhn in 1965 for blazing a trail into new revolutionary times that by comparison "will make Copernicus' confrontations seem 'little league.'" The new science of Extra Sensory Perception was on its way, Woodmansee explained, and Kuhn's insights will surely "help make the transition to a new world view or <u>world views</u> faster and less difficult." A physician promoting "birth control by male continence" sent Kuhn an article he wrote and inscribed it, "Dear Prof Kuhn! I enjoyed your book on the shattering of paradigms. May many more be shattered quickly!!"[39]

What Are the Problems of Modern Life Today?

When The Lowell Institute publicized Kuhn's lectures by asking, "What are the Problems of Research Today?" Kuhn was deeply distressed by the suggestion that he himself was a reformer or planner offering recipes for scientific change. Now he responded politely and usually briefly to these admiring letters from would-be revolutionaries. But he was surely frustrated by their frequent combination of enthusiasm and misunderstanding. Yes, he had explained that revolutions have deep and profound "conversion experiences" at their core, and he implied (and his letter to Lawrence Kubie confirmed his suspicion) that reality itself may be fundamentally malleable and plastic. But Kuhn knew that neither scientists nor historians have special, paradigm-free access to the world or reality, so none can predict when revolution-inducing anomalies will turn up, what they will look like, how future scientists will respond to them, or what new and different paradigm will change the world. It was the study of *history*, Kuhn made clear, that opened a window to these revolutionary dynamics—not anyone's hopes for the future.

As Kuhn himself once put it, *Structure* "is a profoundly conservative book."[40] But its conservative posture was easily obscured in the eyes of those who read it with the forward-looking optimism and intentions of the revolutionary sixties. The immense success of *Structure*, in other words, heralded not only a decisive transformation in the scholarly understanding of science, as Kuhn predicted. It coincided with a transformation within cold war culture—from a conservative fear of revolutions circa 1950, to the progressive embrace of revolutionary logic in both political and personal life circa 1968. As a professional scholar whose days of speaking to the public were behind him, Kuhn did not address this second transformation and may have cared little about it. But as surely as these two transformations would become mixed and confused in the welter of popular culture and the evolving mood of the nation, this mixture would help turn *Structure* into *the* scholarly book about science that mattered, one that spoke to academics as well as politicians, advertisers, inventers, and misunderstood geniuses of all stripes. Even when it was misunderstood, therefore, *Structure* was bound to seem a great and important book. As one David Layton put it when he wrote to congratulate Kuhn for the second edition of *Structure* and the important lesson he took from it: "Keeping an open mind is the main thing—and what you've insisted on."[41]

17

A Revolution and a New Ideology

Officially, at least, it began on the evening of April 30, 1970. President Richard Nixon appeared on American television for twenty-two minutes to announce that he had ordered attacks inside Cambodia (see Figure 17.1). The communist enemy in Vietnam was taking advantage of Cambodia's neutrality, so United States bombers and troops were going in. Pointing to a map of Southeast Asia in the oval office, Nixon offered a string of paradoxes, contradictions, and ad hoc rationalizations to justify the invasion. "This is not an invasion of Cambodia," he said. It was, rather, a difficult but necessary

Figure 17.1. Nixon's address to the nation on the bombing of Cambodia, April 30, 1970. (National Archives, White House Photo Collection, RN-WHPO, 194674)

step to advance the nation's noble crusade for peace and freedom. This was not an expansion of the war, but a way to bring it to an end, to achieve "the just peace we all desire." Besides, Nixon explained, the consequences of inaction would be catastrophic. The war for men's minds in the region would be lost, and the nation would be humiliated. The United States would be seen as "a pitiful, helpless giant" and "the forces of totalitarianism and anarchy will threaten free nations and free institutions throughout the world." He closed by appealing to the so-called Silent Majority, who had elected him in 1968, to maintain order and keep dissenters in check. They must stand against the hippies, the violent radicals, and naive pacifist intellectuals by whose protests the nation's "great universities are being systematically destroyed."

With "Thank you and good night," Nixon blinked off the nation's television screens and ratcheted up antiwar fervor. Americans were increasingly doubtful about the war, especially after the North Vietnamese stunning Tet offensive of early 1968. But Nixon was intent on expanding it. And he was up to his usual tricks. He had campaigned for reelection by promising to end the war, but now he wrapped it in flattering, patriotic language that twisted facts and history. American bombers had already expanded the war into Laos, and the Cambodia story had been leaked to *The New York Times* almost a year before.[1] Though he spoke to the nation's citizens as if they were an informed party to his decision, America under Nixon was no participatory democracy.

Those who had opposed the war from its beginnings under John F. Kennedy and Lyndon Johnson were outraged. But with Nixon's announcement that B-52s loaded with bombs were headed over Cambodian farmers and peasants, it seemed clear to increasing numbers of Americans that something was deeply wrong. Even before Nixon's speech, James Conant had begun to question the goals, methods, and military leadership behind Nixon's expanding war. The organization Women Strike for Peace, along with Elizabeth Moos, now in her early eighties, would convene on Washington repeatedly to brandish signs reading "Impeach the Mad Bomber" (See Figure 17.2).

༄

Since 1963, Kuhn had been teaching at Princeton University. On the evening of Nixon's speech, President Robert Goheen hosted a delegation of ten Soviet intellectuals. They were touring the nation to publicly debate their American counterparts. *The New York Times* described the tour as "an

Figure 17.2. Elizabeth Moos (holding sign) at Women Strike for Peace (WSP) protest in Washington, March 22, 1973. (Image courtesy of the Swarthmore Peace Collection)

extraordinary public discussion of such problems as arms control, pollution, East-West trade and peace-keeping."[2] Goheen invited the Soviets to watch Nixon's address on television in a university lounge along with selected members of the faculty and a group of students, all of whom became upset and angry. Later that night, Goheen brought them to University Chapel, where more than two thousand agitated students and faculty met to debate the issues. By the end of the meeting, they voted to boycott classes in protest of Nixon's escalation.

As the long night wore on, these Russians sat silently and became ever more nervous. As Communists on an American campus, they figured, they might be blamed for the anarchy unfolding around them. The next morning they abruptly left campus before their scheduled events. The director of the Fund for Peace that sponsored them explained that "the Russians preferred to leave town rather than risk becoming involved in the strike movement by holding their scheduled seminars." Some faculty members tried to convince them that "even J. Edgar Hoover wouldn't seriously claim that a dozen Soviet professors could arrive on an American campus and five hours later have the place overturned in a protest demonstration."[3]

But having just seen Nixon, a politician who owed his political career to his fervent anticommunism, point on television to student unrest on the nation's campuses, they evidently felt unsafe. Even official Communists did not want to be tagged as revolutionaries in Nixon's America.

The Committee on Special Research

In the days after Nixon's announcement, student strikes and protests erupted on college campuses. At Kent State University in Ohio, national guardsmen killed four demonstrators.[4] At Princeton, Goheen and thirty-two administrators wrote to Nixon in "deep dismay" over the developments in Cambodia.[5] Leaflets from SDS and other groups criticized not just Nixon's war but Princeton itself, a university that "aids and abets the ever-expanding war in Southeast Asia," by undertaking "war-related research," one of them read. On May 4, the day of the shootings at Kent State, close to four thousand students and faculty filled Jadwin Gym to adopt the proposition that Princeton "as an institution oppose the Cambodian invasion, American foreign policy and domestic oppression and that its facilities and manpower be used to register that opposition."[6] Goheen himself approved new end-of-semester schedules and exam requirements to accommodate students who wished to campaign for the national elections coming up in November.

Goheen's "Princeton Plan" and his solidarity with protestors may explain why Princeton escaped the kind of violence that marked Columbia University and Harvard. (Conant's successor Nathan Pusey called in police to clear University Hall of occupying students the year before.)[7] Princeton faculty responded constructively, as well. On the issue of war-related research, they formed a special committee to investigate the university's involvement in military research. The committee would explore how the university could comply with its new resolution to "refuse to accept any outside funds for research on campus which is directly and specifically related to weapons and weapons systems," and how it could work with other campuses to petition Congress and ensure that research funds come to universities not through the Department of Defense but through civilian agencies such as the departments of Housing, Education, and Welfare and the National Science Foundation.[8]

Kuhn was asked to chair the new committee. He was not eager to do so, given his aversion to publicity as the author of *The Structure of Scientific Revolutions*. But the request obviously had significance. Were he to agree, he would return to the leading role he once played at Elizabeth Moos's school where he defended student strikes against war and publicly denounced American imperialism. He would affirm once again "the dignity of all peoples" as he did in his graduation speech from Harvard. Nixon's expansion of the war seemed to reignite some of the moral urgency he once felt about pacifism and made it difficult for him to say no. Leading this new committee, he wrote at the time, "is a bit of glory I would gladly have declined, but after Cambodia I saw no alternative but to participate when asked."[9]

Until it issued its final report the next year, Kuhn chaired the Special Committee on Sponsored Research and met with students, faculty, and nonvoting members of the administration to wrestle with questions about scientific research at the university, its relationship to the nation, and its sources of funding. The questions proved surprisingly complicated and difficult to answer, however. Kuhn wrote to other universities to ask how they handled these matters. He introduced his committee and its goal "to make certain that the University has not, as an institution, involved itself so deeply with the federal government as to surrender an essential part of its traditional independence and assume functions inappropriate to its nature"; to make sure that the way federal funds come to university research does not "somehow distort the normal evolution of academic research." Yet some responded by questioning the assumptions behind Kuhn's questions. The vice president for research at Cornell acknowledged that sources of funding naturally influence the kind of research undertaken and thought Kuhn was naive to think otherwise. Berkeley president Charles Hitch took exception to Kuhn's special interest in "mission-oriented" funding agencies and said it shone no light on the matter. Both of these respondents denied that military funding distorted scientific research in ways that funding from the National Science Foundation or philanthropic foundations did not.[10]

Princeton faculty also dissented from any simple notion of distorted research. At a meeting that September, a professor of aerospace science insisted that "the selection of research topics is influenced by the sponsor." The ongoing research of a professor in botany altogether stymied the committee when they asked whether the research in question—an Army-funded study of "leaf abscission and senescence"—helped to aid and abet the killing in Southeast Asia. To critics on campus, this was plainly research in defoliants, such as the infamous Agent Orange sprayed by American planes in Vietnam. But the professor defended himself and his research in the *Daily Princetonian* saying that "none of it was concerned with the practical problems of defoliation." The same research project had been supported by the National Science Foundation, he noted.[11]

The committee found no consensus on these issues and reached no definitive recommendations. Faculty disagreed among themselves about whether their research was connected to military purposes, and the committee found redundancy and overlap among sources of funding. Most departments received grants from the Department of Defense as well as NASA, the Atomic Energy Commission, and the National Science Foundation—some or all of which might support the same research projects. Unable to formulate a proposal to free the university from military funding, the committee's final report in 1971 settled on a more symbolic measure: opposing the existence of a classified library on campus and calling for vigilance and greater awareness of how research depends on its sponsors.[12]

In his president's report that year, Goheen congratulated Kuhn and his committee

> on the completion of its investigations it reported to the CPUC [Council of the Princeton University Community] as well as the faculty. The committee found the University's procedures for safeguarding the integrity of sponsored research at Princeton to be essentially sound, and it may be hoped that the committee's detailed and thorough report helped to bring out fuller and more accurate understanding of these matters on campus.[13]

Research on campus would proceed more or less as it had.

Yet Kuhn's philosophy of science was fundamentally challenged. Some twenty years before, inspired by a letter about the sociology of science from Philipp Frank, he insisted that factors "like Government, Church, etc. . . . at this time and place have relatively little impact upon decisions made by professional scientists about problems arising within their own sciences." He made the claim in *Structure*, as well, and it can be seen as one of the book's cornerstones.[14] This image of each scientific community having a sociology or ideology *of its own* led Kuhn to look for and articulate the controlling "predispositions," "prejudices," "ideologies," and finally "paradigms" at work inside them. But the committee's investigations and discussion had not seen this kind of isolation.

In one of many drafts he wrote for the committee to review, Kuhn seemed to admit that research and knowledge could change by subtly "drifting" in one direction or another without being distorted by external forces and without experiencing revolutionary paradigm shifts:

[O]ne may categorically deny "co-option" and still recognize the possibility that the differential availability of external funds for different university activities may gradually and subtly alter the direction of institutional development, producing over time a quite decisive transformation.

Because these effects could not be predicted in advance, and might prove beneficial as well as harmful for scientific progress, Kuhn and the committee did not see any basis for a reformulation of university policy. In these notes, however, Kuhn put *Structure* aside to acknowledge that paradigms and conversion experiences were not the only mechanisms by which research could change and progress be achieved. Kuhn's views were in flux, and he was not alone.

∽

After the construction of the Berlin Wall in 1961, Conant returned to Germany to help organize a new center for education in West Berlin, an outpost on the front lines of the ideological stalemate. By 1965, the Conants were back in New York, but Conant's age, his gradually failing health, and family difficulties (often sparked by his son James's manic depression) brought an end to his quest to improve public education and to teach Americans to understand science. Writing his autobiography, *My Several Lives*, dominated his schedule for the rest of the 1960s. "The last years were difficult ones," Hershberg wrote. After a series of strokes condemned him to dependency on others and only occasional lucidity, Conant died in 1978.[15]

During this last dozen years, Conant carried the burden of his critical, skeptical outlook. At times it became heavy. He never stopped worrying about the nuclear technology he had helped to bring about, even admitting in his autobiography that, from at least one comparative point of view, it had been a mistake: "To my mind, the potentialities for destruction are so awesome as to outweigh by far all the imaginable gains that may accrue in some distant future" of cheap and plentiful atomic energy. As for the many thousands of deaths in Nagasaki and Hiroshima, Conant never voiced any regrets. His daughter claimed to detect some, nonetheless.[16]

When *My Several Lives* appeared, Conant revealed his second thoughts about McCarthy and the McCarthy years. He devoted a chapter to his tense encounter with the senator in 1953 and began that chapter with a blunt assessment: "One can brand Senator Joseph McCarthy as a ruthless, effective demagogue who, like Hitler, used the 'big lie' technique." Deployed effectively, a big lie is a form of mind control, a most effective kind of propaganda. By reporting something so outrageous and extreme, it seems to guarantee its own truth, for no one would believe that an otherwise sensible and credible person would make such a claim. Did Conant mean to suggest that the virus of totalitarianism had in fact taken hold in the United States—as he had warned it might when he lectured on "civil courage" in 1945? As he had hinted when he denounced campus McCarthyism in his final report as Harvard's president? Not quite, but he admitted that one could reasonably see it that way.[17]

Conant could denounce and delegitimize McCarthy like this because the specter of the red menace had faded and a new consensus had formed. As Conant's comparison to Hitler suggested, McCarthy and his brand of anticommunism was now recognized as reckless and ethically corrupt. McCarthy had become McCarthyism, as un-American as the totalitarianism or "the Soviet Philosophy" he railed against. The shift had started about a year after Conant's encounter when the trusted war reporter Edward R. Murrow attacked McCarthy and his tactics on television, a medium that was not flattering to the senator and his exaggerated claims. Weeks later, again on television, the nation watched McCarthy and Roy Cohn implausibly accuse the United States Army and the Central Intelligence Agency of secretly harboring Communists. These hearings led to the famous denouncement from the Army's lawyer Joseph Welch—"At long last, have you left no sense of decency?"—and a burst of applause from the gallery. As Welch stood up to McCarthy's bluster and paranoid accusations, a new public image of McCarthy formed. After being censured by the Senate, he faded from the spotlight he craved and died an alcoholic at the age of forty-eight.

Could Conant have been McCarthy's Joe Welch a year before? Welch seemed to muster the courage he needed when McCarthy began to smear a legal assistant that Welch had brought to Washington for the hearings. McCarthy described this assistant, on camera during widely televised proceedings, as having once belonged to the suspiciously left-leaning National Lawyer's Guild. Welch, that is, did not set out to defend civilized decency in the abstract; he was protecting a young colleague ("Let us not assassinate this lad further, Senator," he said. "You've done enough. Have you no sense of decency, sir, at long last?"). When Conant wrestled face to face with McCarthy over suspicious books in the Information Service libraries, he was a leading educator, the former president of Harvard, and well positioned to defend liberalism and intellectual freedom as principles. He had the convictions and he had defended them before against powerful alumni of Harvard and Massachusetts legislators in the 1930s. He also emerged unscathed from those battles, if not stronger than before.

But Conant was evidently not then prepared to take this stand in 1953, for he himself had bought into this "big lie" that communists and communist ideology threatened the nation—if not with certainty, then with sufficient concern to oppose the civil rights of communist teachers in 1949, to lead the Committee on the Present Danger in the early '50s, and to offer only mild resistance to McCarthy's bullying that day in June. This is why Conant qualified his provocative remark in his autobiography. Yes, he wrote, one can say that McCarthy lied to America as Hitler lied to Germany, but "one must not forget," he continued, the unnerving reality of atomic spies and the shocking revelation that Alger Hiss had lied about his Communist past. These and other factors turned Conant into a true-believing cold warrior, he freely confessed. While he and other anticommunists of the time were *sincere* in their beliefs and arrived at them through good-faith efforts to do what was best for the nation, Conant now understood that McCarthy was just a "great and malignant opportunist" who never even believed his own dire warnings about communists and the communist philosophy.[18]

Conant had moved on from McCarthyism by denouncing the man himself and by skirting questions about who was right and who was wrong about the communist conspiracy and the struggle for men's minds. The distinction that mattered in hindsight was between the majority who were sincere "in their exaggerated fears of Communism" and opportunists like McCarthy who were not. Even that distinction, however, could give way to a note of unity as Conant concluded his chapter on McCarthy: everyone, he wrote, no matter where they stood on the issues, would agree that "those days of inquisition" were difficult and not pleasant.

As for the pressing events of the late 1960s, Conant did not have this luxury of hindsight to reflect on the past and refine (or, if necessary, deflect) the issues raised. But his thoughts about McCarthy's opportunism and lack of integrity may have encouraged his skepticism about the ongoing war in Vietnam. In 1967, as American payloads rained on Southeast Asia, Conant hesitated to lend his name to a public relations project, The Citizens Committee for Peace with Freedom in Vietnam. Not unlike his Committee on the Present Danger years before, it teamed leading citizens, intellectuals, and retired politicians to publicly support the war effort. Conant signed on, but only after asking for reassurance by its chair, former senator Paul Douglas, that repeated bombing of North Vietnamese civilians was at least inching the conflict to an end. "There can be no doubt," Conant told Douglas, that "our policy of bombing the North has given the United States a very black eye throughout the world."[19]

The committee's mission statement rehearsed the familiar view that communists in the north aimed to "impose a government and political system upon their neighbors by internal subversion, insurrections, infiltration, and invasion." Though it compared North Vietnamese leaders to expansionists such as Hitler and Mao, Conant's private notes reveal some sympathy with the student movement and critics of the war who maintained that the conflict had always been a civil war rather than another battle in the expansion of international communism orchestrated from afar. To make matters worse, Conant felt misled by Nixon and his generals. The North's successful Tet offensive of January 1968 surprised Conant as much as the public, for Washington had insisted up to that moment that victory and an end to the bombing of North Vietnam was just weeks or months away. To a man who naturally trusted experts, who assumed they had earned their power and credibility, Tet challenged his understanding of the war and those prosecuting it. He "felt burned by his earlier swallowing—and endorsement—of the Pentagon's Pollyanna projections."[20]

In "our benignly chaotic system of political democracy," he had written in 1946, almost any small town leader or citizen can arrive in a position of leadership and power. There was a boundary between leaders and those who are led, of course, and Conant insisted that in times of war and crisis those boundaries must be strong. Leaders must sometimes simplify and hide complex realities, worries, and doubts from the public. But on pain of becoming like the Kremlin, he believed, American leaders must respect the public and the spectrum of talent and insight it contains. Nixon's Pentagon, Conant feared, was losing this connection and becoming an insular, military community. In an urgent letter to Douglas about Tet, he wrote,

> I was assured that those who understood the military situation were absolutely certain that no further increase in American fighting men was needed. In view of the events of the last two weeks I now question the accuracy of this reply. I am afraid that I am becoming more and more suspicious that the Air Force Generals are in control and, as I know from my World War II experience, they may be extremely dogmatic and even unscrupulous in respect to questions which challenge their premises.

Like Kuhn's experience with the Princeton committee, Conant's expectations were confounded.[21] Kuhn learned that scientists and research communities were less insular and closed, it seemed, than he had assumed when writing *The Structure of Scientific Revolutions*. To Conant, Nixon and his generals now seemed dogmatic and unscrupulous, intolerant of competing premises, and disdainful of counterevidence that the war was not going as well as they believed. They had become, in other words, like Kuhn's "normal" scientists—beholden to a single, shared paradigm and a single point of view. Kuhn was becoming more like Conant in his philosophy of science, and Conant became more Kuhnian in his view of the cold war military establishment.

The Frankenstein Paradigm

They did not correspond much during the last half of the 1960s, and the dedication to Conant, once of obvious significance to Kuhn, did not appear in subsequent editions of *Structure*. Still, Conant's and others' criticisms of the book continued to sink in and affect Kuhn's views. Though he had fervently hoped that the book would make a big splash—"If there are no squeals of outrage," he told Conant just before it was published, "I shall have failed in part of my objective"[22]—Kuhn grew increasingly frustrated and dismayed over the success and attention the book brought. At times, *The Structure of Scientific Revolutions* became a revolution-sized headache for a man who no longer aspired to public notoriety as he did when he was a teenager and who, as a self-described "neurotic," was sometimes emotionally unequipped to handle the attention and critical scrutiny his claims inspired. Feyerabend, Popper, Shapere, and other philosophical critics were not squealing with outrage as Kuhn hoped they would be. They did not see the "traditional epistemological paradigm"[23] in ruins. Instead, they demanded that Kuhn make good on claims that seemed vague, confused, or as Conant had put it, "far too grandiose."

Preparing to revolutionize the world of scholarship, in other words, was more rewarding than the aftermath. Now that the exciting historiographic revolution had taken place, Kuhn found himself in the shoes of a plodding, "normal" scientist of the sort for whom Popper and others felt only pity. Kuhn was surely not mind-controlled by his new image of science, but he was professionally controlled by it. Over and over he was asked to explain, discuss, and account for what he had written, to admit the mistakes and confusions that had come to light in countless critical essays and lectures, and publicly mop up after what he himself sometimes recognized as a sprawling, conceptual mess.

"Sometimes I fear I have fathered a monster," he told his editor at the University of Chicago Press in the early 1980s. And he often wished that those who so admired it "would just go away."[24] Like Mary Shelley's Victor Frankenstein, Kuhn had toiled privately for years to test and formulate his insights. In the late '40s, for example, he decided not to take classes in philosophy and history of science. Instead, he educated himself to unlock the significance of his Aristotle experience by reading psychology, history, sociology, and philosophy on his own. Even Conant, on reading *The Copernican Revolution,* was given no hint that Kuhn was working out a new theory of science that would be unveiled only when *Structure* was finished and finally came to life. Now it roamed the earth on its own, far out of Kuhn's control yet never far from his mind. His friend the anthropologist Clifford Geertz, whose book *The Interpretation of Cultures* enjoys popularity and influence comparable to *Structure,* watched him struggle with his monster for decades. After Kuhn's death from cancer in 1996, Geertz wrote that "he lived, anguished and passionate, in its shadow for nearly thirty-five years."[25]

Kuhn ignored the monster when he could. His book of 1978, *Black Body Theory and the Quantum Discontinuity,* is "scrupulously silent" about *Structure,* he confessed.[26] But in the first several years of its life, Kuhn dutifully took the monster back into his laboratory for modifications and improvements. He described these modification in three essays that appeared by the end of the '60s: His "Second Thoughts on Paradigms," "Reflections on My Critics"—those being Popper, Feyerabend, Lakatos, Watkins, and the others at the Bedford conference in 1965—and the postscript to *Structure's* second edition. The postscript drew on the other two essays to tell readers that the very centerpiece and heart of the book they were reading was to be removed and replaced. Kuhn was finally done with paradigms.[27]

Paradigms had been the answer to the question he had wrestled with ever since the Aristotle experience: What do the members of a scientific community share that explains how they work together effectively, how they understand each other's work and its goals, and why their professional judgments about scientific matters fundamentally agree? His new answer, he explained, was not "a paradigm or a set of paradigms," but a "disciplinary matrix." Realizing he was upsetting a beloved intellectual apple cart, he reassured readers that not everything was changing. Whatever *Structure* described as paradigms, or parts of them, or importantly related to paradigms can be found in a disciplinary matrix. Not that much had changed, he implied, except for this crucial difference: "They are, however, no longer to be discussed as though they were all of a piece."[28]

Conant had urged Kuhn to break up his paradigms into the parts he detected inside them—"(a) a theoretical framework, (b) certain unexpressed premises, (c) a number of unformulated rules of correct experimentation"— and to avoid the magic word *paradigm* that bound them together and made him so uncomfortable. "I think it best," he wrote, "when possible to distinguish each of the three components and the way they change in a revolution." No, that's just impossible, Kuhn had replied, for it would overlook the basic priority paradigms have in the scientific enterprise.[29] But by now Kuhn had decided that Conant was right. Scientific paradigms had parts. They included "symbolic generalizations" such as the symbols and mathematics of theoretical science; "models," like the image of electrical current as fluid or gases as swarms of hard particles; celebrated problem-solutions or exemplars (the primary, original meaning of paradigm, he noted); and scientific "values" that inform scientists' judgments about whether and how a theory is successful.

Introducing "values" was another momentous change. A shared commitment to certain values is surely "deep and constitutive of science," he wrote in the postscript. But how they are applied and the difference they make may often depend on "individual personality and biography that differentiate the members of the group."[30] In the early 1950s, Kuhn paid little attention to individual differences within communities because he aimed to understand what they all shared, what made them act and think in similar, compatible ways—"the single professional ideology" that controlled the inner workings of any community. As he put it in an early draft of *Structure*, this ideology is

> gained initially from popular writings and speeches about science, reinforced and elaborated by teachers and texts, and made very nearly rigid by the rigours of professional intercourse and publication. While the ideology remains stable, it governs both the activity and the evaluation of research, and science appears to proceed by accretion.[31]

He placed this image front and center in "The Function of Dogma in Scientific Research" and in *Structure*'s account of normal science. If values are guides to action and decision making, normal scientists did not need to address them because they were predisposed and guided to act in accordance with the values embedded in their paradigm. Yet in the wake of the Bedford conference where he crossed swords with Popper, if not earlier, Kuhn began to think of scientific communities as groups of autonomous, dissimilar individuals.

"Reflections on My Critics"

Kuhn was clearly changing his mind about some things, but not his confident, combative attitude. His official reply to his critics at Bedford, written in 1969, some four years after the conference, opened with an amusing and self-assured gambit. Looking at what had been written about him by the other authors at the conference, he wrote,

> I am tempted to posit the existence of two Thomas Kuhns. Kuhn$_1$ is the author of this essay and. . . . He also published in 1962 a book called *The Structure of Scientific Revolutions*.

But there is also a Kuhn$_2$. He is

> the author of another book with the same title. It is the one here repeatedly cited by Sir Karl Popper as well as by Professors Feyerabend, Lakatos, Toulmin, and Watkins. That both books bear the same title cannot be altogether accidental, for the views they present often overlap and are, in any case, expressed in the same words. But their central concerns are, I conclude, usually very different.

"I have been misunderstood" is often the first resort of a scholar caught in critical and unfriendly headlights. But Kuhn took this strategy a step farther by suggesting that at least some of his critics were *unable* to understand his views. Much of what was said at the Bedford conference, that is, had all the marks of "partial or incomplete communication—the talking-through-each-other that regularly characterizes discourse between participants in incommensurable points of view." Popper's incomprehension, that is, pointed to flaws in Popper's account of science, not Kuhn's. If it were true, as Popper insisted, that the human mind can put on and take off conceptual frameworks as it pleases, then why, Kuhn asked, were Popper and others having difficulty as they attempted to "step into mine"? Though they denied it, these critics really were imprisoned by their mental frameworks and did not appreciate the rich set of problems this normal condition posed for understanding science and human knowledge.[32]

After this clever opening move, Kuhn defended his claims about incommensurability between scientific languages and theories; his view of conversions at the heart of scientific revolutions; and his theory of normal science that so agitated Popper and Feyerabend. In fact, he returned to claims he had made clearest in "The Function of Dogma" to emphasize that *only* normal science made revolutions possible. Its rigidity and dogmatism reveal "severe trouble spots" or anomalies in current knowledge that make critical, revolutionary science possible—a fact, Kuhn said, that his critics are simply unable to see.[33]

Behind this overt defiance, however, Kuhn had begun to retreat. He could not maintain defiant for long, if only because he was now saying incompatible things—for example, that incommensurability explained why Popper and others could not truly understand his theory; *and* that his and Popper's theories of science were not very different. After having criticized Popper at the conference for altogether ignoring "normal science," he now minimized their differences, writing at one point that "on this set of questions our differences are over nuances." As for *Structure*, the book whose claims he now defended, Kuhn admitted that it was nonetheless deeply flawed. It was organized poorly, he confessed, and he blamed himself for at least some of the misunderstandings surrounding it.[34]

The main problems Kuhn saw in it, however, were fundamental. He now rejected *Structure*'s claim that modern, "normal" science is born with the acquisition of a single, unifying paradigm. "If I were writing my book again," he explained, "I would therefore begin by discussing the community nature of science"—without reference to paradigms, that is—and he would acknowledge disunity and the plurality of different schools within large scientific communities. Kuhn had to think of Conant's and Feyerabend's objections when he wrote

> though scientists are much more nearly unanimous in their commitments than practitioners of, say, philosophy and the arts, there are such things as schools in science, communities which approach the same subject from very different points of view.

In 1961, Feyerabend pressed this point repeatedly. Conant too had complained, "You tend to treat the scientific community far too much as a community with a single point of view."[35]

The New Ideology and the New Normal

Paradigms and the communities they belonged to were separating and disintegrating in Kuhn's thinking. Disciplinary matrices had separable parts, scientific communities contained separate, dissimilar schools of thought, and sometimes autonomous individuals guided by different values. As late as 1965, at the Bedford conference, Kuhn spoke of "ideology" as a system of values that is shared in a community as well as "enforced" by the sociological mechanisms of science.[36] But by the time he finished writing "Reflections" and his postscript, he described normal scientists differently—not as a mob or crowd that thinks and reasons in the same way, and not even as a "group" that shares paradigmatic values, beliefs, and professional norms as a condition of membership. *Structure*'s focus on what unifies communities gave way to a new emphasis on the individual autonomy of scientists. Normal science, he now explained, did not mean that a group of collaborating scientists shared a single "normal mind" or made their decisions in the same way according to "a shared algorithm." Lakatos and these other critics had the wrong idea, Kuhn insisted. Once they saw normal science in terms of "the normal group rather than the normal mind," they would understand that the shared ideology of values that Kuhn theorized would affect scientists differently. They might share values that have the same names ("such values as accuracy, simplicity, scope") but they will understand and apply them differently and idiosyncratically according to their own sensibilities and experiences. Plurality and variety are necessary, Kuhn went on, for it allows a community to "hedge its bets" and move through a revolutionary crisis. If "no one were willing to take the risk and then seek an alternate theory, there would be none of the revolutionary transformations on which scientific development depends."[37]

Kuhn had arrived at this point before in his notes from the late 1950s, just before he'd introduced paradigms into his manuscript. Then conceiving of scientists bound by some kind of sociological consensus, he reasoned that when revolution loomed at least one unique and adventurous person must glimpse the promise of unorthodoxy and a new, rewritten history of the field that would lure scientists into the revolutionary fold. But after his discovery of paradigms, revolutionary scientists no longer only *reasoned* or *thought* their way into accepting the new consensus; they took the leap on faith or were sometimes converted and intellectually relocated to a new world. This innovation, perhaps more than any other in *Structure*, created the world of theoretical problems that these and other critics had found.

By so rejecting paradigms and the image of normal scientists thinking and reasoning in the same ways, Kuhn was returning to liberalism, to individual, conscious choice and intellectual daring as the main engine of science. He still spoke of "the group as a whole," but this had to be understood as a collection of individuals sharing no "group mind" and beholden to no ideology that enforced dogmas or consensus. The kind of professional ideology he now envisioned even cultivated skepticism about the truth and finality of current knowledge: "All scientists must be taught—it is a vital element in their ideology—to be alert for and responsible to theory-breakdown." Whether you call that breakdown "severe anomaly," or—as Kuhn reached out to Popperians—"falsification," it was now essential for the health of science that scientists be aware of these eventualities.

This was not the image of scientists Kuhn had in mind in the early 1950s when he theorized about unconscious predispositions that guided scientific thinking; or the powerful social norms within scientific communities that he theorized in response to Philipp Frank's invitation to collaborate in sociology of science. Nor was it the image behind the concept of "theory as ideology" he introduced to Charles Morris in the summer of 1953, or behind his debate with H. Bentley Glass about dogmatism in professional research. Even as he agreed to stop using the term *dogma*, Kuhn professed that paradigms and the dogmatic tendencies that scientific education cultivates were essential: "I doubt that science will get on without them," he told Glass. By the end of the 1960s, however, Kuhn changed his mind. Not just science, but the history and philosophy of science, should get on just fine without paradigms and their powerful hold on the scientific mind that he once believed was necessary for scientific progress.[38]

Kuhn had finally accepted Conant's and Feyerabend's criticisms; their insistence that, whatever philosophers and historians may wish to say about it, science and knowledge always struggle to move ahead any way they can. Sometimes through revolutionary breaks, or something like it, and sometimes not; sometimes because of pioneering individuals and sometimes because of groups; sometimes because one theory replaces another, and sometimes as competing theories exist side by side; sometimes insulated from social, economic, and practical realities, and sometimes in concert with them. In 1970, Kuhn made Conant's forward motion his own by undergoing a kind of revolution. It was not, however, a paradigm shift. He moved ahead by circling back to where he had started, in politics and in philosophy of science.

Epilogue

Writing and Rewriting History

With *Structure* behind him, Kuhn's thinking is often said to have taken a "linguistic turn." He put aside his former interests in the psychological and sociological dynamics of scientific change to adopt in the 1980s a narrower, philosophical focus on language, semantics, "lexicons," and "lexical systems" (as he called them). As these chapters suggest, however, this was a return to Kuhn's early theoretical explorations of the shifting "meaning systems" that seemed to lie beneath his Aristotle experience. Another candidate for revision is Kuhn's own account of where his theories of normal science and paradigms came from. When narrating the development of his ideas and terminology, his recollections usually began with his apprenticeship with Conant and his momentous Aristotle experience; and then moved to the end of the 1950s, when he replaced the "consensus" of normal science with its "paradigms."[1] His recollections jump over the 1950s and his proposals to analyze science on the basis of predispositions that filter and reduce experience and theory-as-ideology that inhibits the creativity of the scientific mind.

When hot-button political terms or images of mental captivity or ideological conversion bubbled up in reviews or critical debates about *Structure* in the 1960s, Kuhn was quick to imply that these political overtones were spurious and that readers had supplied them. In his "Reflections on My Critics," for example, faced with Lakatos's impressions of mobs and Watkins's reference to paradigms as infectious agents, he admitted errors. But they were mainly sins of omission—as if a lack of clarity or flaws in the book's organization had allowed readers to concoct bizarre, incredible interpretations of what Thomas Kuhn$_1$ was trying to say. It was Lakatos and other critics, not Thomas Kuhn$_1$, who wished to explain science on the basis of the "normal mind" known to modern psychology. Thomas Kuhn$_1$ could not be responsible for Kuhn$_2$ who, according to these critics, claimed that

355

a paradigm was "like a quasi-mystical entity or property that, like charisma, transforms those infected by it."[2]

As time went on, however, Kuhn came closer to admitting that when crafting *Structure* he was indeed fascinated by something like a "group mind" or collective "normal mind" operating within scientific communities. He confessed in the 1980s and early '90s to mistakenly supposing that "a group is somehow an individual writ large" that undergoes conversions and gestalt switches in understanding and perception. That use of the gestalt switch "now seems to me mistaken," he admitted, for "revolutions should be described not in terms of group experience but in terms of the varied experiences of individual group members."[3] On a different occasion he called this "a damaging category mistake, one of which I was repeatedly guilty in *Structure*." It creates misleading oversimplifications, as if there exists a "group mind (or group interest)"—such as *Structure*'s early suggestion that "it is sometimes just its reception of a paradigm that transforms a group previously interested merely in the study of nature into a profession or, at least, a discipline." Still, the problem was larger than *Structure*, Kuhn seemed to believe, because contemporary scholarship as a whole lacks effective tools and terminology with which to understand groups of scientists: "We badly need to learn ways of understanding and describing groups that do not rely on concepts and terms we apply unproblematically to individuals," he told his fellow scholars.[4]

When Kuhn first began in the 1950s to outline the book that would become *Structure*, there was no lack of ways to understand and describe groups and individuals. In the partisan, geopolitically divided world he and Conant knew, it was widely understood by the public, their elected leaders, and the majority of public intellectuals that the world's future would be determined by the struggle for men's minds and the single ideology left standing victorious. Where one stood in this battle between the incompatible ideologies of liberalism and totalitarianism, what groups of partisans one belonged to, determined one's political and social identity, which professions, careers, and personal futures were open, and which were closed. For Communists, Sidney Hook insisted—and Conant came to agree—the individual and the collective were united by conspiratorial logic. Communist professors could not be trusted to teach, the reasoning went, because their political affiliations compelled them to betray principles of intellectual freedom, to indoctrinate their students, and so advance Moscow's conspiracy. In cold war politics, in other words, ideology was destiny—both for individuals and collectives. To understand properly the history of science, Kuhn reasoned, one must acknowledge a similar subordination of the individual to the theory, school of thought, and worldview into which he or she was initiated. Even for those pioneers whose careers transcended the normality of a single paradigm and passed through a scientific revolution, paradigms were any scientist's destiny.

When Kuhn remarked on the subject of historiography and the philosophy of history—a field that examines the character of historical understanding, its sources, and what justifications are available for claims about events that exist only in the past—some of his observations were tailor-made to address this lacuna of the 1950s in his recollections. Above all, he emphasized, historical writing is creative. Like the "creative science" he described in his Lowell Lectures and in his early notes for *Structure*, an effective, successful historian does not impart understanding by collecting facts and stringing them into stories. The task is to choose certain facts and to arrange them in ways that strike the historian—and then his or her readers—as familiar, plausible, and credible at a basic, primitive level:

> If history is explanatory . . . it is because the reader who says, "Now I know what happened," is simultaneously saying, "Now it makes sense; now I understand; what was for me previously a mere list of facts has fallen into a recognizable pattern."[5]

Historical understanding, he added, depends on "the primitive recognition that the pieces fit to form a familiar, if previously unseen, product."[5]

What, then, were the cultural and sociological patterns available and familiar to Kuhn himself when his Aristotle experience first suggested to him the insights he would elaborate in *Structure*? Born in 1922, his life began soon after the Russian civil war had ended and the image of political revolution as a sudden, winner-take-all transformation guided by philosophy and ideology became widespread in the United States. Inspiring for some Americans, but terrifying for most, this specter of revolution cultivated an emphasis on patriotism and loyalty that came to the fore during the postwar red scare, partly because of controversial figures who defied the nation's political mood, such as Harlow Shapley and Alger Hiss (whose case especially agitated Conant) and rebels in Kuhn's orbit, including William Remington, Elizabeth Moos, Robert Gorham Davis, and Wendell Furry. There was in addition Kuhn's longstanding interest in propaganda, illustrated in his essay "The Crisis in Democracy" and, before that, in plays and assemblies held at the Hessian Hills School. There was the phenomenon of conversion illustrated by Conant's swings of opinion and conviction as well as those of other intellectuals, such as Karl Popper, Arthur Koestler, and Sidney Hook, whose transformations blurred boundaries between political, scientific, and philosophical beliefs and values. And there were Kuhn's own crises and conversions—in politics, as his pacifism clashed with his interventionism; and in his view of science, as Aristotle transported him to a new way of understanding nature that had been long forgotten and misunderstood.

These interests, events, and individuals—all framed within this era's obsession with the determinative powers of ideology—provided taken-for-granted patterns that Kuhn used to organize his growing knowledge of science's history, to make sense of his shocking Aristotle experience, to theorize those ways by which the scientific mind consciously and unconsciously engages the world, and to elaborate *Structure*'s fundamental point: scientific change is like political change. They are primarily revolutionary, rooted in human sociology, and driven by the considerable power of ideas over the mind to follow the same narrative structure and, sometimes, to create entirely new, different worlds of experience.

And yet, in the wake of *Structure* and its success, Kuhn minimized the significance of politics to his theorizing. In 1973, for example, he received a letter from a Mr. Kenneth Pietrzak of Hartford, Connecticut, who had been told that Kuhn had published an essay on politics. He understood that Kuhn was a historian of science and not an expert on politics; but he reasoned that since paradigms so illuminated the workings of science, they might also illuminate the workings of politics. So he wrote to ask if Kuhn had in fact written such an article and, if so, where it might be available for him to read. Kuhn replied briefly to say he had written on the topic in *Structure* (presumably the first pages of "The Nature and Necessity of Scientific Revolutions") but he dismissed this as merely "a few scattered remarks." He also downplayed the "bit of information" on this subject in *Structure*'s postscript and then signed off: "Beyond that there is nothing. I cannot imagine what your informant had in mind."[6]

A similar unease with politics appears within Kuhn's remarks about Ludwik Fleck, whose *Genesis and Development of a Scientific Fact* he praised in *Structure* as an inspiration. In the late 1970s, before writing a foreword to its English translation, Kuhn reread it and applauded the newly translated text as a "brilliant and largely unexploited resource" for contemporary intellectuals. But he recoiled at Fleck's theorizing about "group minds" in science, his signature concept of the *Denkkollectiv*—"the to me unknown and yet vaguely repulsive perspective of a sociology of the collective mind,"

Kuhn wrote.[7] However repulsive he may have found Fleck's concept, it was not unknown to Kuhn twenty years before when he first encountered it. As a teenager, he had extolled the positive, constructive powers of propaganda and proposed a "ministry of propaganda" to cultivate a powerful and uniform pro-democracy consensus in the United States. The theorizing that led him finally to his conception of paradigms usually took collectivism for granted and aimed to explain how and why groups of scientists see and understand nature, despite its unbounded complexity, in similar ways. Ironically, phrases and passages in *Structure* that emphasize collectivism and uniformity agitated Conant, H. Bentley Glass, Paul Feyerabend, and others. Before Kuhn recoiled at Fleck in the 1970s, in other words, these readers had recoiled at collectivist aspects of Kuhn's theories that they found troubling and repulsive.

By the 1980s and early 1990s, while never acknowledging an original connection between *Structure* and American politics, Kuhn paid increasing critical attention to those parts of *Structure* that most resonated with the struggle for men's minds and had once agitated his politically attuned critics. As shown in the last chapter, he described it once as a monstrous and largely mistaken picture of scientific communities as cohesive groups governed by a single "group mind" (or possibly, he told the psychoanalyst Kubie, a "group-unconscious") not too unlike the *thought collectives* that later made him wince when rereading Fleck. As for "The Function of Dogma in Scientific Research," in which the hot-button word *dogma* and an emphasis on the narrowness of scientific training brought his theory of normal science close to popular notions of brainwashing and mental captivity, Kuhn was reported to be "adamant" that the essay be excluded from *The Road Since Structure*. Published after Kuhn's death, this collection of essays explore his "later attempts to rethink and extend his own 'revolutionary' hypotheses."[8] The essay that most strongly documents the earlier, more overtly political aspects of Kuhn's theorizing was—emphatically—not to be included, rethought, or extended.

Like other scholars who lived through the anxieties and inquisitions of the early cold war, Kuhn may have been averse to recognizing the politics of paradigms because of the tensions and anxieties of the years in question. If exploring the methodology of science is difficult when one wishes to examine "the science in which we happen to believe," as Kuhn told his audience at the Boston Public Library,[9] then it may have been difficult or impossible for him to retrace the paths of his own thought through ideas and concepts that he once found intriguing and compelling, but which later came to seem less credible, unprofessional, even "vaguely repulsive." So it is, he once remarked, that "no one is an expert on the events of his or her own life."[10] Especially, it would seem, when the events in question belonged to the turbulent ideological times into which Kuhn was born, which reached a crescendo as *The Structure of Scientific Revolutions* was commissioned, and which did not fully relax until the cold war ended in the last decade of Kuhn's life—roughly thirty years after *Structure* was published.

∽

Except for these reflections about Kuhn in his subsequent career, I chose to end this study in the year 1970 because at that point, as far as the author of *The Structure of Scientific Revolutions* was concerned, the story of paradigms was over. Other theoretical directions, more philosophical and historical, and less sociological and political, marked Kuhn's future intellectual path. At least some readers, I suspect, will find this decision puzzling for it seems to leave aside trends and developments that cemented *Structure*'s reputation. As one reader of this manuscript put it, "What on earth happened to make it *legitimately* famous and influential?"[11] The question is worth unpacking for what it reveals about the past and the present.

One answer is simple: too much happened for any book to recount. To examine how *Structure*'s ideas were received, criticized, and appropriated would be to explore the histories of philosophy, history of science, sociology, anthropology, literary criticism, business, law, the self-help industry, leftist politics and other endeavors from 1962 onward. Standing at the feet of these mountains, most scholars demur, as did I. Studies of *Structure*'s enormous influence are thus likely to remain metahistorical and based on the frequencies of words and phrases in digitized libraries or citation analyses that chart where and how frequently writers in different fields referred to *Structure* in their own writings.[12] Studies like this confirm that *Structure* became famous and influential, but they do not explain why and whether—as this reader took care to emphasize—those reasons are *legitimate*.

From at least one point of view, however—Thomas Kuhn's—there is reason to wonder whether *Structure*'s fame and influence were legitimate. In public and in private, he criticized the book (some of its central claims, its theory of paradigms, and its sociological approach) and in its second edition he announced "a new version of the book" in which he would attempt to rectify these problems.[13] He also continued to believe that lay readers and world-class scholars—those who liked the book and those who did not—often misunderstood what it said, or what it was trying to say.

Together with his personal, complicated relationship to it, *Structure*'s ubiquity and popularity in postwar culture suggests that it can be seen not only as a work of scholarship but as a work of modern art. At least part of its compelling appeal and influence rests in questions such as, "What exactly does it mean?" and "Why is it so important?"

What kind of legitimacy did this reader have in mind as a foundation for *Structure*'s fame and influence? Most professional historians of science and intellectual historians see the past as an interconnected and evolving network of ideas, perspectives, and theories. On this view and despite Kuhn's misgivings, *Structure* deserves its status as an influential classic because it can be situated among *other* important and influential texts, the celebrated authors who wrote them, and the pathways—sometimes personal relationships, sometimes through books and articles—over which reasoned arguments and compelling evidence travel from scholar to scholar, from teachers to students, and from the past to the present. In Kuhn's case, that means his teachers and colleagues at Harvard, more distant scholars he admired (such as those mentioned and cited in *Structure*), as well as classic figures from decades and centuries before. Most important for any book's reputation, of course, are those scholars who first received *Structure*, such as Karl Popper and his followers who criticized it and Richard Rorty who championed Kuhn's ideas in his widely read *Philosophy and the Mirror of Nature* of 1979.

There is no place for politics in this image of intellectual history—and for good reason. The agendas of politics are subjective, often passionate, inconsistent, and poorly reasoned. Its struggles are often local, transient, and quickly forgotten (or regretted) from one generation to the next. There is wisdom in this approach, therefore, as it guides historians toward ideas, discoveries, and intellectual values of genuine, enduring value and consequence. To borrow again from Kuhn's theory of historical understanding, it provides a pattern for historical understanding that renders the unfamiliar familiar and produces the "aha" of recognition and understanding. Aristotle corrected and built upon his teacher, Plato. Isaac Newton stood on the shoulders of Copernicus, Kepler, and Galileo. Thomas Kuhn, it is often said, corrected the logical empiricists before him and inspired many who came after, thus earning his place in the pantheon.

At the same time, however, this approach to intellectual history can be a trap. As Conant's, Feyerabend's, Glass's, and Kuhn's own evolving concerns about *Structure* demonstrate, standards of scholarly legitimacy and importance are not necessarily shared by scholars in the same field. More importantly, they have historical lives of their own that are readily obscured by ideals of timeless, unchanging standards of importance and legitimacy.[14] As Kuhn might have put it in the early 1950s, this belief that the origins and significance of great, classic texts lay primarily in their relations to other great, classic texts is a prior assumption—a "predisposition" or possibly "unconscious prejudice" that lacks prior justification but nonetheless encourages a researcher "to ignore or discard certain portions of experience in formulating or verifying his theories."[15] To borrow Kuhn's distinction between "textbook" and "creative" science, it belongs to "textbook" historiography, which serves a number of disciplinary and professional functions, but tends to oversimplify, if not overlook, the different and complex processes by which a text became an influential classic. In the case of *Structure*, this book argues, cold war sensibilities that firmly divide politics from scholarship have led us to overlook the formative roles of politics in both Conant's and Kuhn's thinking about science and, consequently, in *Structure* itself. The fact that the struggle for men's minds is no longer a framework of academic life and experience may indeed make that cold war struggle seem an unlikely—possibly illegitimate—context for understanding how and why *Structure* became so famous and influential. But unfamiliarity constitutes neither an argument nor evidence against the claim that it was once a regular and, in *Structure*'s case, consequential feature of scholarly life.

Once we place this assumption aside, we can begin to see how *Structure*'s fame and influence, on the one hand, and those sensibilities that recognized it as important, revealing, and legitimate, on the other, grew together as parts of a larger story within American postwar intellectual history. In this story, politics is not evanescent, but productive and transformative. In schematic form, it begins with the coincidence of three things in the late 1940s: Kuhn's intellectual ambitions, Conant's determination to teach the nation how to understand science, and the souring relations between the United States and the Soviet Union. This geopolitical context sustained the militarization of the nation that followed its entry into World War II and sparked momentous convergences and compromises between political and intellectual life, some of which were detailed in these chapters.[16] These included institutional compromises, as the liberal arts curriculum collided with officer training on university campuses, goldfish-swallowing students joined ranks of uniformed student-soldiers, and FBI agents arrived in the 1950s to interview professors, students, and department secretaries about subversives in their midst. They included cultural and civic compromises, as public fear and intense, personal dislike of communism led scholars such as Sidney Hook and administrators like Conant to discard long-held principles of intellectual freedom and declare that some of their colleagues, precisely because of their political beliefs and associations, should no longer be permitted to teach. And they include intellectual convergences, as diplomatic crises raised fundamental questions about the mind and its susceptibilities to ideological control—questions that supported careers for pundits and psychologists who studied brainwashing and for political scientists and philosophers who became experts in the study of totalitarianism or Sovietology.

Structure itself can profitably be understood as one fruit of these convergences, as Kuhn borrowed from this era's political concerns and its fascination with ideology to theorize the nature of science and scientific change. They set the stage for Kuhn's shocking Aristotle experience, the ways he interpreted it, and for the early, often politically sensitive readings of *Structure* from H. Bentley Glass, Feyerabend, Conant, and others. Yet the political circumstances that supported and surrounded *Structure*'s origin and its early reception did not dissipate after *Structure*'s publication in 1962. For the many audiences that would later read it, the cold war continued as a shared cultural and political backdrop, a reservoir of shared assumptions and sensibilities, against which it grew in popularity and influence.

Crucially, this political background helped make *Structure* a great and rewarding book to read. Kuhn led his readers to discover novel connections between his case studies, on the one hand, and what the ongoing ideological struggle had already taught most Americans to understand about the nature of revolutionary change, the faculties and susceptibilities of the mind and its perceptions, and the powers of language, sociology, and ideology itself to shape experience. As for the "conversion experiences" that Kuhn identified with scientific revolutions, they were longstanding features of American culture that the cold war's brainwashing controversy had revived in a new, urgently politicized form. Chapter by chapter, many of Kuhn's readers felt themselves transformed and enlightened—rescued, perhaps, from the faulty image of science by which they had been possessed. Despite the ambiguities and mistakes that he later acknowledged, Kuhn's voice was comparably enthusiastic, clear, and confident. It reflected the "extraordinarily exciting time" he had when its concepts, terminology, and organization finally came together in his first complete draft. Even when the book consisted of half-written chapters that went off track, and when Kuhn placed them aside in the 1950s for other projects and commitments, writing *Structure*, he later said, was "what I really wanted to be doing."[17] It showed.

Second, by absorbing key elements of the politics of the 1950s and yet transposing them into a different domain of inquiry (history of science, not politics) and into a different terminology ("paradigms" and not "ideology"), *Structure* became a transitional text that moved its readers from the academy of the 1950s to that of the 1960s and after. For scholars intimidated, if not professionally damaged, by the campus inquisitions of the late 1940s and 1950s, *Structure*'s view of closed, autonomous communities as engines of intellectual progress—properly isolated from political and social concerns and forces—was likely to be welcomed. For the philosophers whom Kuhn most hoped to reach and impress with *Structure*, the effects of this ongoing disengagement from politics were in some cases transformative. Besides the predictable decline of Marxism as a philosophical specialty, research in social theory and values moved aside to make room for growing interest in logic and linguistic analysis. Even new conceptions of reason and rationality, scholars have recently argued, became familiar in papers, books, and lecture halls as the discipline accommodated itself to the now-enduring norms of cold war life that were reflected in *Structure*'s picture of relatively closed scientific communities.[18] For historians and philosophers of science, for whom *Structure* became and remained a founding professional text, Kuhn posed provocative questions that had all the urgency and fascination of the early cold war ("Do Russians really live in a different, inscrutable mental world?") but found a scholarly life of their own well removed from the controversies and suspicions that had surrounded political debate in the '50s ("Did Galileo and Aristotle really see and understand motions so differently?"). Understood as "paradigm shifts" and not changes in "theory-as-ideology," scholars could debate the nature and implications of scientific revolutions without even mentioning political revolutions; or puzzles about rationality and "incommensurability" without trespassing into cold war debates about diplomacy and ideology.[19]

As these chapters show, Kuhn himself transitioned from a politically engaged intellectual of the 1930s and early '40s to a comparatively unpolitical, professional scholar of this postwar world. At the Hessian Hills School, his political and intellectual development went hand in hand with his growing confidence in "reason," as he later explained to Robert Gorham Davis. In secondary school and college, he wrote essays on democracy, his personal struggles as a (former) pacifist, and the quandaries of nihilism. But by the early 1950s, Kuhn began to embrace the emerging postwar norms of professionalism and political disengagement. To Ralph Lowell and his audience at the Boston Public Library, he firmly and repeatedly distinguished himself from those philosophers and sociologists of science who presumed to engage voters and politicians about the future of scientific research. To Mr. Pietrzak, twenty years later, he emphasized "there is nothing" in his writings of substantive political interest. And when he recounted the story of paradigms from his Aristotle experience to *Structure* thirty years later, he never mentioned his view that "theory-as-ideology" was the key to understanding scientific change. His overt and frequent appeal to political concepts for understanding science evidently came to seem distant and unconnected—perhaps even illegitimate—to his sensibilities as a world-famous scholar.

For those outside the academy, finally, this movement away from the anxious 1950s also helped to connect *Structure* to the new cultural moods of the 1960s. By abstracting the powers of ideology and revolution from politics and applying them to science's history, Kuhn laid a foundation for adapting them in yet other ways and directions. As it happened, *Structure* arrived as American students collectively resolved to move past the repressive, omnipresent specter of communist revolution toward constructive, progressive revolutions in civil rights, feminism, and democracy itself. Not unlike Kuhn's own teenage transvaluation of propaganda as a force for cultural and democratic progress, this generation now transvalued the specter of revolution they had been taught to fear as a force they could embrace for positive, progressive change. As Kuhn's files of letters from readers show, not only scholars but professionals and basement inventors embraced *Structure* as a guide or justification for their own, revolutionary aspirations.

The politics of paradigms, it turns out, are larger and more extensive than Thomas Kuhn's famous book. They include as well these political, institutional, and cultural developments that helped make *Structure* famous and influential. Acknowledging them does not mean *Structure* cannot be read in abstraction from its time and place, as a text that transcends politics and joins the great historical conversations of philosophy, sociology, psychology, and other fields. The creative purposes to which scholars put texts generally are not and ought not be constrained by their origins. Yet that distinction may be little comfort to those who nonetheless uphold *Structure* as a classic reflection of the academy and all that is noble in contemporary thinking about science and knowledge. As one reader predicted, to those who expect to see Kuhn's famous claims about revolutions, paradigms, and conversion experiences portrayed in gleaming, marble white and untouched by political controversy—much less by characters like Sidney Hook, Joseph McCarthy, or even James Bryant Conant at the height of his cold warrior paranoia—these chapters may be "profoundly uncongenial."

If so, I recommend taking to heart Paul Feyerabend's incisive commentary on *Structure*'s first draft. The past, Feyerabend told Kuhn, is merely "a series of accidents combined with struggle for power, etc. etc." in which scholars find meaning by imposing some kind of rational order that suits their aims and purposes. Feyerabend accused Kuhn of turning this situation around and claiming to see the past itself essentially ordered and arranged into the communities, crises, and revolutions from which he drew his philosophical conclusions. That was the tell, Feyerabend exclaimed, for "anybody who wants to derive reason from history is therefore bound to cheat at one place or another!"[20]

Kuhn did not believe he had cheated. But he knew that something about Feyerabend's chaotic, anarchic picture of the past was correct. "An apparently arbitrary element, compounded of personal and historical accident," he wrote at the beginning of *Structure*, "is always a formative ingredient of the beliefs espoused by a given scientific community at a given time."[21] If that reasonable observation extends to intellectual communities, to Kuhn himself—an ambitious, creative scholar who happened to begin theorizing about science in the first, frosty depths of the cold war—and to ourselves, for whom *Structure* remains a towering and legitimate classic, we can see its proximity to the cold war in a different, more productive and congenial light. Not as a text whose legitimacy is threatened or tarnished, but as a paradigm of a different sort—one that shows how classic works in philosophy, history, and literature are often connected in accidental but nonetheless vital ways to the struggles of their times.[22] If that reduces them, the problem belongs not to these great books, but to the stories we tell ourselves about them.

Acknowledgments

Many scholars, colleagues, and friends contributed to the ideas in this book and the methods I used to organize and express them. They include Naomi Oreskes, John Krige, and the other participants of a 2010 workshop on cold war history of science at Caltech; Elena Aronova and fellow conferees who met at the Alfred Krupp Wissenschaftskolleg in Greifswald, Germany, in March 2012; Don Howard and Martin Carrier, whose workshop on the social relevance of philosophy of science at the University of Bielefeld in 2012 I was privileged to attend; and Robert J. Richards, Lorraine Daston, and the University of Chicago Press, whose celebration of the fifty-year anniversary of *Structure* provided more opportunity to refine and adjust this revisionist picture of Kuhn and his famous book. I also thank audiences at Notre Dame, Butler University, the University of Cincinnati, University of British Columbia, the University of Western Ontario, and the Hungarian Academy of Sciences, as well as those many individuals who have answered questions and offered perspective along the way, including Juan Mayoral, Chris Hamlin, Steve Fuller, Joel Isaac, John McCumber, Simone Turchetti, Karen Darling, Andrew Jewett, Jeff Dean, Geert Somsen, George Smith, Patrick Slaney, Chris Haufe, Dawne Moon, David Ramsay Steele, Matteo Collodel, Gerald Holton, Michael Schorner, and Paul Hoyningen-Huene. Other helpful supporters, critics, and skeptical sounding boards include Thomas Uebel, Alan Richardson, Elizabeth Nemeth, Ron Giere, John Beatty, Phil Mirowski, Heather Douglas, Janet Kouraney, Audra Wolfe, Hans-Joachim Dahms, Don Brosnan, John Rossi (who kindly shared archival documents), and John Huss. I am also grateful to Bob Ross, former leader of the Students for Democratic Society, for fielding questions I posed about *Structure* in the 1960s, and Edgar Schein for his recollections and references to his research in brainwashing and coercive persuasion in the 1950s.

    A different kind of thanks goes to James Hershberg, whose monumental biography of James Bryant Conant supports these chapters (and which I have relied on throughout as authoritative), Christopher Phelps, whose *Young Sidney Hook* illuminates the radicalized world of New York philosophy in the 1920s and '30s like no other, Timothy Melley, whose paper "Brainwashed! Conspiracy Theory and Ideology in the Postwar United States" nudged me toward my own aha experience connecting Kuhn, Conant, and the early cold war, and Sarah Bridger who shared with me her then-unpublished research into the Kuhn committee at Princeton that helped to inspire this book's concluding chapter.

I am particularly indebted to Gary Hardcastle, Ádám Tuboly, and K. Brad Wray who went above and beyond to read and comment on entire drafts of this book, sometimes repeatedly, and to Randy Auxier and John Shook, in whose series American Philosophy and Cultural Thought this manuscript found a perfect home at SUNY Press. I also thank editors Andrew Kenyon and Rafael Chaiken and two anonymous readers who provided support and constructive suggestions, as well as Kathleen League for preparing the index.

I thank Theodore Conant for permission to research and quote from the unpublished letters of James Bryant Conant, as well as the Harvard University Archives for providing scans of images held in their collections and for permission to reproduce these images and to quote from documents held in their collections. I also thank Nora Murphy and the staff at the Institute Archives and Special Collections at MIT for help locating archival documents and permission to quote from Thomas Kuhn's papers. In their own ways, Sarah Rothermel and Raffaello LaMantia also made this book possible by graciously providing me with guestrooms while I visited these archival facilities in Cambridge. For assistance searching documents at the Hoover Institution archives I thank Ron Basich. For locating other images and arranging or extending permission to reprint them, I am grateful to Dr. Manfred Lube at the University Library Klagenfurt, Austria; Adam Jaenke at the Cleveland Public Library; Kathi Neal at the University of California, Berkeley; Janet Bunde and Celeste Leigh Brewer at the New York University Archives; Marianna Apostolakis and Nanci Young at Smith College; Wendy E. Chmielewski and Mary Beth Sigado at the Swarthmore College Peace Collection; Christina Moretta at the San Francisco Public Library; Aaron Lisec at Southern Illinois University; Dorothy Pezanowski at the Croton Historical Society; Julia Gardner at the University of Chicago Library; Stephanie Stewart at the Hoover Institution Archives; Ruth Cahir at British Pathé Ltd.; James Stimpert at Johns Hopkins University; Scott Drawe at Chicago Public Library; Jeff Bridgers at the Prints and Photographs Division as well as the staff at the Newspaper and Current Periodical Reading Room of the Library of Congress.

Abbreviations

FRUS. U.S. State Department, *Foreign Relations of the United States* series. Washington, D.C., U.S. Government Printing Office.

Objectives. Harvard University, Committee on the Objectives of a General Education in a Free Society, *The Objectives of a General Education in a Free Society* (Cambridge: Harvard University Press, 1945).

RSS. Thomas S. Kuhn, *The Road Since Structure,* ed. James Conant and John Haugeland (Chicago: University of Chicago Press, 2000).

Structure. Thomas S. Kuhn, *The Structure of Scientific Revolutions* (Chicago: University of Chicago Press, 4th edition, 2012).

TSK-MIT. Thomas S. Kuhn Papers, Massachusetts Institute of Technology, Institute Archives and Special Collections, Manuscript Collection MC240.

UCPR. University of Chicago Press Records, Department of Special Collections, Regenstein Library, University of Chicago.

Notes

Preface

1. Daniel Bell, *The End of Ideology: On the Exhaustion of Political Ideas in the Fifties* (Glencoe, IL: Free Press, 1960), 15.

Introduction

1. On McCarthy, his Wheeling speech, and his crusade, see David Oshinsky, *A Conspiracy So Immense: The World of Joe McCarthy* (New York: MacMillan, 1983), esp. ch. 7 and 109; on the Canwell Commission investigating communist subversion at the University of Washington, see Ellen Schrecker, *No Ivory Tower: McCarthyism and the Universities* (New York: Oxford University Press, 1986), esp. 94–112; on young William F. Buckley Jr. at Yale, see Sigmund Diamond, *Compromised Campus: The Collaboration of Universities with the Intelligence Community, 1945–1955* (New York: Oxford University Press, 1992), ch. 7; and on Buckley's *God and Man at Yale* (Washington, DC: Regnery, 1951), see 166–78; on Conant's presidency of Harvard and its reputation as "The Kremlin on the Charles," see James Hershberg's definitive biography, *James B. Conant: Harvard to Hiroshima and the Making of the Nuclear Age* (New York: Knopf, 1993), 626.

2. Hershberg, *James B. Conant*, 632, 635.

3. Quoted by Edward Ranzal, "McCarthy Charges 'Mess' at Harvard," *New York Times*, Nov. 6. 195.

4. Quoted in Ellen Schrecker, *The Age of McCarthyism: A Brief History with Documents* (New York: Bedford/St. Martins, 2002), 239; James B. Conant, *My Several Lives: Memoirs of a Social Inventor* (New York: Harper and Row, 1970), 578; on the Hiss affair, a classic account remains Whittaker Chambers, *Witness* (Washington, DC: Regnery, 1952).

5. On Eisenhower and McCarthy, see Hershberg, *James B. Conant*, 651; on the note about McCarthy, and being on McCarthy's "list," see Conant, *My Several Lives*, 564.

6. European newspaper reports of the Cohn-Schine tour are summarized and quoted in USIA files, "German Press and Radio Comments on Cohn and Schine," dated 4-22-53, and 4-24-53. FRUS, 1952–1954, v. 1, pt. 2, General: Economic and Political Matters, Doc. 215. Available at history.state.gov.

7. Telegram from John Foster Dulles to Conant and other diplomats, April 3, 1953. FRUS, 1952–1954, v. 1, pt. 2, General: Economic and Political Matters, Doc. 206. Available at history.state.gov.

8. "Adenauer, Conant Visit Here Today," *Crimson*, April 17, 1953; also Hershberg, *James B. Conant*, 658. On "junketeering gumshoes," see Hershberg, *James B. Conant*, 656 and Nicholas Von Hoffman, *Citizen Cohn* (New York: Doubleday, 1988), 147. Cohn's phone call is mentioned in "German Press and Radio Comments on Cohn and Schine" (FRUS, 1952–1954, v. 1, pt. 2, General: Economic and Political Matters, Doc. 215). Theodore Kaghan, quoted in "McCarthy Blamed by Kaghan in Bonn," *New York Times*, May 13, 1953.

9. Conant, *My Several Lives*, 565; on Eisenhower's remarks, see C. P. Trussell, "Some Books Literally Burned After Inquiry, Dulles Reports," *New York Times*, June 16, 1953.

10. "President Denies Clemency to Rosenbergs in Spy Case," *New York Times*, Feb 12, 1953.

11. Anthony Leviero, "Eisenhower Backed on Book Ban Talk," *New York Times*, June 17, 1953.

12. Ibid. On McCarthy's history as a boxer, see Oshinsky, *A Conspiracy So Immense*, 10, 13.

13. This and the following quotations from the Conant-McCarthy encounter are given in Conant, *My Several Lives*, 561–75.

14. McCarthy quoted in Schrecker, *The Age of McCarthyism*, 240; on whether McCarthy claimed there were "two hundred and five" or "fifty seven" or some other number of hidden Communists, see Oshinsky, 109–12.

15. Kuhn, *The Structure of Scientific Revolutions* (Chicago: University of Chicago Press, fourth edition, 2012), (hereafter, *Structure*), 111. A. R. Hall's *The Scientific Revolutions, 1500–1800* (Longmans, Green) was first published in 1954. A history of historians' treatment of scientific revolutions is I. B. Cohen, *Revolution in Science* (Cambridge: Harvard University Press) 1985.

16. Kuhn, *Structure*, 111.

17. Stephen R. Covey, *The Seven Habits of Highly Effective People* (New York: Free Press, 1989); *Adbusters*, Aug. 5, 2013; see also "An Open Letter to the Ecological Economics Movement," *Adbusters*, July 16, 2009.

18. Kuhn to Charles Morris, Oct. 6, 1953, Thomas S. Kuhn Papers, MIT Institute Archives and Special Collections, Manuscript Collection-MC240 (hereafter, TSK-MIT), box 25, folder 53.

19. The summer session is posted in *The Mississippi Valley Historical Review* 38, no. 1 (June 1951): 162.

20. Sidney Hook, *Out of Step* (New York: Carroll and Graf, 1987); Schrecker, *No Ivory Tower*; on the practice of "naming names," see Victor Navasky, *Naming Names* (New York: Viking Press, 1980); on anticommunism's effects on academic philosophy, see John McCumber, *Time in the Ditch* (Evanston: Northwestern University Press, 2001) and *The Philosophy Scare* (Chicago: University of Chicago Press, 2016); and George Reisch, *How the Cold War Transformed Philosophy of Science: To the Icy Slopes of Logic* (New York: Cambridge University Press, 2005).

21. Kuhn, *Structure*, 167.

22. Ibid., 1.

23. See, e.g., George Reisch, "Did Kuhn Kill Logical Empiricism?" *Philosophy of Science* 58, no. 2 (June 1991); Michael Friedman, "Remarks on the History of Science and the History of Philosophy," in *World Changes: Thomas Kuhn and the Philosophy of Science*, ed. P. Horwich (Cambridge: MIT Press, 1992), 37–54; Guy Axtell, "In the Tracks of the Historicist Movement: Re-assessing the Carnap-Kuhn Connection," *Studies in History and Philosophy of Science* 24, no. 1 (1993): 119–46; Gurol Irzik and Teo Grünberg, "Carnap and Kuhn: Arch Enemies or Close Allies?" *British Journal for the Philosophy of Science* 46, no. 3 (Sept. 1995): 285–307.

24. On Kuhn's relations to Kant, see, for example, Michael Friedman, "Kant, Kuhn, and the Rationality of Science," *Philosophy of Science* 69, no. 2 (2002): 171–90; to Kant and Hegel, see Bird, *Thomas Kuhn* (Princeton: Princeton University Press, 2000) and Hoyningen-Huene, *Reconstructing Scientific Revolutions* (Chicago: University of Chicago Press, 1993); to Talcott Parsons, see Joel Isaac, *Working Knowledge: Making the Human Sciences from Parsons to Kuhn* (Cambridge: Harvard University Press, 2012). On Kuhn's relationship to David Hume and other influences, a recent helpful summary is Juan V. Mayoral, "Five Decades of *Structure*: A Retrospective View," *Theoria* 27, no. 3 (2013): 261–80. On Kuhn's relationship to psychology, see David Kaiser, "Thomas Kuhn and the Psychology of Scientific Revolutions," in *Kuhn's* Structure *at Fifty*, ed. Richards and Daston (Chicago: University of Chicago Press, 2016), 71–95, as well as Maurice Mandelbaum, "A Note on Thomas Kuhn's *The Structure of Scientific Revolutions*," *The Monist* 60, no. 4 (1977): 445–52, 450.

25. K. Brad Wray, *Kuhn's Evolutionary Social Epistemology* (Cambridge: Cambridge University Press, 2011); Alexander Bird, *Thomas Kuhn*, viii, ix, 2–3.

26. Hoyningen-Huene, *Reconstructing Scientific Revolutions*, 14–19, 258.

27. John Earman, "Carnap, Kuhn, and the Philosophy of Scientific Methodology," in *World Changes*, ed. Paul Horwich (Cambridge: MIT Press, 1993), 9–36, 19. For Kuhn's confession to writing this purple prose, see Kuhn, "Afterwords," in the same volume, p. 314, or in Kuhn, *The Road Since Structure* (hereafter, *RSS*), ed. James Conant and John Haugland (Chicago: University of Chicago Press, 2000), 228.

28. David Hollinger, *Science, Jews, and Secular Culture* (Princeton: Princeton University Press, 1996), 169, 170; Peter Novick, *That Noble Dream* (Cambridge University Press, 1988), 530. Mary Jo Nye's *Michael Polanyi and His Generation* (Chicago: University of Chicago Press, 2011), acknowledges political meanings and

values latent in Kuhn's *Structure* (see, e.g., ch. 7), but treats them as peripheral to the book's intellectual core: "It is tempting to argue that it is precisely the absence of an impassioned and explicit political agenda" in *Structure*, Nye writes, that explains its academic and popular success.

29. Isaac, *Working Knowledge*.

30. Ibid., 65; George Homans, *Sentiments and Activities* (Glencoe, IL: Free Press, 1962), 4; see also Homans's *Coming to My Senses: The Autobiography of a Sociologist* (New Brunswick, NJ: Transaction Books, 1984), 102–103 and, for his relationship to Kuhn, 47. Homans's personal investment at Harvard in Pareto's political theorizing qualifies Isaac's generalization that "it was Pareto's theory of scientific knowledge, not his critique of social democracy that mattered to Henderson and his associates" (64). Another associate, the historian Crane Brinton, reported that Henderson and his fellow devotees of Pareto were likely to be called "the Pareto Cult" and that Henderson was known to attack "people he called 'liberals'" and their faith in "the great traditions of American democracy" (Barbara S. Heyl, "The Harvard 'Pareto Circle,'" in *Talcott Parsons: Critical Assessments Vol. 1*, ed. P. Hamilton [Routledge, 1992], 29–49, see p. 30, where Heyl discusses Brinton and Homans). Scholars who have acknowledged the well-known politics of Henderson and his Pareto Circle include Alvin Gouldner, *The Coming Crisis of Western Sociology* (New York: Avon, 1971), 149; Novick, *That Noble Dream*, 243; Samuel Bloom, *The Word as Scalpel* (New York: Oxford University Press, 2002), 149; Steve Fuller, *Thomas Kuhn: A Philosophical History for Our Time* (Chicago: University of Chicago Press, 2000), 164–66.

31. Isaac, *Working Knowledge*, 26.

32. Jamie Cohen-Cole, "Instituting the Science of Mind: Intellectual Economies and Disciplinary Exchange at Harvard's Center for Cognitive Studies," *British Journal for the History of Science* 40, no. 4 (2007): 567–97, 570. See also Cohen-Cole's *The Open Mind: Cold War Politics and the Sciences of Human Nature* (Chicago: University of Chicago Press, 2014), 150, 158; and Jamie Cohen-Cole, "The Creative American: Cold War Salons, Social Science, and the Cure for Modern Society," *ISIS* 100 (2009): 219–62, 235–36.

33. Steve Fuller, *Thomas Kuhn: A Philosophical History for Our Time* (Chicago: University of Chicago Press, 2000), 5, 6.

34. Conant to Kuhn, June 5, 1961, TSK-MIT, box 25, folder 53.

35. Marx Wartofsky described Kuhn's "glorious explosion" in *Models: Representation and the Scientific Understanding* (Dordrecht: Reidel, 1979), 131. Kuhn mentions these other works throughout *Structure* (e.g., 3, 113, 190).

36. Kuhn, *Structure*, 92–94.

37. Conant, *My Several Lives*, 579.

38. See Kuhn's writings on historiography, such as "The Relations between the History and the Philosophy of Science," in *The Essential Tension* (Chicago: University of Chicago Press, 1977), 3–20, esp. 16, as well as Kuhn's Lowell Lectures, discussed in chapter 9.

Chapter 1. Progress and Revolution in the Suburbs of New York

1. Elisabeth Cushman, "Wisdom on War Flows from Students' Mouths," *The Daily Argus* (Mt. Vernon, NY), Nov. 18, 1935. The story appeared also in other area papers on the same date.

2. A draft of Kuhn's speech "Report on the Student Strike" remains in his archived papers, TSK-MIT, Box 1, folder 2. Additional details about these events are given in Katherine Moos Campbell, *An Experiment in Education: The Hessian Hills School, 1925–1952*, PhD Dissertation, Boston University (1984), 172–73, 194–95, 199–200. See also Jensine Andresen, "Crisis and Kuhn," *Isis* 90, supplement: S43–S67, S45.

3. Quoted in Moos Campbell, *An Experiment in Education*, 199.

4. The Kuhn family appears in Stephen Birmingham's engaging account *"Our Crowd": The Great Jewish Families of New York* (New York: Dell, 1967).

5. Ibid., 109–10. "Mrs. Simon Kuhn, 83, A Cincinnati Leader," *New York Times*, April 23, 1952.

6. James A. Marcum, *Thomas Kuhn's Revolution: An Historical Philosophy of Science* (New York: Continuum, 2005), 3. In 1941, the Kuhns rented the penthouse at 230 E 50th Street, an apartment earlier admired in 1928 by *New Yorker* columnist Stephanie Zimbalist ("Streetscapes/Readers' Questions," *New York Times*, Sept. 7, 2003). Karl Hufbauer, "From Student of Physics to Historian of Science: T. S. Kuhn's Education and Early Career," *Physical Perspectives* 14 (2012): 423; "Topics: A Hobby Headquarters," *New York Times*, April 17, 1959; "Mrs. Simon Kuhn, 83, A Cincinnati Leader," *New York Times*, April 23, 1952; on Jewish Board of Guardians, "Child-Aid Leaders to Convene May 2," *New York Times*, April 22, 1935; on Junior Achievement, "Brooklyn Boys Create, Manage Manufacturing Concerns that Flourish in Big League Style, *Brooklyn Daily Eagle*, Nov. 20, 1932, B11; and "Junior Craft Clubs Open Their Exhibit at Rumsey Studio," *New York Sun*, Jan. 26, 1933, 20.

7. Dewey, *The School and Society* (Chicago: University of Chicago Press, 1900), 44. A concise account of Dewey's education theory remains his *Experience and Education* (New York: Simon and Schuster, 1938).

8. Parker, "Obstacles in Our Way," *Journal of Education* 15 (May 18, 1882): 315; Caldwell, quoted in Elmer Davis, "Four-Year-Old Educational Experiment," *New York Times*, August 28, 1921.

9. "Progressive Schools Cut 'Three R' Drills," *New York Times*, May 26, 1929; "Teaching Three R's Held Time Waster," *New York Times*, March 11, 1928.

10. Quotations from Counts's lectures are taken from "Dare the School Build a New Social Order" (New York: John Day, 1932) in which the three lectures are collected, pp. 11, 31.

11. M. Ilin [pseud.], *New Russia's Primer: The Story of the Five-Year Plan*, trans. George S. Counts and Nucia P. Lodge (New York: Houghton Mifflin, 1931); George S. Counts, *The Soviet Challenge to America* (New York: John Day, 1931).

12. Counts, "Dare the School Build a New Social Order?" 28.

13. Quoted in Lawrence Cremin, *Transformation of the School* (New York: Vintage, 1961), 260. Moos Campbell, *An Experiment in Education*, 147.

14. Ibid., 74–76. Leslie Fishbein, *Rebels in Bohemia: The Radicals of* The Masses, *1911–1917* (Chapel Hill: University of North Carolina Press, 1982). See, for example, the dance-centered radicalism of Isadora Duncan, 55.

15. Moos Campbell, *An Experiment in Education*, 141–43. "Smith, Wagner, and Thomas Write Views on Disarmament for Schoolboy Editor, 14," *New York Times*, Feb. 3, 1932.

16. *Progressive Education* 9, no. 4 (April 1932): 260. Quoted in Moos Campbell, *An Experiment in Education*, 148.

17. Quoted in Moos Campbell, *An Experiment in Education*, 158, 153, 154.

18. Ibid., 120ff; "Hessian Hills School Opens $1,000,000 Drive: Lincoln Steffens, Floyd Dell and Stuart Chase Laud Progressive Education Aims at Dinner," *New York Times*, May 5, 1931; "Tradition Ignored in School Design," *New York Times*, May 3, 1931; Edward Alden Jewell, "Art," *New York Times*, May 12, 1931;"Benefit for Hessian Hills School," *New York Times*, May 25, 1933; "Croton School Sponsors Rivera Lecture on Art," *Ossining Citizen Register*, May 13, 1933.

19. Moos Campbell, *An Experiment in Education*, 156, 159.

20. Quoted in ibid., 349.

21. Ibid., 238, note 28; Andresen, "Crisis and Kuhn," S48.

22. Christine K. Erickson, "'I Have Not Had One Fact Disproven': Elizabeth Dilling's Crusade against Communism in the 1930s," *Journal of American Studies* 36 (3): 473–89.

23. The title page lists her address as 545 Essex Rd. Kenilworth, Illinois, and 53 W. Jackson Blvd, Chicago.

24. Elizabeth Dilling, *The Red Network* (Elizabeth Dilling, 1934), 9, 3.

25. Ibid., 48, 49.

26. Quoted in "Atomic Education Urged by Einstein," *New York Times*, May 25, 1946.

27. Einstein's correspondence with Mrs. Simon Kuhn is published by the Oregon State University Special Collections & Archives Research Center (scarc.library.oregonstate.edu) and consists of eight letters from her to Einstein and others in regard to Einstein's appeal and written between May and November 1946.

28. Dilling, *Red Network*, 216–17.

29. Quoted in Cremin, *Transformation*, 263, 264. See also Counts et al., "A Call to the Teachers of the Nation" (New York: John Day, 1933), 17.

30. Grace L. H. Brosseau, "Annual Message of the President General," *Proceedings of the Thirty-Eighth Continental Congress*, National Society of the DAR (1929), 11.

31. "Dr. Counts Assails 'Liberty's Enemies,'" *New York Times*, Feb. 26, 1936.

32. Cremin, *Transformation*, 264.

33. Ilin, *New Russia's Primer*, vii, viii.

34. *"I Want to Be Like Stalin" From the Russian Text on Pedagogy by B.P. Yesipov and N.K. Goncharov*, trans. George S. Counts and Nucia P. Lodge (New York: John Day, 1947), 18, 21, 26, 2, 3. As early as the late '30s, Counts had criticized communists within the American Federation of Teachers. See Gerald Gutek, *George S. Counts and American Civilization* (Mercer University Press, 1982) 10, 144.

35. A highly readable account of the nation's debates over intervention is Lynn Olson, *Those Angry Days* (New York: Random House, 2013).

36. Moos Campbell, *An Experiment in Education*, 211, 220; Gary May, *Un-American Activities: The Trials of William Remington* (New York: Oxford University Press, 1994), 60.

37. Moos Campbell, *An Experiment in Education*, 253–55.

38. Ibid., 256, 280.

39. Ibid., 283, 286–88.

40. Additional history of the film, its origins, and reception among critics and fairgoers is examined by Craig Kridel, "Towards an Understanding of Progressive Education and 'School': Lee Dick's 1939 Documentary Film on the Hessian Hills School," unpublished.

41. May, *Un-American Activities*, 46, 51, 60, 52.

42. Ibid., 57.

43. Ibid., 62, 68.

44. Quoted in ibid., 84.

45. Ibid., 138.

46. "Soviet Spy Lists Reds in Key U.S. Jobs," Associated Press (*Miami Daily News*, Aug. 1, 1948); May, *Un-American Activities*, 99, 105. On Bentley, see C. P. Trussell, "Woman Links Spies to U.S. War Offices and White House," *New York Times*, July 31, 1948, and Kalman Seigel, "Ex-Wife Identifies Remington as Red," *New York Times*, Dec. 27, 1950.

47. May, *Un-American Activities*, 129, 290.

48. Elizabeth Moos, "A Woman's Place Is in the Factory," *New China*, Spring 1975, 28; "Educator Reports on Second China Visit," *Herald Statesman* (Yonkers, NY), March 29, 1976, 16.

49. "'World Peace' Plea Is Circulated Here," *New York Times*, July 14, 1950; "Moscow Decrees 'Peace' Sabotage," *New York Times*, July 22, 1950. On Moos's arrest, "Foreign Aid Charge Denied by Mrs. Moos," *New York Times*, April 3, 1951. "Du Bois Acquitted in Foreign Agent Trial," *Indianapolis Recorder*, Nov. 24, 1951, 1; "Ex-Kin of Remington Seized on Foreign Agent Charge," *Brooklyn Eagle*, March 30, 1951.

50. Herbert A. Philbrick, *I Led 3 Lives* (New York: Grosset and Dunlap, 1952), 245.

51. May, *Un-American Activities*, 5–7; "The Nation: Two in Lewisburg," *New York Times*, Nov. 28, 1954. Some believed Remington was killed during a cell

robbery, while columnist Drew Pearson trumpeted that it was all about homosexuality (See Pearson's syndicated column "Washington Merry-Go-Round," Dec. 4, 1954: "Murder Highlights Prison Vice").

52. "Hearings before the Committee on Un-American Activities, House of Representatives," Dec. 11–13, 1962 (Washington, DC: U.S. Government Printing Office, 1963), 2157; Amy Swerdlow, *Women Strike for Peace: Traditional Motherhood and Radical Politics in the 1960s* (Chicago: University of Chicago Press, 1993), 98, 107.

53. Cremin, *Transformation*, 268–69.

54. Moos Campbell, *An Experiment in Education*, 319, 322.

55. Ibid., 326.

56. Quoted in Andresen, "Crisis and Kuhn," S52, S46; see also Kuhn, *RSS*, 256–59; Moos Campbell, *An Experiment in Education*, 195.

57. Quoted in Moos Campbell, *An Experiment in Education*, 350; Kuhn, *RSS*, 256, 257.

Chapter 2. War and General Education at Harvard

1. Conant, *My Several Lives*, 52.

2. Hershberg, *James Bryant Conant*, 45–48. Another helpful source of information and perspective on Conant's life, values, and personality is Jennet Conant's biography *Man of the Hour: James B. Conant, Warrior Scientist* (New York: Simon and Schuster, 2017). Jennet Conant is Conant's granddaughter.

3. "We need a man and mind of distinction," the lawyer Felix Frankfurter wrote to a fellow member of the Harvard Corporation. "A distinguished chemist is not enough." Quoted in Jennet Conant, *Man of the Hour*, ch. 8, see also ch. 9.

4. Hershberg, *James Bryant Conant*, 56, 59, 78.

5. Ibid., 57, 80–81.

6. Ibid., 61, 62. Jennet Conant, *Man of the Hour*, chs. 6, 7.

7. John Desmond Bernal, *The Social Function of Science* (New York: Macmillan, 1939).

8. Fifth International Congress for the Unity of Science, Harvard University, Sept. 3–9, 1939, brochure and schedule, University of Chicago Press Papers, University of Chicago Library Special Collections (hereafter, UCPR), Box 346 Folder 2.

9. Conant's intervention is documented in letters between Frank and Harlow Shapley. See Shapley to Frank, April 3, 1939; April 10, 1939; Frank to Shapley, April 13, 1939, all in Harlow Shapley Papers, Harvard University Archives. For more on Frank's relationship to Conant, see my "Pragmatic Engagements: Philipp Frank and James Bryant Conant on Science, Education, and Democracy," *Studies in East European Thought* 69: 227–44.

10. Conant, *Modern Science and Modern Man* (New York: Columbia University Press, 1952), 22, See also Justin Biddle, "Putting Pragmatism to Work in the

Cold War: Science, Technology and Politics in the Writings of James B. Conant," *Studies in History and Philosophy of Science* 42 (2011): 552–61.

11. Hershberg, *James B. Conant*, 169, 170.

12. "Dr. Conant Urges Universal Military Service," *New York Times*, June 13, 1940.

13. "Pacifists Plan to Display Feelings in Rallies Fixed for Armistice Weekend," *Crimson*, Nov. 9, 1939. See also Gerald M. Rosberg, "War Protest at Harvard Is Not New; Pacifists Got Support in '16 and '41," *Crimson*, June 16, 1966.

14. Lynn Olson, *Those Angry Days*, ch. 15.

15. "American Interests Jeopardized if U.S. Intervenes in Europe's War, McKay Warns," *Crimson*, Nov. 24, 1939; "Lead, Kindly White," *Crimson*, Oct. 25, 1940.

16. "Split Keeps HSU From Joining Walkout Called by National Organization on Over 100 Campus," *Crimson*, Oct. 11, 1940,

17. "Conant Urges Aid to Allies at Once," *New York Times*, May 30, 1940; "President Roosevelt's Message to Congress on the State of the Union," *New York Times*, Jan. 7, 1941.

18. "President Roosevelt's Message to Congress on the State of the Union," *New York Times*, Jan. 7, 1941.

19. Hershberg, *James B. Conant*, 116. "Text of President's Baccalaureate Address," *Crimson*, June 21, 1937.

20. "The President's Message," *Crimson*, Jan. 7, 1941.

21. "Conant Sets Value of Colleges in War," *New York Times*, Jan. 15, 1942; Lewis M. Steel, "College Life During World War II Based on Country's Military Needs," *Crimson*, Dec. 7, 1956. For a more detailed account of Harvard's militarization and its relations to research being done on campus, see Isaac, *Working Knowledge*, ch. 5.

22. Hershberg, *James B. Conant*, 310.

23. "University Mobilized Rapidly in '42, Was Naval Training Camp by '43," *Crimson*, Feb. 7, 1951; "Colonel Jay Reviews ROTC Regiment of 600," *Crimson*, Dec. 13, 1941.

24. "Pacifist Group Holds Fast Amid Quickening War Fever," *Crimson*, Aug. 19, 1942. The paper also printed letters to the editor from Quakers and the Harvard Pacifist Association ("The Mail," April 23, 1942). On the decline of the Student Union, see Martin McLaughlin, "Prewar Student Activism: A Profile," in *American Students Organize: Founding the National Student Association after World War II* (Praeger, 2006) 15–18, 16.

25. "College Days Few, Harvard is Told," *New York Times*, Oct. 7. 1942. On Buck's work with Conant, see Hershberg, *James B. Conant*, 172–73. Buck quoted in Steel, "College Life."

26. "Thus Far and No Farther," *Crimson*, Feb. 12, 1941.

27. "E for Effort," *Crimson*, April 30, 1942. Kuhn is identified as author of this and other editorials by Karl Hufbauer, "From Student of Physics to Historian of

Science," 425, 426. Kuhn's election to editorial chairman is announced in "Dan Fenn Jr., Only 19, Elected Harvard Crimson President," *The Boston Globe*, May 26, 1942.

28. "Forecast for '47," *Crimson*, Oct. 7, 1942.

29. Kuhn, "The War and My Crisis," TSK-MIT, Box 1, folder 3.

30. Kuhn wrote that advocates of competing schools of thought "will inevitably talk through each other when debating the relative merits of their respective paradigms." *Structure*, 109.

31. See Schrecker, *No Ivory Tower*, 195; "Communist Methods of Infiltration (Education)," Hearings before the Committee on Un-American Activities, House of Representatives, Feb. 25, 26, and 27, 1953 (Washington, DC: Government Printing Office), 3–45.

32. "Willkie Lifts His Voice for Liberal Arts Study," *New York Times*, Jan 17, 1943.

33. Conant, "No Retreat for the Liberal Arts," *New York Times*, Feb. 21, 1943.

34. Ibid.

35. Conant, *My Several Lives*, ch. 27; Harvard University, Committee on the Objectives of a General Education in a Free Society, *The Objectives of a General Education in a Free Society* (Cambridge: Harvard University Press, 1945), 31 (hereafter, "*Objectives*").

36. *Objectives*, 30.

37. Conant, "Introduction," *Objectives*, viii.

38. *Objectives*, 227.

39. Ibid., 77.

40. Conant, *On Understanding Science* (New Haven: Yale University Press, 1947), 25.

41. Kuhn, *RSS*, 275. See also Interview of Paul Buck by Katherine Sopka on 1977 March 2, Niels Bohr Library & Archives, American Institute of Physics, College Park, MD USA, www.aip.org/history-programs/niels-bohr-library/oral-histories/32154.

42. "Subjective View. Thomas S. Kuhn, on Behalf of the Recent Students, Reflects Undergraduate Attitude," *Harvard Alumni Bulletin*, Sept. 22, 1945, 29–30.

Chapter 3. History of Science in a Divided World

1. "Student Opinion Wins First Victory in Tutorial See-Saw," *Crimson*, April 27, 1946; "Poll of Students Uphold General Education Plan," Jan. 22, 1947.

2. Wright to Kuhn, Feb. 3, 1947, TSK-MIT, B14 F10.

3. "Prescription for Cambridge," *Harvard Alumni Bulletin*, Sept. 22, 1945, 29.

4. Hufbauer, "From Student of Physics," 427–29; Joel Isaac, "Kuhn's Education: Wittgenstein, Pedagogy, and the Road to *Structure*," *Modern Intellectual History* 9, no. 1 (2012): 89–107, 98; Kuhn, *RSS*, 273, 274.

5. Kuhn, *RSS*, 274, 275.

6. Hershberg, *James B. Conant*, 29–30, 31. On Conant's Quaker-inspired pacifism, see Jennet Conant, *Man of the Hour*, ch. 4; on Conant's membership in the Signet Society, ch. 3.

7. Hershberg, *James B. Conant*, 79. On Kuhn's and Conant's view of George Sarton, the historian of science at Harvard, see Kuhn, *RSS*, 275 and Hershberg *James B. Conant*, 407–408.

8. Kuhn, *RSS*, 275. Other young faculty involved in science teaching within the General Education program included Gerald Holton and I. B. Cohen.

9. One fruit of this collaboration was James Bryant Conant, ed., *Harvard Case Studies in Experimental Science*, 2 vols. (Cambridge: Harvard University Press), 1957.

10. "Conant to Teach New Course in Philosophy," *Crimson*, April 9, 1952; Kuhn to Conant, Oct. 8, 1950; Conant to Kuhn, Oct. 11, 1950, Harvard University Archives, Records of James B. Conant, Box 402. Conant returned at Kuhn's invitation to *Natural Sciences 4* to lecture on Pasteur in the spring of 1952. See Conant to Kuhn, November 14, 1951 and March 11, 1952, Box 433.

11. Kuhn, "Objectives of a General Education Course in the Physical Sciences," May 1947, TSK-MIT, box 1 folder 4.

12. On these concerns about Truman's qualifications, see for example David McCullough's biography, *Truman* (New York: Simon and Schuster, 1992), 431–32.

13. Conant, *On Understanding Science*, 2–5.

14. Ibid., 16.

15. Ibid., 14, 15; Karl Pearson, *The Grammar of Science* (Mineola, NY: Dover, 2004).

16. Conant, *On Understanding Science*, 24, 25.

17. Ibid., 18, 91, 36.

18. Ibid., 17.

19. Conant did not invent this theory of conceptual schemes. Like many others at Harvard who would cross paths with Kuhn, he inherited it most likely from his wife's uncle, the biochemist L. J. Henderson, who in turn distilled conceptual schemes from the writings of the Italian sociologist and economist Vilfredo Pareto. Henderson led widely attended seminars on Pareto, instituted courses to expose undergraduates to Pareto's ideas, and made Pareto and conceptual schemes a staple of conversation and debate within the prestigious Society of Fellows that Kuhn would later join. See Isaac, *Working Knowledge*, 68–74.

20. Conant, *On Understanding Science*, 25, 31, 106, 64. Conant made the same point earlier in his essay "Science and the National Welfare," *The Journal of Higher Education* 15, no. 8 (1944): 399–406, 406. Recent interest in Conant's historiography and philosophy of science and his many influences on Kuhn is illustrated by Justin Biddle, "Putting Pragmatism to Work in the Cold War"; K. Brad Wray, "The Influence of James B. Conant on Kuhn's *Structure of Scientific Revolutions*," HOPOS 6, no. 1 (2016): 1–23; and Christopher Hamlin, "The Pedagogical Roots

of the History of Science: Revisiting the Vision of James Bryant Conant," *ISIS* 107, no. 2 (2016): 282–308.

21. Hershberg, *James B. Conant*, chs. 14, 15.

22. Conant, *Education in a Divided World: The Function of the Public Schools in our Unique Society* (New York: Greenwood Press, 1948), ix.

23. Ibid., 1, 6, 10, 7.

24. Ibid., 17–18, 30.

25. Ibid., 36–37.

26. Kuhn, "The War and My Crisis," TSK-MIT, Box 1, folder 3.

27. Some of Kuhn's presentations at these events are examined in chapter 13.

28. On impressions of Conant's personality, see Hershberg, *James B. Conant*, 9, 736. On Kuhn's comparisons of his father to Conant, see John Forrester, "On Kuhn's Case: Psychoanalysis and the Paradigm," *Critical Inquiry* 33 (Summer 2007): 782–819, 785, 816.

29. On Conant and the Society of Fellows, *RSS*, 276, 284. For Kuhn's recollections of Quine's influence on him, see *RSS*, 279–80, and also Isaac, *Working Knowledge*. Kuhn's Lowell Lectures, discussed in chapter 9, evidence Kuhn's particular interests in Quine's influential essays "Truth By Convention" (in *Philosophical Essays for A. N. Whitehead*, ed. O. H. Lee [New York: Longmans, 1936], 90–124) and "On What There Is" (*Review of Metaphysics* 2, no. 5 [Sept. 1948]: 21–38).

30. George Kennan, Telegram to George Marshall ("Long Telegram"), February 22, 1946. Harry S. Truman Administration File, Elsey Papers, www.trumanlibrary.org.

31. Kennan, writing as Mr. X., "The Sources of Soviet Conduct," *Foreign Affairs* 25 (July 1947): 566–78, 580–82.

32. Conant, *Education in a Divided World*, 26, 27–28.

33. Quoted in Lawrence E. Davies, "Conant Sees Peace Under Atom Pact," *New York Times*, Sept. 9, 1947.

34. Conant, *Education in a Divided World*, 4, 21.

35. Hershberg, *James B. Conant*, 338–39, 370–72, 480–81, 483–84, 495; Kennan, Long Telegram, 17. For a more detailed analysis of Kuhn's *Structure* in the context of Kennan's and Conant's diplomacy, see my "Telegrams and Paradigms: On Cold War Geopolitics and *The Structure of Scientific Revolutions*," in *Science Studies during the Cold War and Beyond: Paradigms Defected*, ed. Elena Aronova and Simone Turchetti (London: Palgrave, 2016), 23–53.

36. James Burnham, *The Coming Defeat of Communism* (New York: John Day, 1950), 226.

37. Conant, *Education in a Divided World*, 26, 27–28.

38. Hershberg, *James. B. Conant*, 328. On Gromyko's behavior, see "This Week's Events," *New York Times*, March 31, 1946; "Russians Blame Byrnes," *New York Times*, March 29, 1946; "Days of Seclusion Ended by Gromyko," *New York Times*, April 5, 1946.

39. Hershberg, *James B. Conant*, 845, n. 81. For a survey of the growth of Sovietology during the cold war that discusses the Russian Research Center (in ch. 2) as well as Columbia University's Russian Institute, see David C. Engerman, *Know Your Enemy: The Rise and Fall of America's Soviet Experts* (New York: Oxford University Press, 2009).

40. Diamond, *Compromised Campus*, 66.

41. See, e.g., Barrington Moore Jr., *Terror and Progress USSR* (Cambridge: Harvard University Press, 1954); Alex Inkeles, *Public Opinion in Soviet Russia: A Study in Mass Persuasion* (Cambridge: Harvard University Press, 1950); David Dallin, "The Making of the Russian Mind," *New York Times*, July 2, 1950.

42. On Conant's involvement with the new center, see Hershberg, *James B. Conant*, 412; Diamond, *Compromised Campus*, 66–68; and Engerman, *Know Your Enemy*, 47–48; David Halberstam, "Russian Center Studies Soviet Social System," *Crimson*, Oct. 9, 1953.

43. Kuhn, *Structure*, 109, 4.

44. Kennan, "Russia, the Atom and the West," Reith Lecture no. 2, "The Soviet Mind and World Realities," broadcast Nov. 17, 1957. Text available at bbc.co.uk and in George Kennan, *Russia, the Atom, and the West* (London: Oxford University Press, 1958).

45. Kuhn, *Structure*, 111, 102.

Chapter 4. The Cold War Conversions of Thomas S. Kuhn aand James Bryant Conant

1. Aristotle, *The Physics*, two vols., trans. Philip H. Wicksteed and Francis M. Cornford (New York: G. P. Putnam, 1929), 200b 26, 224 25. If not the Wicksteed and Conford translation, Kuhn may have read the popular translation by R. P. Hardie and R. K. Gaye (Oxford: Clarendon Press, 1930).

2. Ibid., 216a15.

3. "What Are Scientific Revolutions?" in Kuhn, *RSS*, 16–17.

4. Ibid., 18.

5. Conant, *On Understanding Science*, 20, 21.

6. Ibid., 21.

7. Ibid., 20–21; *Education and a Divided World*, 123; *Science and Common Sense* (New Haven: Yale University Press, 1951), 38; "Science and the National Welfare," 400; *Modern Science and Modern Man*, 13; *Two Modes of Thought* (New York: Trident, 1964), 15.

8. *Objectives*, 221, 159.

9. "A Historian Views the Philosophy of Science," lecture notes dated 1957, TSK-MIT, box 3, folder 10. Kuhn refers to the event as a "revelation" in his fore-

word to Ludwik Fleck, *The Genesis and Development of a Scientific Fact* (Chicago: University of Chicago Press, 1979), vii–viii.

10. "I had wanted to write *The Structure of Scientific Revolutions* ever since the Aristotle experience," Kuhn later explained. "That's why I had gotten into history of science—I didn't know quite what it was going to look like, but I knew the noncumulativeness; and I knew about what I took revolutions to be . . . but that was what I really wanted to be doing." See Kuhn, *RSS*, 292.

11. Quoted in Hershberg, *James B. Conant*, 112–13.

12. Ibid., 113, 114.

13. Ibid., 115, 119. See also Conant, *My Several Lives*, 228.

14. Hershberg, *James B. Conant*, 149.

15. Ibid., 475–77.

16. Ibid., 8.

17. For a classic account of American religiosity and revivalism, see Richard Hofstadter, *Anti-Intellectualism in American Life* (New York: Vintage, 1962), esp. ch. 3.

18. William James, *The Varieties of Religious Experience: A Study in Human Nature* (New York: Collier, 1961), 299–300.

19. Hershberg, *James B. Conant*, 236.

20. Ibid., 88–89.

21. "Student Council Urges Repeal of Teachers Oath Bill in Resolution," *Crimson*, March 10, 1936; "Free Inquiry of Dogma," *Atlantic Monthly*, April 1935, 436–42, quoted in Hershberg, *James B. Conant*, 89.

22. "Conant Letter Urges Faculty to Sign Oath, but Criticizes Bill," *Crimson*, Oct. 8, 1935.

23. Conant, *My Several Lives*, 561; Hershberg, *James B. Conant*, 426, 428.

24. Hershberg, *James B. Conant*, 428–29; also William L. Marbury, "The Hiss-Chambers Libel Suit," *Maryland Law Review* 41, no. 1 (1981): 75–102, 85. See also Chambers's memoir *Witness*.

25. Hershberg, *James B. Conant*, 464, 464–65.

26. Ibid., 476.

27. "Chain Reaction," *New York Times*, Feb. 12, 1950.

28. Hershberg, *James B. Conant*, 483.

29. Ibid., 493–94.

30. Ibid., 496. Conant, *My Several Lives*, 508–509.

31. Hershberg, *James B. Conant*, 495. Conant, *My Several Lives*, 509.

32. Hershberg, *James B. Conant*, 506, 483, 499.

33. "The Present Danger," *New York Times* editorial, December 22, 1952.

34. Hershberg, *James B. Conant*, 509, 512, 500, 502.

35. "Barnes Fails to See Cause for Alarm," *Crimson*, Feb. 6, 1948.

36. "Conant Advises Cut in University's Enrollment," *Crimson*, Jan 22, 1948; Hershberg, *James B. Conant*, 414.

37. Hershberg, *James B. Conant*, 415.

38. Ibid., 609, 610.
39. Ibid., 433, 431.
40. Conant, quoted in "Text of Conant's Speech," *The Crimson*, June 23, 1949.

Chapter 5. Sidney Hook and the Anticommunist Inquisition

1. "Rally No Red Front, Shapley States," *Crimson*, March 24, 1949.
2. Hook to Rabbi Louis Newman, quoted in Edward Shapiro, ed., *Letters of Sidney Hook: Democracy, Communism, and the Cold War* (Armonk, NY: A. E. Sharpe, 1995), 124.
3. Hook, *Out of Step* (New York: Carroll and Graf, 1987), 388.
4. Neil Jumonville, *Critical Crossings: The New York Intellectuals in Postwar America* (Berkeley: University of California Press, 1990), 54–55.
5. Quoted in *Letters of Sidney Hook*, 120.
6. "Sponsors of the World Peace Conference," *New York Times*, March 24, 1949.
7. Carnap to Hook, March 24, 1949, Rudolf Carnap Papers, Archives of Scientific Philosophy, University of Pittsburgh, ASP RC088-38-13. Others to whom Hook appealed unsuccessfully were Algernon Black, head of the Ethical Culture Society in New York, and author Thomas Mann (see Shapiro, ed. *Letters of Sidney Hook*, pp. 120–28).
8. Hook, *Out of Step*, 383.
9. On Counts's participation with Hook, Charles Grutzner, "Pickets to Harass Cultural Meeting; Delegates Arrive," *New York Times*, March 24, 1949; William R. Conklin, "Soviet is Attacked at Counter Rally," *New York Times*, March 27, 1949. Hook later told the historian John Rossi that the AIF "was such a spontaneously organized thing run exclusively by friends and members of our families" (Hook to Rossi, March 1, 1982).
10. Hook, *Out of Step*, 391.
11. John Rossi, in his article "Farewell to Fellow Traveling: The Waldorf Peace Conference of March 1949," *Continuity* 10 (1985): 1–31, notes that letters on Hook's behalf had arrived and now exist in Shapley's archived papers. It was most likely, Rossi concludes, "Shapley wanted to keep Hook off the program at any cost" (17 n52).
12. Shapley's letter is quoted in *Out of Step*, 390–91. On Hook's exchange with Shipler, see *Letters of Sidney Hook*, 123.
13. Hook, *Out of Step*, 391.
14. Hook to Shapley, March 23, 1949, titled "Copy of a letter delivered by hand." Copy in Rudolf Carnap Papers, Archives of Scientific Philosophy, University of Pittsburgh (ASP RC 088-38-08).
15. "There is no evidence," John Rossi wrote, "that the NCASP acted in response to Soviet Orders"; "Farewell to Fellow Traveling," 4.

16. Hook, *Out of Step*, 391; Carnap to Hook, March 24, 1949, ASP RC 088-38-13.

17. "Hook Cites Letters in Shapley Dispute," *New York Times*, March 27, 1949.

18. "Hook Confronts Shapley in Latter's Hotel Room," *New York Times*, March 26, 1949.

19. Hook, *Out of Step*, 392. Hook later wrote that coverage of the encounter was sympathetic in the *Herald Tribune* but misleading in the *New York Times*, which portrayed him as "a wild man bursting in on an innocent Shapley." Yet the resulting front-page *Tribune* report also portrayed Hook as a surprise invader who "burst into the hotel room of Dr. Shapley" (Mac R. Johnson, "Hook Invades Shapley's Room to Ask Apology," *New York Herald Tribune*, March 26, 1949).

20. Quoted in "Panel Discussions of the Cultural Conference Delegates Cover a Wide Range of Subjects," *New York Times*, March 27, 1949.

21. "Dupes and Fellow Travellers Dress up Communist Fronts," *Life*, April 4, 1949, 39–43.

22. William L. Laurence, "'One-World' Ideal Is Seen in Science," *New York Times*, Sept. 24, 1946.

23. To the end of his life, Kennan believed his long telegram was misunderstood and that "containment" did not require military buildup. See Kennan's memoir, *At a Century's Ending: Reflections 1982–1995* (New York: Norton, 1996), 115.

24. Sidney Hook, "Should Communists Be Permitted to Teach?" *New York Times*, Feb. 27, 1949.

25. Burnham quoted in Daniel Kelly, *James Burnham and the Struggle for the World* (Wilmington, DE: ISI Books, 2002), 38–39.

26. Hook, "Should Communists Be Permitted to Teach?"

27. Sidney Hook, *Heresy, Yes—Conspiracy, No* (New York: John Day, 1953), 180.

28. Ibid., 252.

29. Hook, "Should Communists Be Permitted to Teach?"

30. Russell B. Porter, "Reds Here Shift in Stand on War," *New York Times*, July 27, 1941.

31. "Harvard's United Front," *Crimson*, Dec. 14, 1939.

32. "Youth Talks Back," *Crimson*, Feb. 28, 1940.

33. Kuhn, "The War and My Crisis."

34. Sidney Hook, "The Philosophy of Dialectical Materialism. I." *Journal of Philosophy* 25, no. 5 (Mar. 1, 1928): 113–14, 114; "The Philosophy of Dialectical Materialism. II." *Journal of Philosophy* 25, no. 6 (Mar. 15, 1928): 141–55, 142; Schrecker, *No Ivory Tower*, 31. On Burnham's career, see Kelley, *James Burnham and the Struggle for the World*.

35. Sidney Hook, *Out of Step* (New York: Carroll and Graf, 1987), 229–32, 233–47; Harold Kirker and Burleigh Taylor Wilkins, "Beard, Becker and the Trotsky Inquiry," *American Quarterly* 13, no. 4 (Winter 1961): 516–25. For a detailed account

of Hook's involvement in the Trotsky Committee and the aftermath of the Dewey Commission, see Christopher Phelps, *Young Sidney Hook* (Ithaca: Cornell University Press, 2005), 155–71. The Dewey Commission report appeared as Dewey et al., *Not Guilty: Report of the Commission of Inquiry into the Charges Made Against Leon Trotsky in the Moscow Trials* (New York: Harper and Brothers, 1938).

36. Howe, quoted in Neil Jumonville, "Polemics, Open Discussion, and Tolerance," in *Sidney Hook Reconsidered* (Amherst, NY: Prometheus Books, 2004), 225–44, 228.

37. Arnold Forster, quoted in Navasky, *Naming Names*, 112.

38. Bund leader Fritz Kuhn—no relation to Thomas—spent the early 1940s in a series of internment camps built for enemy aliens in Texas and New Mexico. See "Bundist Round-Up Nets 72 More Here," *New York Times*, July 9, 1942; "Mass Bund Trials Now in Prospect," *New York Times*, Oct. 21, 1942; "Kuhn is Ordered Deported to Reich," *New York Times*, May 19, 1945.

39. Hook to Carnap, March 20, 1949, Rudolf Carnap Papers, Archives of Scientific Philosophy, University of Pittsburgh, ASP RC088-38-10.

40. Christopher Phelps, "Why Wouldn't Sidney Hook Permit the Republication of His Best Book?" *Historical Materialism* 11, no. 4 (2003): 305–15, 311, 315; see also Thomas J. Main, "Fearless Sidney Hook," *Policy Review*, no. 120 (Aug. 1, 2003).

41. Sidney Hook, "Conant on Politics and Education," *New York Times*, Oct. 24, 1948.

42. Hook to Conant, April 11, 1949. This and other letters between Hook and Conant, unless otherwise noted, are contained in the Sidney Hook Papers, Stanford University, box 10, folder 6.

43. Hook's use of ellipses and other devices to alter quotations taken from Party literature was denounced in the article "Civil Liberties and the Philosopher of the Cold War," *The New International* 19 (1953): 184–227. The authors were Julius Falk and Gordon Haskell, but Falk was a pseudonym of Julius Jacobsen. See Phelps, *Young Sidney Hook*, 228, n. 44.

44. Hook, "International Communism," *Dartmouth Alumni Magazine* (March, 1949): 13–20.

45. Hook to Conant, April 7, 1949; Conant to Hook, April 11, 1949; Sidney Hook papers, Hoover Institution Archives, Stanford, CA.

46. Conant, quoted in *The Crimson*, June 23, 1949.

47. Jennet Conant, *Man of the Hour*, ch. 20; Benjamin Fine, "Conant Sees Peril to U.S. Education," *New York Times*, April 8, 1952.

48. Hook to Conant, April 28, 1952; Conant to Hook, May 5, 1952.

49. Walter R. Story, to the Editor, *New York Times*, March 15, 1953; William L. Maier, to the Editor, *New York Times*, March 15, 1953.

50. Hook, to the Editor, *New York Times*, March 15, 1953.

51. Hook to Conant, Sept. 23, 1961, Papers of James B. Conant, Harvard University, UAI 15.898, H–J Box 130; Conant to Hook, Oct. 16, 1961; Ralph Blumenthal, "Vietnam Backers Urged to 'Shout,'" *New York Times*, Nov. 29, 1965.

52. Hook, "Conant on Politics and Education."
53. Shapiro, ed. *Letters of Sidney Hook*, 236–40, 242–43.
54. Hook, *Heresy, Yes—Conspiracy, No*, 246, 250.
55. Alexander Meiklejohn, "Should Communists Be Allowed to Teach?" *New York Times*, March 27, 1949.
56. Victor Lowe, "A Resurgence of 'Vicious Intellectualism,'" *Journal of Philosophy* 48, no. 14 (July 5, 1951): 435–47, 441.
57. Hook, *Heresy, Yes—Conspiracy, No*, 236; Hook, "Mindless Empiricism," *Journal of Philosophy* 49, no. 4 (Feb. 14, 1952): 89–100, 91, 98.
58. Fred Schwarz, *Communism: Diagnosis and Treatment* (Los Angeles: World Vision, Inc., n.d.).
59. Conklin, "Soviet Is Attacked."
60. "A Word from Harlow Shapley," in *Speaking of Peace* (New York: National Council of the Arts, Sciences, and Professions, 1949), vii.
61. Conklin, "Soviet Is Attacked."
62. Grutzner, "Pickets to Harass Cultural Meeting." Coverage in the *New York Herald Tribune* by Seymour Freidin included headlines "'Peace' Rally Opens at Waldorf . . ." (March 26); "Russians at 'Peace' Rally Assail U.S . . ." (March 27); and "Red 'Peace' Group Here . . ." (March 24).
63. As of June 24, 2015, the newsreel can be viewed on youtube.com and at britishpathe.com.
64. Grutzner, "Pickets to Harass Cultural Meeting"; John Fisher, "NY Rally Part of Red Move to Snarl Artists and Scientists," *Nashua* (N.H.) *Telegraph*, April 7, 1949, 16.
65. Hook to Carnap, March 20, 1949.
66. *Life*, "Dupes and Fellow Travellers Dress up Communist Fronts."
67. "Science: Stargazer," *Time*, Mon. Aug. 29, 1949.
68. Hershberg, *James B. Conant*, 617, 618.
69. Kuhn, "International Morale and a United States Declaration of War," dated Oct 19, 1941, TSK-MIT, box 1, folder 3. In this essay Kuhn correctly assumed, about seven weeks before the bombing of Pearl Harbor, that only "military action" could unite Americans behind a new war effort.
70. Schrecker, *No Ivory Tower*, 194, 195.
71. "Communist Methods of Infiltration (Education)," quotes from 26, 29, 31 (see also 16), 38. Davis and others protesting the conference appeared in the *New York Times* March 25, 1949, under "200 Sponsors Join Culture Unit Foes."
72. The other resisters were Helen Deane Markham in the medical school and Leon Kamin in the social relations department. Jenner also called two students to testify, the twins Jonathan and David Lubell. Hufbauer, "From Student of Physics to Historian of Science," 423–24; Schrecker, *No Ivory Tower*, 197, 200–202; "Communist Methods of Infiltration (Education)," 37, 66.
73. Schrecker, *No Ivory Tower*, 197, 200–202. "Furry, Named in Probe, Still Teaching at Harvard," *Boston Globe*, Feb. 26, 1953.

74. Ibid., 197, 203.
75. Edward Ranzal, "McCarthy Charges 'Mess' at Harvard."
76. See Hershberg, *James B. Conant*, 632; John H. Fenton, "Pusey Is 'Unaware' of Reds on Faculty," *New York Times*, Nov. 10, 1953.
77. Kuhn's recollections about exactly when and where Morris approached him remained vague (*RSS*, 292; Kuhn, personal correspondence with author, Aug. 13, 1993). On Morris visiting Harvard, see "East, West View of Life Merging, Lecturer Asserts," *Crimson*, Feb. 7, 1952.

Chapter 6. Brainwashing, or The Structure of Philosophical Revolutions

1. Quoted in David Caute, *The Fellow Travellers: Intellectual Friends of Communism* (New Haven: Yale University Press, 1973), 20.
2. Philbrick, *I Led 3 Lives*, 38, 118.
3. "How to Identify an American Communist," *LOOK* Magazine, vol. 11, no. 5, March 4, 1947; "Spotting Communists," *New York Times* editorial, June 14, 1955.
4. J. Edgar Hoover, *On Communism* (New York: Random House, 1969), 62–63.
5. J. Edgar Hoover, *Masters of Deceit* (New York: Henry Holt, 1958), 319.
6. Executive Sessions of The Senate Permanent Subcommittee on Investigations of the Committee on Government Operations, v. 2 (Washington, DC: Government Printing Office, 1953), 1248.
7. Ibid., v. 4, 2961, 3167, 2968, 3141.
8. For a recent examination of the Russell controversy, see Thomas Weidlich, *Appointment Denied: The Inquisition of Bertrand Russell* (Amherst, NY: Prometheus Books, 2000); Plato, *Apology*, 18b–c.
9. John McCumber, *Time in the Ditch*, 26.
10. Abbott Gleason, *Totalitarianism* (New York: Oxford University Press, 1995), 100–101.
11. Ibid., 102, 100.
12. George Reisch, *How the Cold War Transformed Philosophy of Science*, 246.
13. Edward Hunter, *Brainwashing in Red China: The Calculated Destruction of Men's Minds* (New York: Vanguard Press, 1951) and *Brainwashing: The Story of Men Who Defied It* (New York: Farrar, Strauss and Cudahy, 1956); Stuart Lillico, "Black-Out in Red China," review of Hunter, *Brainwashing in Red China*, *New York Times*, Oct. 21, 1951; Anthony Leviero, "Thinking to Order," review of Hunter, *Brainwashing: The Story of Men Who Defied It*, *New York Times*, May 20, 1956.
14. Anthony Leviero, "New Code Orders P.O.W.'s to Resist in 'Brainwashing,'" *New York Times*, Aug. 18, 1955.
15. Eugene Kinkead, *In Every War But One* (New York: Norton, 1959).

16. Richard Hofstadter "The Paranoid Style in American Politics," in *The Paranoid Style in American Politics and other essays* (Vintage reprint, 2008), 32 (orig: *Harper's*, November 1964, 77–86).

17. Gleason, *Totalitarianism*, 99.

18. Joost Meerloo, "Pavlov's Dog and Communist Brainwashers," *New York Times*, May 9, 1954; Hunter, *Brainwashing in Red China*, 7–9; Leviero, "Thinking to Order."

19. Joost Meerloo, "Pavlov's Dog and Communist Brainwashers," *New York Times*, May 9, 1954.

20. Gleason, *Totalitarianism*, 93–95, 103–107; Conant, "Text of President's Baccalaureate Address"; "No Retreat for the Liberal Arts."

21. George Orwell, *1984* (New York: Knopf, 1987), 257.

22. Hannah Arendt, *The Origins of Totalitarianism* (New York: Harcourt, 1985), 325. On Arendt's influence shaping the West's understanding of totalitarianism, see Gleason, *Totalitarianism*, ch. 6. The SA in this quote refers to Hitler's *Sturmabteilung*, commonly known as his paramilitary "Brown Shirts."

23. See Arendt's Preface to part III, note 1, in the 1966 edition.

24. A useful survey of brainwashing in films is Susan Carruthers, "Redeeming the Captives: Hollywood and the Brainwashing of America's Prisoners of War in Korea," *Film History* 10, no. 3, "The Cold War and the Movies," (1998): 275–94. On advertising, see Vance Packard, *The Hidden Persuaders* (New York: McKay, 1957); Norman Vincent Peale, *The Power of Positive Thinking* (New York: Prentice-Hall, 1952).

25. S. L. A. Marshall, "P.O.W.'s in North Korea: The Way They Were and Why," review of *In Every War But One*, by Eugene Kinkead, *New York Times*, Feb. 22, 1959; Edward Hunter, Letter to the Editor, *New York Times*, March 29, 1959.

26. "Communist Psychological Warfare (Brainwashing). Consultation with Edward Hunter," Committee on Un-American Activities (Washington, DC: Government Printing Office, 1958), 12–13.

27. Theodore Chen, *Thought Reform of the Chinese Intellectuals* (Hong Kong: Oxford University Press, 1960); Robert Jay Lifton, *Thought Reform and the Psychology of Totalism: A Study of "Brainwashing" in China* (Chapel Hill: University of North Carolina Press, 1989) (orig. New York: Norton, 1961); Edgar Schein, Inge Schneier, and Curtis H. Barker, *Coercive Persuasion: A Psycho-Social Analysis of the 'Brainwashing' of American Civilian Prisoners by the Chinese Communists* (New York: Norton, 1961); Albert D. Biderman and Herbert Zimmer, *The Manipulation of Human Behavior* (New York: John Wiley and Sons, 1961). Two review articles are E. F. O'Doherty, "Brainwashing," *Studies: An Irish Quarterly Review* 52, no. 205 (Spring 1963): 1–15; and Robert Waelder, "Demoralization and Re-education," *World Politics* 14, no. 2 (Jan. 1962): 375–85.

28. Meerloo, "Pavlov's Dog and Communist Brainwashers"; Chen, *Thought Reform*, 71; Schein, "Brainwashing and Totalitarianization in Modern Society," *World Politics* 11, no. 3 (April 1959): 430–441; "2 Challenge View on Brainwashing," *New York Times*, Sept. 22, 1956.

29. Biderman, "The Image of 'Brainwashing,'" *Public Opinion Quarterly* 26, no. 2 (1962): 547–63, 552, 558; Bauer, quoted in *Science News Letter* 70, 1 (Sept. 15, 1956): 167.
30. "Communist Psychological Warfare (Brainwashing)," 16, 17.
31. Ibid., 10.
32. Lifton, *Thought Reform*, 5, 66, 13.
33. Schein, "Brainwashing and Totalitarianization," 435, 437–38.
34. Schein, "From Brainwashing to Organizational Therapy," *Organization Studies* 27, no. 2 (2006): 287–301, 291.
35. Schein, "Brainwashing and Totalitarianization," 437.
36. Schein, Schneier, and Barker, *Coercive Persuasion*, 282. Quoted in Chen, review of *Coercive Persuasion*, 175.
37. Kuhn, *Structure*, 57.
38. Ibid., 94, 198.
39. Ibid., 77, 113; Conant, *On Understanding Science*, 37, see also 84, 89.
40. Kuhn, *Structure*, 122.
41. Ibid., 5.
42. Ibid., 88, 18, 5.
43. Ibid., 88, 89.
44. Ibid., 65.
45. Ibid., 63, 64.
46. Ibid., 1, 137–38.
47. Orwell, *1984*, 41; Kuhn, *Structure*, 167.
48. Conant, *On Understanding Science*, 103, 74, 15, 80.
49. Ibid., 86, 89, 95.
50. Bernard Barber, "Resistance by Scientists to Scientific Discovery," *Science* 134, no. 1 (Sept. 1961): 596–602.

Chapter 7. The Necessary Dangers of Consensus and Unity

1. Conant, quoted in "'My Dear Sir:' A Sealed Letter from the University Archives Reaches Drew Faust on the Occasion of Her Inauguration," *Harvard University Library Letters*, v. 2, no. 5 (Fall 2007/Winter 2008): 5. Jennet Conant noted that "a habit of a lifetime disinclined him from expressing his feelings and private fears" (*Man of the Hour*, ch. 16).
2. Hershberg, *James B. Conant*, 520.
3. Ibid., 47.
4. Kuhn, *RSS*, 259–60.
5. On Greenbaum's civic service alongside Samuel L. Kuhn, "Child-Aid Leaders to Convene May 2," *New York Times*, April 22, 1935; and "Mrs. Borg Heads Board," *New York Times*, Jan. 13, 1939; "Edward S. Greenbaum, Lawyer, Is Dead at 80," *New York Times*, June 13, 1970; *Red Channels*, 121, 273.

6. On Ernst and Hoover, see Curt Gentry, *J. Edgar Hoover: The Man and His Secrets* (New York: Norton, 2001), 232–34, and Harrison Salisbury, "The Strange Correspondence of Morris Ernst and J. Edgar Hoover," *The Nation*, Dec. 1, 1984. On Greenbaum's national reputation, see "Man in the News" feature, "Legal Perfectionist: Edward S. Greenbaum," *New York Times*, Jan 27, 1957.

7. "Text of Group's Statement on Present Peril," *New York Times*, Dec. 13, 1950; Paul Isherwood, "A Failed Elite: The Committee on the Present Danger and the Great Debate of 1951," MA thesis, Ohio University, March 2009, 87, 94.

8. "Edward S. Greenbaum, Lawyer, Is Dead at 80."

9. "Footprints of the Trojan Horse: Some Methods Used by Foreign Agents within the United States," Citizenship Educational Service, New York, special edition printed in 1942. See also "5th Column Methods Bared in Pamphlet," *New York Times*, Sept. 20, 1940; Wendy L. Wall, *Inventing 'The American Way': The Politics of Consensus from the New Deal to the Civil Rights Movement* (New York: Oxford University Press, 2008), 135–37.

10. Kuhn, "The Crisis in Democracy," TSK-MIT, box 1, folder 3.

11. Ibid.

12. Ibid.

13. "Thomas Mann to Lecture on the Problem of Freedom," *Daily Princetonian* 65, no. 11 (Feb. 13, 1940), 4; Thomas Mann, "The Problem of Freedom," *Vital Speeches of the Day* 5, no. 18 (July 1, 1939), 547–550, 549.

14. Kuhn, "The Crisis in Democracy," TSK-MIT, box 1, folder 3.

15. Edward S. Greenbaum to Kuhn, Sept. 20, 1940, TSK-MIT, box 1, folder 3.

16. Kuhn, *RSS*, 269.

17. Ibid., 271, 272. For a more detailed account of Kuhn's wartime work as a radar technician, see Peter Galison, "Practice All the Way Down," in *Kuhn's Structure at Fifty*, 42–69.

18. Kuhn, *Structure*, 27–28; *RSS*, 273, 274.

19. Kuhn, *RSS*, 270, 271.

20. Conant, *General Education in a Free Society*, 40; Conant to John Boyer, Sept. 12, 1952, James B. Conant Personal Papers, Harvard University Archives. See also Hershberg, *James B. Conant*, 578.

21. Conant, *General Education in a Free Society*, 40–41.

22. Ibid., 76, 32.

23. "Text of President's Baccalaureate Address," *Crimson*, June 21, 1937.

24. "Excerpts from Conant Valedictory Address," *Crimson*, Jan. 11, 1943. On the "overflow" crowd expected, see "Free Society Conant Topic," *Crimson*, Jan. 7, 1943.

25. Conant, "Civil Courage," in A. Craig Baird, ed. *Representative American Speeches, 1945–1946* (New York: H. W. Wilson, 1946), 223–28, 224.

26. Ibid., 224.

27. Ibid., 225.

28. "Excerpts from Conant Valedictory Address," *Crimson*.

29. Sigmund Diamond's persistent research, described in his *Compromised Campus*, 40–47, documents that Buck and Conant approved of information about Harvard faculty and the Russian Research Center being made secretly available to Hoover's FBI in the late 1940s and early '50s. On this issue, see also Hershberg, *James B. Conant*, 624.

30. *Objectives*, 248.

31. Conant, *My Several Lives*, 515.

32. Hershberg, *James B. Conant*, 520–21.

33. "Conant Discusses Role of Scholars," *New York Times*, Jan. 26, 1953.

Chapter 8. The Language, Psychology, and Psychoanalysis of Scientific "Reorientations"

1. Kuhn, "What Are Scientific Revolutions?" 16.

2. Ibid., 19, 18.

3. Ibid., 15.

4. William James, *Principles of Psychology v. 1* (New York: Henry Holt, 1890), 488.

5. In a notebook titled "Notes and Ideas," (TSK-MIT, box 1, folder 7) an entry "reading to 3-31-49" lists "Quine, Mathematical Logic; Tarski: Introduction to Logic; Woodger: Techniques of Theory Construction; Bloomfield: Linguistic Aspects of Science"—the last two of which were monographs in the International Encyclopedia of Unified Science to which Kuhn would later contribute *Structure*— "Dewey: Reconstruction in Philosophy; Kant: Critique of Pure Reason (thru Judgement), Mill: Logic; Russell: The Scientific Outlook; James: Essays in Pragmatism; [Feigl and Sellars, ed.] Readings in Philosophical Analysis, Ayer: Language, Truth and Logic, J.M.W. Sullivan, The Limitations of Science, Wiener: Cybernetics, [Eric Ashby]: Scientist in Russia."

6. James, *The Varieties of Religious Experience*, 120, 121, 122. Kuhn, notecard on James, William, The Varieties of Religious Experience, dated 12/30/43, TSK-MIT, box 8.

7. Kuhn, "The Metaphysical Possibilities of Physics," TSK-MIT, box 1, folder 3, 4–5. Kuhn's essay is undated but marked for Mr. Davis's class, English A-1, which his "The War and My Crisis" dates to the fall of 1941.

8. Kuhn, "The Metaphysical Possibilities of Physics," 5.

9. Henri Poincaré, *Science and Method* (New York: Science Press, 1905), 10.

10. Untitled document in a folder titled by Kuhn "Incomplete Memos and Ideas, 1949," TSK-MIT, box 1, folder 6, 1. Kuhn mentions Whorf in *Structure*, xli.

11. Untitled document in "Incomplete Memos and Ideas, 1949," 2.

12. Kuhn, *Structure*, 119.

13. Untitled document in "Incomplete Memos and Ideas, 1949," 3.
14. Ibid., 3, 4, 5–6.
15. Ibid., 4, 6, 8–9. Kuhn's view of science as something more than an essentially "logical" construction reflects the widespread interest at Harvard in conceptual schemes, understood as partly unconscious guides to action (see Isaac, *Working Knowledge*, ch. 2).
16. Untitled document in "Incomplete Memos and Ideas, 1949," 5.
17. Kuhn, "A Function for Thought Experiments," in *The Essential Tension*, 240–65, 243. For more on Kuhn's reading of Piaget and other psychologists, see David Kaiser, "Thomas Kuhn and the Psychology of Scientific Revolutions" and Peter Galison, "Practice All the Way Down."
18. Kuhn, "Notes and Ideas" notebook (TSK-MIT, box 1, folder 7).
19. Conant, *On Understanding Science*, 13, 25.
20. On Dewey and Piaget as the masterminds behind what is perceived by some as America's educational decline, see Kieran Egan, *Getting It Wrong from the Beginning: Our Progressivist Inheritance from Herbert Spencer, John Dewey, and Jean Piaget* (New Haven: Yale University Press, 2004).
21. Kuhn, notecard for Piaget, J: *Les Notions de Mouvement et de Vitesse Chez L'enfant*, TSK-MIT, box 8.
22. Kuhn, "A Function for Thought Experiments," 245.
23. Kuhn, "Notes and Ideas" notebook (TSK-MIT, box 1, folder 7).
24. Kuhn, *RSS*, 280; see also Andresen, "Crisis and Kuhn"; Hufbauer, "From Student of Physics to Historian of Science," 429, 463; and John Forrester, "On Kuhn's Case: Psychoanalysis and the Paradigm."
25. Robert D. Richardson, *William James: In the Maelstrom of American Modernism* (New York: Houghton Mifflin, 2007), ch. 90.
26. Horney thanked Minette Kuhn in *Our Inner Conflicts: Toward a Constructive Theory of Neurosis* (New York: Norton, 1945), 9. See also Andresen, "Crisis and Kuhn," S48; Fishbein, *Rebels in Bohemia*; Forrester, "On Kuhn's Case," 787; Kuhn, *RSS*, 280. For more on Kuhn's early interests in psychoanalysis and neglect of that interest by scholarly commentators, see my "On the Couch with Freud and Kuhn," *Metascience*, vol. 27, no. 1 (March 2018): 37–46.
27. Kuhn, *RSS*, 280.
28. Kuhn's letter to Radó is in "Incomplete Memos and Ideas, 1949," TSK-MIT, box 1 folder 6.
29. Jennet Conant, *Man of the Hour*, chs. 10, 22.
30. Kubie addressed frequent emotional problems in fledgling scientists in his two part essay, "Some Unsolved Problems of the Scientific Career," *American Scientist* 41, no. 4: 596–613; 42, no. 1: 104–12. Kuhn's letter to Kubie, which refers to their discussion at Kuhn's father's recent birthday party, is dated March 17, 1955, TSK-MIT, box 25, folder 53, 3.
31. Kuhn to Kubie, March 17, 1955, 4; Kuhn, *Structure*, ch. 11, "The Invisibility of Revolutions."

32. See the first footnote in the chapter "Normal Science as Puzzle Solving."
33. Kuhn to Professor David Owen, Jan. 6, 1951, TSK-MIT, box 3 folder 10.
34. Kuhn, "Notes and Ideas" notebook.
35. Kuhn's outline for "THE BOOK" is in his folder "Incomplete Memos and Ideas, 1949," TSK-MIT, box 1, folder 6.

Chapter 9. "Attention Senator McCarthy"

1. Kuhn's alarm and his conversations with the Lowell Institute officers are reconstructed from Kuhn to Ralph Lowell, Feb. 17, 1950 [sic, actually 1951] and Feb. 20, 1951; William H. Lawrence to Kuhn, Feb. 17, 1951, all in TSK-MIT, box 3, folder 10.
2. Kalman Seigel, "College Freedoms Being Stifled By Student's Fear of Red Label," *New York Times*, May 10, 1951.
3. William S. Fairfield, "FBI's Activities Spread Fear at Yale," *Crimson*, June 4, 1949.
4. On this elite, see Birmingham, *Our Crowd*. On Jewish intellectuals in New York broadly, a very readable account remains Jumonville, *Critical Crossings*; see also David Hollinger's collection of essays, *Science, Jews, and Secular Culture*.
5. Erickson, "'I have Not Had One Fact Disproven,'" 487.
6. Edward Weeks, *The Lowells and their Institute* (Boston: Little Brown, 1966), 112–15, 125, 126–27, 133–38, 142–43; "Public Lectures in the City of Boston under The Lowell Institute, Program for 1950–1951," advertisement, in TSK-MIT, box 3, folder 10.
7. Kuhn to Ralph Lowell, March 19 1950, TSK-MIT, box 3 folder 10. The titles were, in order: (1) Introduction: Textbook Science and Creative Science, (2) The Foundation of Dynamics, (3) The Prevalence of Atoms, (4) "The Principle of Plenitude": Subtle Fluids and Physical Fields, (5) Evidence and Explanation, (6) Coherence and Scientific Vision, (7) The Role of Formalism, and (8) Canons of Constructive Research.
8. Reisch, *How the Cold War Transformed Philosophy of Science*, 59, 104; on Malisoff, see ch. 5.
9. Otto Neurath, "Encyclopedia as 'Model,'" in *Philosophical Papers 1913–1946*, ed. R. S. Cohen and Marie Neurath (Boston: Reidel 1983), 145–58, 153.
10. Kaempffert, "Science in Review," *New York Times*, Oct. 1, 1944; Neurath, "Personal Life and Class Struggle," in *Empiricism and Sociology*, ed. R. S. Cohen and Marie Neurath (Boston: Reidel, 1973), 294–95.
11. George Reisch, "Planning Science: Otto Neurath and the International Encyclopedia of Unified Science," *British Journal for the History of Science* 27 (1994): 153–75; Mary Jo Nye, *Michael Polanyi and His Generation*; Friedrich Hayek, *The Road to Serfdom* (London: Routledge, 1944).

12. Conant, *On Understanding Science*, 63, 64, 107. Kuhn later mentions Conant's attraction to Polanyi's writings in Kuhn, *RSS*, 294.

13. Otto Neurath, "After Six Years," *Synthese* 5 (1946): 77–82.

14. Horace Kallen, "The Meanings of 'Unity' Among the Sciences," *Educational Administration and Supervision* 26, no. 2 (1940): 81–97, 92, 91.

15. Kallen, "The Significance of the Unity of Science Movement: Reply," *Philosophy and Phenomenological Research* 6, no. 4 (June 1946): 515–26, 519, 518–19.

16. Kallen, "The Meanings of Unity," 83, 82.

17. See Marie Neurath's recollections in Neurath, *Empiricism and Sociology*, 62.

18. Vannevar Bush, *Science: The Endless Frontier* (Washington, DC: National Science Foundation, 1945), 22, 34.

19. Kaempffert, "For a Hierarchy of Scientists," review of Bush, *Endless Horizons* (Public Affairs Press, 1946), *New York Times*, March 17, 1946; "Science in Review: Organization and Planning Held Necessary to Close Gaps in Our Knowledge," *New York Times*, Sept. 9, 1945. For a detailed survey of the long-running debates over funding the NSF, see Jessica Wang, *American Scientists in an Age of Anxiety* (Chapel Hill: University of North Carolina Press, 1999).

20. "Research for Defence," *New York Times*, July 21, 1945.

21. Conant, Letter to the Editor ("National Research Argued"), *New York Times*, Aug. 13, 1945; Kaempffert, "Science in Review: Further Arguments in Favor of Research Organized on a National Scale," *New York Times*, Aug. 19, 1945.

22. Roosevelt to Bush, Nov. 19, 1944, reprinted in Bush, *Science: The Endless Frontier*, 3–4; Harry S. Truman, Letter to Congress, printed in *New York Times*, Sept. 7, 1945.

23. The letter is reprinted in full in "Truman Aid Asked for Magnuson Bill," *New York Times*, Nov. 27, 1945. On Truman's preference for a "straightline administration," see "Truman on Research Bill," *New York Times*, Oct. 30, 1945.

24. Conant, "Science and Politics in the Twentieth Century," *Foreign Affairs* 28, no. 2 (Jan. 1950): 189–202.

25. Warren Weaver to Editor, *New York Times*, "Free Science Sought," *New York Times*, Sept. 2, 1945.

26. Kaempffert, "Toward Bridging the Gaps Between the Sciences," *New York Times*, Aug. 7, 1938.

27. Kaempffert, "Science in Review: Organization and Planning Held Necessary to Close Gaps in Our Knowledge"; Weaver, "Free Science Sought."

28. William H. Lawrence to Kuhn, Feb. 17, 1951.

29. Kuhn to Lowell, Feb. 20, 1951; Kuhn to Owen, Jan. 6, 1951, TSK-MIT, box 3 folder 10.

30. Kuhn, "Introduction: Textbook Science and Creative Science," manuscript. All quotations from Kuhn's Lowell Lectures are taken from his typed and often hand-edited manuscripts in TSK-MIT, box 3, folders 10–11. Quotations on pages 3, 13, 4.

31. Kuhn, lecture V ("Evidence and Explanation"), hand-numbered pages 15–16.

32. Ibid., hand-numbered pages, 19, 21, 25, 26, 31.

33. Kuhn, "Introduction: Textbook Science and Creative Science," 4–5.

34. Conant, *On Understanding Science*, 21; Kuhn, lecture V ("Evidence and Explanation"), hand-numbered pages 32–33.

35. Kuhn, "Introduction: Textbook Science and Creative Science." These remarks are contained on two consecutive pages labeled "6."

36. Ibid., 7, 8–9.

37. Ibid., 9.

38. Ibid., 7.

39. Kuhn, lecture V ("Evidence and Explanation"), hand-numbered page 44 (upper left) and 43 (upper right).

40. Kuhn, lecture VIII ("Canons of Constructive Research"), page VIII-1-1. Kuhn put a check mark ("indicates read in toto") next to Woodger's monograph in his "notes and ideas" notebook. Joseph Woodger, *The Technique of Theory Construction* (Chicago: University of Chicago Press, 1939).

41. Ibid., hand-numbered 7, 14.

42. Ibid., hand-numbered 25, 35–36.

43. Kuhn's surviving lecture notes reminded him to draw these circles at this point in his discussion. Alfred Korzybski, *Science and Sanity* (Lakeville, CT: International Non-Aristotelian Library, 1933); Stuart Chase, *The Tyranny of Words* (New York: Harcourt Brace, 1938); S. I. Hayakawa, *Language in Action* (New York: Harcourt Brace, 1941).

44. See Reisch, *How the Cold War Transformed Philosophy of Science*, 69. In his autobiography, W. V. O. Quine, whom Kuhn knew at Harvard, discusses Korzybski's project and its dubious allure for Harvard's students in the late 1930s. See *The Time of My Life* (Cambridge: MIT Press, 1985), 139–40.

45. Kuhn, lecture VIII ("Canons of Constructive Research"), hand-numbered pages 21, 22. Later in 1951, precisely this complaint would be leveled at Sidney Hook by the philosopher Victor Lowe. See Victor Lowe, "A Resurgence of Vicious Intellectualism," and Hook's rejoinder, "Mindless Empiricism," *Journal of Philosophy* 49, no. 4 (Feb. 14, 1952): 89–100. The debate between Lowe and Hook was watched closely by young philosophers according to Richard Rorty's recollections. See his "Afterword," in M. J. Cotter, ed., *Sidney Hook Reconsidered* (Amherst, NY: Prometheus Books, 2004), 281–86, 285, 286.

46. Kuhn, lecture VIII ("Canons of Constructive Research"), hand-numbered pages 23, 24, 28, 33. In fact, Kuhn misread logical empiricist research such as Woodger's. Its aim was to clarify ordinary scientific language by reconstructing it within formal language. That aim did not extend to replacing either scientific language or the language of the "layman," as Kuhn presupposed in his lecture VII, with formal reconstructions like Woodger's. For their part, Neurath, Carnap, and other logical

empiricists were clear that scientific language necessarily rested on ordinary language, what Neurath called "jargon" and Carnap called the "thing language," that could be refined but not replaced.

47. Ibid., hand-numbered page 6. The page is type-numbered as VII-1-6 but appears to belong as VIII-1-6. On the makeup of audiences for the Lowell Lectures, Edward Weeks pointed out (in *The Lowells and Their Institute*, 162) that "in addition to those who were attracted to a particular subject, the Lectures had at that time a small but loyal legion of listeners who attended out of habit, and most of whom were elderly spinsters." Weeks wrote of the 1930s but I have seen no evidence that Kuhn's audience was likely to have been much different.

48. Kuhn, *RSS*, 289. Marcum, *Thomas Kuhn's Revolution*, 13–14.

Chapter 10. Ideology and Revolution in the *International Encyclopedia of Unified Sciences*

1. On Shils's "feeler" about Kuhn coming to the university's Committee on Social Thought, see Kuhn to Shils, Sept. 13, 1963, in TSK-MIT, box 14, folder 23.

2. Louis Wirth, Preface to Karl Mannheim, *Ideology and Utopia: An Introduction to the Sociology of Knowledge*, trans. Louis Wirth and Edward Shils (New York: Harcourt, Brace, 1954 [orig. 1936]), xxv–xxvi.

3. Mannheim, *Ideology and Utopia*, 54, 67.

4. Otto Neurath, *Modern Man in the Making* (New York: Alfred A. Knopf, 1939). For additional biographical information about Neurath, see his *Empiricism and Sociology*, 1–79.

5. Frank, "Introductory Remarks," in *Contributions to the Analysis and Synthesis of Knowledge, Proceedings of the American Academy of Arts and Sciences* 80, no. 1 (1951): 5–8, 7, 8; Conant to John Boyer, Sept. 12, 1952, James B. Conant Personal Papers, Harvard University Archives; Hershberg, *James B. Conant*, 578.

6. Conant, "Greetings to the National Conference of the Institute for the Unity of Science. Boston, Massachusetts: April, 1950." *Proceedings of the American Academy of Arts and Sciences* 80, no. 1 (1951): 9–13, 9, 10.

7. Ibid., 13.

8. Frank's unfinished manuscript is held in Philipp Frank Papers, Harvard University Archives. Frank, "The Variety of Reasons for the Acceptance of Scientific Theories," in *The Validation of Scientific Theories*, ed. Frank (Boston: Beacon Press, 1957), 3–18, 14–16, 18.

9. Frank to Kuhn, Dec. 2, 1952, in TSK-MIT, box 25, folder 53.

10. Frank, Merton, Nagel, "Research Project in the Sociology of Science," 3 pp., in TSK-MIT, box 25, folder 53. Mannheim wrote, "The emergence and the crystallization of actual thought," is influenced by "extra-theoretical factors of the most diverse sort. These may be called, in contradistinction to purely theoretical

factors, existential factors." Mannheim, *Ideology and Utopia*, 240 (see also 34, 35, 210 239, 243, 261). On Merton and Mannheim, see David Kaiser, "A Mannheim for All Seasons: Bloor, Merton, and the Roots of the Sociology of Scientific Knowledge," *Science in Context* 11, no. 1 (1998): 51–87, 70.

11. Frank, Merton, Nagel, "Research Project in the Sociology of Science."

12. Untitled document ("Dear Professor Frank"), TSK-MIT, box 25, folder 53.

13. Kuhn, Lowell Lecture V, hand-numbered page 44–45 (upper left) and 43–44 (upper right).

14. Untitled document ("Dear Professor Frank"), TSK-MIT, box 25, folder 53.

15. Philipp Frank, "Report of the Institute for the Unity of Science to the Rockefeller Foundation for the Period of July 1952–1953," Rockefeller Foundation Papers, Record Group 1, Box 35, Folder 285, Rockefeller Archive Center, Tarrytown, NY. On Frank's declining professional reputation and Hoover's investigation of Frank, see Reisch, *How the Cold War Transformed Philosophy of Science*, 308–309, 268–71.

16. Internal Press memo, 1949, UCPR Box 346, Folder 5. On the postwar history of the encyclopedia, see Reisch, *How the Cold War Transformed Philosophy of Science*, ch. 14.

17. Neurath to Morris, Nov. 12, 1936, Unity of Science Movement Papers, Regenstein Library, University of Chicago, Department of Special Collections, box 2, folder 6. The portrayal of Kuhn in Isaac's *Working Knowledge* (chapters 1 and 6) emphasizes the interdisciplinary character of Kuhn's education, research, and teaching at Harvard during these years.

18. Morris to Kuhn, July 27, 1953; Kuhn to Morris, July 31, 1953, in TSK-MIT, box 25, folder 53; Quine, "Ontology and Ideology," *Philosophical Studies* 2, no. 1 (Jan. 1951): 11–15, quote on 14. For more on Quine's theories of ideology, see in particular the work of Lieven Decock, including *Trading Ontology for Ideology* (The Netherlands: Kluwer, 2002).

19. United States. House of Representatives. Select Committee to Investigate Tax-Exempt Foundations and Comparable Organizations. Final report (Washington, DC: U.S. Government Printing Office, 1953). The report faulted this encyclopedia for allowing known communists to contribute articles because they lack "objectivity" and "have a way of bringing things political into almost any subject." Quoted in "Excerpts from Reports on Congressional Investigation of Tax-Exempt Foundations," *New York Times*, Dec. 20, 1954.

20. On the FBI investigations of Malisoff, Frank, and Carnap, see Reisch, *How the Cold War Transformed Philosophy of Science*, chs. 13, 15.

21. Kuhn to Morris, Oct. 6, 1953, TSK-MIT, box 25, folder 53. On Carnap's support for clemency for the Rosenbergs, see Reisch, *How the Cold War Transformed Philosophy of Science*, 271–72. For Carnap's reaction to Kuhn's title (he suggested simply "Revolutions in Science" or "Scientific Revolutions") see Morris to Kuhn, Oct. 6, 1953, TSK-MIT, box 25, folder 53. Kuhn disliked Carnap's suggestions and told Morris that they obscured his aim to explore the historical, as opposed

to logical, structure of scientific revolutions. See Kuhn to Morris, Oct. 25, 1953, Charles Morris Papers, Center for American Thought, Indiana University Purdue University Indianapolis.

22. Kuhn to Morris, Oct. 6, 1953, TSK-MIT, box 25, folder 53.

23. Ibid.

24. Kuhn, "Afterwords," in *RSS*, 227; see also "A Discussion with Thomas S. Kuhn," in *RSS*, 306.

25. Kuhn, *Structure*, 169–71; Otto Neurath, "The New Encyclopedia of Scientific Empiricism," in *Philosophical Papers 1913–1946*, 189–199, 195. On Neurath's reflexive ideal of using science to understand science, see the writings of Thomas Uebel, for example, his *Empiricism at the Crossroads* (Chicago: Open Court, 2007) and "The Enlightenment Ambitions of Epistemic Utopianism," in *Origins of Logical Empiricism*, ed. R. N. Giere and A. W. Richardson (Minneapolis: University of Minnesota Press, 1996), 91–112.

26. For an analysis of Neurath's proposal, see my "Epistemologist, Economist . . . and Censor? On Otto Neurath's Infamous *Index Verborum Prohibitorum*," *Perspectives on Science* 5, no. 3 (1997): 452–80. On Neurath's "clots," see his contribution to the encyclopedia, "Foundations of the Social Sciences," *International Encyclopedia of Unified Science*, v. 2, no. 1 (Chicago: University of Chicago Press, 1944), 18.

27. James, "The Pragmatic Conception of Truth," in *Pragmatism* (Indianapolis: Hackett, 1981), 92.

28. Kuhn described his future monograph as "primarily a work of synthesis" that would make use of "a variety of fields not normally treated together" in his application for a Guggenheim Fellowship in the early 1950s. "Plans for Research," TSK-MIT, box 25, folder 53.

29. Kuhn, "The War and My Crisis," TSK-MIT, box 1, folder 3.

30. Kuhn to Morris, Oct. 6, 1953, TSK-MIT, box 25, folder 53.

Chapter 11. Progress, Ideology, and "Writing History Backwards"

1. Kuhn, "Notes Toward Unity of Science Monograph," hand-labeled "A," box 25, folder 53.

2. Kuhn to Morris, Dec. 17, 1959, TSK-MIT, box 25, folder 53.

3. Kuhn's letter suggests that he enclosed manuscripts for what became "The Essential Tension: Tradition and Innovation in Scientific Research" and "The Function of Measurement in Modern Physical Science."

4. Kuhn to Morris, Dec. 17, 1959, TSK-MIT, box 25, folder 53.

5. Kuhn to Morris, July 11, 1958, TSK-MIT, box 25, folder 53. A copy of this letter also exists in the Charles Morris Papers, Center for American Thought, IUPUI.

6. Kuhn Lowell Lectures manuscript, Lecture 8, hand-numbered page 38.

7. Kuhn, "Notes Toward Unity of Science Monograph," hand-labeled "A," TSK-MIT, box 25, folder 53.

8. Kuhn, "The Structure of Scientific Revolutions, Block Outline #1," TSK-MIT, box 25, folder 53.

9. "Discoveries as Revolutionary," TSK-MIT, box 4, folder 2, 37, 38.

10. "Discoveries as Revolutionary," TSK-MIT, box 4, folder 2, 37. *Structure*, 137.

11. Kuhn, *Structure*, 166.

12. Kuhn, "Foreword," in Ludwik Fleck, *Genesis and Development of a Scientific Fact* (Chicago: University of Chicago Press, 1979, vii–xi, viii–ix.

13. Fleck, *Genesis and Development*, 96, 98.

14. Ibid., 86–87.

15. If, Schein wrote, "one could persuade the peasant to the acceptance of communist premises, the peasant would by definition become a Communist regardless of actual social origins." Schein, "Brainwashing" (Center for international Studies, MIT, December 1960), 3–4.

16. Kuhn, "Foreword," in Fleck, *Genesis and Development*, vii, viii, ix.

17. Kuhn, *Structure*, xli.

18. Dick Pels, *Autonomy and Reflexivity in the Social Theory of Knowledge* (Liverpool: Liverpool University Press, 2003), 60; Thaddeus J. Trenn, preface to Fleck, *Genesis and Development*, xiii–xix, xv.

19. Fleck, *Genesis and Development*, 95. For a recent interpretation that reconciles the revolutionary and evolutionary aspects of Kuhn's philosophy of science, see K. Brad Wray, *Kuhn's Evolutionary Social Epistemology*, esp. ch. 8.

20. Crane Brinton, *The Anatomy of Revolution* (New York: Random House, 1956), 35.

21. Kuhn, notecard on Crane Brinton, *French Revolutionary Legislation on Illegitimacy 1789–1804* (Cambridge: Harvard University Press, 1936), TSK-MIT, box 8.

22. May, *Un-American Activities*, 50.

23. Eugen Rosenstock-Huessy, *Die Europäischen Revolutionen: Volkscharaktere und Staatenbildung* (Jena: E. Diederichs, 1931); *Out of Revolution: Autobiography of Western Man* (Providence, RI: Berg, 1993), 13.

24. The notes read: Some general characteristics of revolutions:

p. 5) "Wenn wir aber, in diesem Buche von Revolution reden so meinen wir nur eine solche, die ein für allemal ein neues Lebensprinzip in die Weltgeschichte hat einführen wollen, also ein Totalumwälzung."

p. 18) "Die Geschichte nach Rückwärts ändert eben ihren Inhalt mit jedem neuen Revolutionaren Ursprung des Volkerlebens nach Vorwärts." And more on the way each revolutionary movement starts calendar anew and destroys the past.

p. 23f.) "Revolution ist . . . Sprechen einer bis dahin unerhörten Sprach, mit anderen Worten Auftauschen einer anderen Logik" etc. Much more of the same on conceptual discontinuities in revolutions.

p. 25) "Das Zeitalter der Revolution schliesst niemals mit einer vollstandigen Rasur des alten Menschen, sondern mit einem neuen Bund."

p. 27) "Die Empörung ist . . . nicht zu Ende mit dem Ende der Revolution. Aber sie ist es in personeller und menschlicher hinsicht."

pp. 29–30) "Die Zeitafel der Revolutionen bedarf daher der Erganzung durch eine solche Demutigungszeiten des Emporen. In diesen Zeiten scheinen die Errungenschaften des Revolution sinnloss geworden zu sein. . . . Aber es scheint nur so. . . . Die Zeit der Demutingung . . . erscheint als Prufungzeit";

Kuhn, notecard on Rosenstock-Huessy, *Die Europäischen Revolutionen*, TSK-MIT, box 25, folder 53.

25. Kuhn, *Structure*, 93; emphasis added, 94.

26. "Conant Claims German Nazism Dead," *Crimson*, January 25, 1956; Conant, *Germany and Freedom: A Personal Appraisal* (Cambridge: Harvard University Press, 1958), 10, 14–15.

27. Conant, *Germany and Freedom*, 14–15.

28. Ibid., 15.

29. Kuhn, *Structure*, 136, 135.

30. Kuhn to Morris, Dec. 17, 1959, TSK-MIT, box 25, folder 53.

31. Morris to Kuhn, Aug. 5, 1956; Morris to Kuhn, Jan. 4, 1960, TSK-MIT, box 25, folder 53.

32. Kuhn to Morris, Feb. 13, 1960, TSK-MIT, box 25, folder 53.

33. Ibid. On Berkeley, see Seth Rosenfeld, *Subversives: The FBI's War on Student Radicals and Reagan's Rise to Power* (New York: Farrar Strauss and Giroux, 2012); on Oppenheimer and Conant's efforts to defend him from political persecution, see, for example, Hershberg, *James B. Conant*, 676–82, and Dean Gordon, "Excerpts From Testimony of Leading Witnesses in Security Hearings on Oppenheimer," *New York Times*, June 17, 1954; on Condon, see *Testimony of Dr. Edward U. Condon: Hearing before the Committee on Un-American Activities, House of Representatives, Eighty-second Congress, second session*. September 5, 1952 (Washington, DC: Government Printing Office, 1952), William L. Laurence, "Scientist Decries Curb on Condon, Physicist a Victim of Rumor and Anxiety," *New York Times*, Dec. 29, 1954, and Philip M. Morse, "Edward Uhler Condon, 1902–1974" (Washington, DC: National Academy of Sciences, 1976), 139–40. For a revealing analysis of this climate's effects on the University of California, Los Angeles, see McCumber, *The Philosophy Scare*.

34. Kuhn, "new outline," TSK-MIT, box 4, folder 2.
35. Kuhn, *Structure*, 163, 20.
36. Kuhn, "new outline," TSK-MIT, box 4, folder 2.
37. Kuhn, note page beginning "Pattern of Evolution," TSK-MIT, box 25, folder 53.
38. Robert Post, "A Very Special Relationship: SHOT and the Smithsonian's Museum of History and Technology," *Technology and Culture* 42 (2001): 401–435, see 402; Kranzberg to Kuhn, TSK-MIT, box 4, folder 4.
39. Kuhn, note page beginning "Pattern of Evolution," TSK-MIT, box 25, folder 53.
40. Kuhn, *Structure*, 161.

Chapter 12. From "Ideology" and "Consensus" to Paradigmania

1. Kuhn, "The Structure of Scientific Revolutions" (draft of "Chapter 1—Discoveries as Revolutionary"), TSK-MIT, box 4, folder 2, 1–2.
2. Ibid., 3–5.
3. Kuhn, "II. The Nature of Scientific Consensus," draft ms., TSK-MIT, box 4 folder 4, 38–39.
4. Ibid., 43, 44.
5. Ibid., 44, 45–47.
6. Ibid., 50, 48.
7. Ibid., unnumbered pages at end of draft.
8. Kuhn, *Structure*, 10–11.
9. Ibid.
10. Kuhn, "The Essential Tension: Tradition and Innovation in Scientific Research," in *The Essential Tension* (Chicago: University of Chicago Press, 1977), 225–39, 231, 235. On others besides Kuhn using the term *paradigm* in the late 1950s, see K. Brad Wray, "Kuhn and the Discovery of Paradigms," *Philosophy of the Social Sciences* 41, no. 3 (2011): 380–97. See also Wray's *Kuhn's Evolutionary Social Epistemology*, ch. 4. Kuhn's puzzle about covering a chessboard with dominoes appears at the end of his fifth Lowell Lecture and the beginning of his sixth. It was evidently adapted from the "mutilated chessboard problem" posed by the philosopher Max Black in his book *Critical Thinking: An Introduction to Logic and Scientific Method* (New York: Prentice-Hall, 1946, 157). Black's puzzle was later popularized by Martin Gardner in his long-running column in *Scientific American*, "Mathematical Games."
11. Kuhn, "The Structure of Scientific Revolutions" (draft of "Chapter 1—Discoveries as Revolutionary"), TSK-MIT, box 4, folder 2, 33–34, emphasis in original.

12. Kuhn, note page beginning "Pattern of Evolution," TSK-MIT, box 25, folder 53.
13. Kuhn, *Structure*, 88.
14. Kuhn to Morris, Oct. 13, 1960, TSK-MIT, box 25, folder 53.
15. Kuhn to Morris, Dec. 17, 1959, TSK-MIT, box 25, folder 53.
16. Kuhn, "Plans for Research," TSK-MIT, box 25, folder 53. *Structure*, xii–xiii.
17. Kuhn, "The Structure of Scientific Revolutions" (draft of "Chapter 1—Discoveries as Revolutionary"), TSK-MIT, box 4, folder 2, 37.
18. Ibid., folder 5, 39, 41–42, 23.
19. Ibid., 43.
20. Ibid., 123; *Structure*, 122.
21. *Structure*, 122–23. Kuhn's notes on James's *Varieties* is in TSK-MIT, box 8.
22. Kuhn, "The Structure of Scientific Revolutions," draft, TSK-MIT, box 4, folder 5, 156; *Structure*, 149.
23. Arthur Koestler, "Arthur Koestler" in *The God that Failed*, ed. Richard Crossman (New York: Columbia University Press, 2001; orig. London: Hamilton, 1950), 23.
24. Koestler, "Arthur Koestler," 23.
25. Kuhn, "The Structure of Scientific Revolutions," draft, TSK-MIT, box 4, folder 5, 157; *Structure*, 150.
26. Kuhn, "The Structure of Scientific Revolutions," draft, TSK-MIT, box 4, folder 5, 23.
27. Kuhn to Morris, Oct. 13, 1960, TSK-MIT, box 25, folder 53.
28. Morris to Kuhn, Oct. 23, 1960, TSK-MIT, box 25, folder 53.
29. Morris to Kuhn, Apr. 14, 1961, Kuhn to Morris, April 16, 1961, both TSK-MIT, box 25, folder 53.
30. Kuhn to Carroll G. Bowen, Jun 18, 1961; Noyes to Kuhn, June 20, 1961; Nagel to Kuhn, June 4, 1961, TSK-MIT, box 25, folder 53.
31. Nagel to Kuhn, June 4, 1961; Bernard Barber to Kuhn, May 8, 1961, TSK-MIT, box 25, folder 53.
32. Kuhn to Carroll G. Bowen, Jun 18, 1961, TSK-MIT, box 25, folder 53.
33. These debates are examined in chapter 15.
34. Norwood Hanson, "Report on Dr. T. Kuhn's *The Structure of Scientific Revolutions*," UCPR, box 278, folder 4.
35. Popper's use of falsifiability to distinguish genuine science from pseudo-science is presented in his *The Logic of Scientific Discovery* (New York: Basic Books, 1959).
36. Norwood Hanson, "Report on Dr. T. Kuhn's *The Structure of Scientific Revolutions*."
37. Ibid.
38. Hanson to Kuhn, telegram, TSK-MIT, box 25, folder 53.

39. Kuhn, *RSS*, 300–301.
40. Kuhn to Bowen, June 29, 1961, TSK-MIT, box 25, folder 53.

Chapter 13. "If Mr. Kuhn Is Right . . ."

1. William J. Jorden, "Soviet Fires Earth Satellite Into Space," *New York Times*, October 5, 1957; John W. Finney, "U.S. Missile Experts Shaken By Sputnik," *New York Times*, October 13, 1957.
2. See, for example, James B. Conant, "Wanted: American Radicals," *Atlantic Monthly* (May 1943), 41–45; James B. Conant, "Science and the National Welfare," 454; James B. Conant, *Education in a Divided World*; James B. Conant, *Modern Science and Modern Man*; James B. Conant, *The Citadel of Learning* (New Haven: Yale, 1956). A helpful account of post-Sputnik education reforms involving H. Bentley Glass is John Rudolph, *Scientists in the Classroom* (New York: Palgrave McMillan, 2002).
3. Arthur H. Compton to Kuhn, May 2, 1949, TSK-MIT, box 3, folder 33.
4. "The Sciences in the Harvard General Education Program," TSK-MIT, box 3, folder 33.
5. Lecture notes beginning "II. Qualifications," TSK-MIT, box 3, folder 33. The first two pages of these fourteen pages of notes, which may have contained a date, appear to have been lost. I take them to be later than the short lecture "The Sciences in the Harvard General Education Program" because of their length and the author's tone.
6. Ibid., 13.
7. Ibid., 10, 14.
8. Kuhn, "II. The Nature of Consensus in Science," draft ms., TSK-MIT, box 4, folder 4, unnumbered pages at end of draft.
9. "Can the Layman Know Science?" TSK-MIT, box 3, folder 33, 1.
10. Ibid.
11. Ibid., 2.
12. Ibid., 3, 4, 5.
13. Ibid., 5, 6.
14. Ibid., 9.
15. Ibid., 10; Thomas S. Kuhn, *The Copernican Revolution* (Cambridge: Harvard University Press, 1957), 76.
16. Kuhn, *RSS*, 292.
17. Kuhn, *The Copernican Revolution*, 40, 75.
18. Kuhn, "Sputnik & American Public Mind," notecards, Dec. 5, 1957, TSK-MIT, box 3, folder 12.

19. Ibid.

20. Thomas S. Kuhn, "The Function of Dogma in Scientific Research," in *Scientific Change: Historical Studies in the Intellectual, Social, and Technical Conditions for Scientific Discovery and Technical Invention, from Antiquity to the Present*, ed. A. C. Crombie (New York: Basic Books, 1963), 347–69, 347.

21. Ibid., 347–48.

22. Ibid., 362, 363.

23. Ibid., 352, 353, 360, 363.

24. Ibid., 349.

25. "Commentary by A. Rupert Hall," in Crombie, ed., *Scientific Change*, 370–75, 374.

26. "Discussion: S. E. Toulmin," in Crombie, ed. *Scientific Change*, 382–84, 383. One participant who liked the paper was Michael Polanyi, whose writings on the "tacit" and "personal" features of scientific knowledge bear a similarity to Kuhn's that remains controversial. See, e.g., Martin Moleski, S.J., "Polanyi vs. Kuhn: Worldviews Apart" and Struan Jacobs, "Michael Polanyi and Thomas Kuhn: Priority and Credit," both in *Tradition and Discovery: The Polanyi Society Periodical* 33, no. 2 (2006/2007): 8–24, 25–36; Struan Jacobs, "Thomas Kuhn's Memory," *Intellectual History Review* 19 (2009): 83–101.

27. No in-depth biography of Glass exists, but useful resources include Audra Wolfe, "The Organization Man and the Archive: A Look at the Bentley Glass Papers," *Journal of the History of Biology* 44 (2011): 147–51; and Frank C. Erk, "H. Bentley Glass," *Proceedings of the American Philosophical Society* 153 (2009): 327–39. On Glass's activity as a public intellectual, see, for example, Bentley Glass, "Academic Freedom and Tenure in the Quest for National Security," *Bulletin of the Atomic Scientists* 12, no. 6 (June 1956): 221–23, 226; Bentley Glass, "Liberal Education in a Scientific Age," *Bulletin of the Atomic Scientists* 14, no. 11 (Nov. 1958): 346–53, and his book *Progress or Catastrophe: The Nature of Biological Science and Its Impact on Human Society* (New York: Praeger, 1985).

28. H. Bentley Glass, "The Establishment of Modern Genetical Theory as an Example of the Interaction of Different Models, Techniques, and Inferences," in *Scientific Change*, ed. A. C. Crombie, 521–41, 541.

29. Glass, "The Establishment of Modern Genetical Theory," 541.

30. Kuhn, "The Function of Dogma in Scientific Research," 351, 363, 369, 350. See also Thomas Kuhn, "The Essential Tension: Tradition and Innovation in Scientific Research," esp. 229.

31. "Discussion: H. Bentley Glass," in *Scientific Change*, ed. A. C. Crombie, 381–82.

32. Ibid., 382.

33. "Discussion: T. S. Kuhn," in *Scientific Change*, ed. A. C. Crombie, 386–95, quoted from 391.

34. "Discussion: H. Bentley Glass," 381.

35. "Discussion: T.S. Kuhn," 390, 392.

36. See Kuhn, *RSS*, 2, note 1. In Kuhn's papers, there exists a request from a publisher in 1991 to reprint "The Function of Dogma in Scientific Research" that Kuhn declined while offering instead his early essay, "The Essential Tension" (TSK-MIT, box 20, folder 23).

37. In the fourth edition of *Structure*, these can be found on the following pages: "binding" (12, 40, 46), "commitment," (11, 25, 28, 40) "rigid" (20, 49, 64, 165), "accepts without question" (48), "relatively inflexible [theoretical] box" (24), "take for granted" (20, 30, 102), "assurance," (42, 151) "confidence in their paradigms" (18, 25, 164). Variants of these phrases in Kuhn's postscript of 1969 are not included here.

38. Kuhn, "The Function of Dogma," 349; *Structure*, 165.

Chapter 14. The Magic of Paradigms

1. Hufbauer, "From Student of Physics to Historian of Science," 452–53.

2. See Kuhn to Conant, April 22, 1961, TSK-MIT, box 25, folder 53. Kuhn's interest in psychoanalysis and his relationship to Conant as a father-like figure is explored by John Forrester, "On Kuhn's Case: Psychoanalysis and the Paradigm," 785, where Kuhn compares Conant to his father as being "the brightest person I had ever known."

3. Conant to Kuhn, June 2, 1961, TSK-MIT, box 25, folder 53.

4. Conant to Kuhn, June 5, 1961, TSK-MIT, box 25, folder 53. A succinct comparison and analysis of Kuhn's and Conant's views of science is K. Brad Wray, "The Influence of James B. Conant on Kuhn's *Structure of Scientific Revolutions*."

5. These annotations and those described below are in Kuhn's handwriting on the letter from Conant.

6. Kuhn, *Structure*, 4; Conant to Kuhn, June 5, 1961, TSK-MIT, box 25, folder 53.

7. Kuhn to Conant, June 29, 1961, TSK-MIT, box 25, folder 53. The quotations appear on pages 6 and 1.

8. Conant, *The Citadel of Learning* (New Haven: Yale University Press, 1956), 3. Three years before, Conant offered the same surprising observation in *Foreign Affairs*. Writing on "Science and Politics in the Twentieth Century," he noted that issues in physics and philosophy and other "matters that would not incite even the passing interest of ninety-nine politicians out of a hundred in the western democracies are treated [in the pages of *Pravda*] as of deep significance." "Science and Politics in the Twentieth Century,"191.

9. Conant, *The Citadel of Learning*, 3, 4, 5, 6.

10. Conant, "Science and Politics in the Twentieth Century," 201, 202.

11. Conant, "Science and the National Welfare," 400.
12. Conant, *The Citadel of Learning*, 8, 67–68.
13. Conant to Kuhn, Dec. 7, 1956, TSK-MIT, box 25, folder 46; Conant, foreword to Kuhn, *The Copernican Revolution*, xiii, xviii.
14. Kuhn, "The Structure of Scientific Revolutions," manuscript draft, TSK-MIT, box 4, folder 5, 1.
15. Isaac, *Working Knowledge*, 68–69.
16. Conant, *On Understanding Science*, 23, 22; Conant to Kuhn, June 5, 1961.
17. Conant to Kuhn, June 5, 1961.
18. Kuhn to Conant, June 29, 1961.
19. Ibid.
20. Ibid.; Kuhn, *Structure*, 113, 121 [Draft manuscript, 112, 121].
21. Kuhn to Conant, June 29, 1961.
22. Conant to Kuhn, Oct. 11, 1950, Harvard University Archives, Records of James B. Conant, Box 402.
23. Conant to Kuhn, July 11, 1961, TSK-MIT, box 25, folder 53.
24. Kuhn to Conant, Aug. 5, 1961, TSK-MIT, box 25, folder 53.
25. On Conant's saying about the adventurous turtle, see Hershberg (1993, 89); Conant to Kuhn, Aug. 14, 1961, TSK-MIT, box 25, folder 53.
26. Conant to Kuhn, Dec. 19, 1962, TSK-MIT, Box 4, folder 8.
27. Kuhn, *Structure*, 46.
28. Ibid., 160, 162–63, 166.
29. Ibid., 169; "Structure" (manuscript), 175; Conant to Kuhn, Dec. 19, 1962, TSK-MIT, box 4, folder 8. Conant's aversion to "truth" as a helpful concept in understanding science's history can also be traced to L. J. Henderson (see Isaac, *Working Knowledge*, 69), but it also expresses his commitment to James's pragmatism.
30. Conant to Kuhn, Dec. 19, 1962, TSK-MIT, box 4, folder 8.
31. Conant to Kuhn, July 29, 1963, TSK-MIT, box 4 folder 8.
32. Conant, *Two Modes of Thought*, 13.
33. Conant cites and quotes from *Structure* when first introducing it on p. 13. In a footnote on p. 17, he refers the reader to it again (along with Butterfield's *Origins of Modern Science* and the *Harvard Case Studies in Experimental Science*) in support of his claim (familiar to his readers since *On Understanding Science*) that science's history cannot be understood as an accumulation of gathered facts.
34. Conant, *Two Modes of Thought*, xxxi; Kuhn, "The Essential Tension: Tradition and Innovation in Scientific Research," 226. In his letter to Conant of June 29, 1961, Kuhn suggests that Conant was familiar with this essay.
35. Conant, *Two Modes of Thought*, 13–14, 15, 24–25, 30.
36. Ibid., 82.
37. Milosz, "The Happiness Pill," *Partisan Review* 18, no. 5 (Sept.–Oct. 1951): 540–46: Conant, *Two Modes of Thought*, 86.
38. Conant, *Two Modes of Thought*, 86, 87.

Chapter 15. Spies, Prisons, Mobs, Bandwagons, and Beasts

1. Leslie R. Groves, *Now It Can Be Told: The Story of the Manhattan Project* (Boston: Da Capo Press, 1983), 144.

2. U.S. Congress, Joint Committee on Atomic Energy, *Soviet Atomic Espionage* (Washington, DC: Government Printing Office, April, 1951), 19–20.

3. Douglas Little, "Red Scare, 1936: Anti-Bolshevism and the Origins of British Non-Interventionism in the Spanish Civil War," *Journal of Contemporary History* 23, no. 2: 291–311; Groves, *Now it Can be Told*, 144.

4. "Cohn and Schine Get Headlines in London's Papers," AP news, *Milwaukee Journal*, March 13, 1954, 2.

5. "British Press Pokes Fun at McCarthy Investigators," *New York Times*, April 26, 1953.

6. Mary Hesse reviewed *Structure* in *Isis* 54 (June 1963): 286–87, and applauded it as a revolutionary breath of fresh air. Another British exception was Margaret Masterman, who participated in the conference described below, and applauded Kuhn's "paradigms" despite the vagueness and ambiguity she found in them (See *Criticism and the Growth of Knowledge*, ed. I. Lakatos and A. Musgrave, [Cambridge: Cambridge University Press, 1970], 59–89).

7. Kuhn, "The Structure of Scientific Revolutions," manuscript draft, TSK-MIT, box 4, folder 5, 1.

8. The review appears beneath the section "Books" written by Anthropologist Marshall D. Sahlins; but Sahlins (personal correspondence) denies being the author of the several "Short Reviews" that follow (*Scientific American*, May 1964, 142, 144).

9. Dudley Shapere, "The Structure of Scientific Revolutions" (review), *The Philosophical Review* 73, no. 3 (July 1964): 383–94, 385.

10. Ibid., 386; Kuhn, *Structure*, 11.

11. Dudley Shapere, "The Structure of Scientific Revolutions," 393, 388.

12. Even for Kuhn's friend and colleague Edward Shils, who commissioned a review of *Structure* in the magazine *Minerva* (an outgrowth of the Congresses for Cultural Freedom) and to whom the political valences of *Structure* might have been evident, *Structure* seemed to have little to do with either the geopolitics of freedom, which had guided Conant's perceptions of the book, or "the planning and administration of science," the avoidance of which had shaped Kuhn's theorizing at least since his Lowell Lectures. Shils commissioned the philosopher Stephen Toulmin to explore these connections, but the review either never materialized or never appeared. See Elena Aronova, "The Congress for Cultural Freedom, 'Minerva,' and the Quest for Instituting 'Science Studies' in the Age of Cold War," *Minerva* 50, no. 3 (2012): 307–37, 329.

13. Hershberg, *James B. Conant*, 511, 660, 661. On CIA interest and research in mind-control, see also John Marks, *The Search for the 'Manchurian Candidate': The CIA and Mind Control* (New York: New York Times Books, 1979);

and Allan Needell, " 'Truth Is Our Weapon': Project TROY, Political Warfare, and Government-Academic Relations in the National Security State," *Diplomatic History* 17 (1993): 399–420.

14. Grace Richards Conant, "The Cold War of the Mind: Regimentation in East Germany," *Modern Age* 5, no. 2 (Spring 1961): 117–24, 121, 124.

15. H. V. Stopes-Roe, "The Structure of Scientific Revolutions by Thomas S. Kuhn," *The British Journal for the Philosophy of Science* 15, no. 58 (Aug. 1964): 158–61.

16. Ibid. Quotations from Kuhn appear in *Structure* on 165 and 138.

17. Popper, "Unended Quest: An Intellectual Biography," in *The Philosophy of Karl Popper* (La Salle, IL: Open Court, 1976), §8.

18. Karl Popper, *Poverty of Historicism* (Boston: Beacon, 1957); *The Open Society and Its Enemies: vol I, The Spell of Plato*" (Princeton: Princeton University Press, 1971), 200, 201.

19. Matteo Collodel, "A Note on Kuhn and Feyerabend at Berkeley (1956–1964)," unpublished.

20. Two letters are transcribed by Paul Hoyningen-Huene in his "Two Letters of Paul Feyerabend to Thomas S. Kuhn on a Draft of *The Structure of Scientific Revolutions*," *Studies in History and Philosophy of Science* 26, no. 3, (1995): 353–87, and an additional two letters are in his "More Letters by Paul Feyerabend to Thomas S. Kuhn on *Proto-Structure*," *Studies in History and Philosophy of Science* 37 (2006): 610–32. What Kuhn wrote to Feyerabend, however, can only be gleaned from Feyerabend's four letters, as Kuhn's letters are currently lost.

21. Hoyningen-Huene, "Two Letters," 367.

22. Ibid., 355.

23. Ibid., 365, 367–68.

24. Kuhn, *Structure*, 42, 41; Hoyningen-Huene, "Two Letters," 363.

25. Hoyningen-Huene, "More Letters," 614.

26. Ibid., 360.

27. Kuhn, *RSS*, 310.

28. Imre Lakatos, "Falsification and the Methodology of Scientific Research Programmes," in *Criticism and the Growth of Knowledge*, ed. I. Lakatos and A. Musgrave (Cambridge: Cambridge University Press, 1970), 91–195, 178; emphasis in original. On Lakatos's background, see "Lakatos, Popper, and Feyerabend—Some Personal Reminiscences," by Donald Gillies (unpublished remarks), and Jancis Long, "The Unforgiven: Imre Lakatos' Life in Hungary," in *Appraising Lakatos,* ed. G. Kampis, L. Kvasz, and M. Stöltzner (New York: Springer, 2002), 263–302.

29. Paul Feyerabend, "Consolations for the Specialist," in Lakatos and Musgrave eds., *Criticism and the Growth of Knowledge*, 197–230, 200. On Feyerabend's passion for operatic singing and philosophical pronouncements, see his autobiography, *Killing Time* (Chicago: University of Chicago Press, 1995), 31–35. On the conference and its continuing resonance for philosophers of science, see Steve Fuller, *Popper vs. Kuhn:*

The Struggle for the Soul of Science (New York: Columbia University Press, 2004).

30. Kuhn, "Logic of Discovery or Psychology of Research?" in Lakatos and Musgrave eds., *Criticism and the Growth of Knowledge*, 1–23, 6.

31. Karl Popper, "Normal Science and Its Dangers," in Lakatos and Musgrave, eds., *Criticism and the Growth of Knowledge*, 51–58, 53, 52.

32. Ibid., 51, 56.

33. Stephen Toulmin, "Does the Distinction between Normal and Revolutionary Science Hold Water?," in Lakatos and Musgrave, eds., *Criticism and the Growth of Knowledge*, 39–48, 41. For a later, more exhaustive survey of scientific and political revolutions, see I. Bernard Cohen, *Revolution in Science*. Cohen adds that besides taking different shapes and sizes in the two domains, assessments of what counts as revolutionary in both science and politics "differs greatly from one age to the next" (xiii).

34. Toulmin, "Does the Distinction between Normal and Revolutionary Science Hold Water?," 43–44, 43.

35. J. W. N. Watkins, "Against Normal Science," in Lakatos and Musgrave, eds. *Criticism and the Growth of Knowledge*, 25–37, 26, 30, 37, 34.

36. Ibid., 28, 33; Kuhn, *Structure*, 165. For an account of how Kuhn's *Structure* was received in East and West Germany, see Michael Schorner's manuscript "The Reception of Kuhn's *Structure of Scientific Revolutions* in West and East Germany." Schorner shows that *Structure* was seen through Marxist-Leninist lenses in the German Democratic Republic as early as 1965. In general, however, the book's influence and stature began to grow in both Germanies in the mid-1970s.

37. Hoover, "The Communist Menace," in "Bills to Curb or Outlaw the Communist Party of the United States," part 2, United States House Committee on Un-American Activities, 80th Congress, 1st sess., March 26, 1947 (Washington, DC: U.S. Government Printing Office), 33–50 (voicesofdemocracy.umd.edu). Acheson quoted in Brian Crozier, "Why the Cold War?," in *The Collapse of Communism*, ed. Lee Edwards (Stanford, CA: Hoover Institute, 2000), 141–58, 153.

38. "Conant Urges Marshall Plan, UMT to Halt Soviet Advance," *Crimson*, March 25, 1948; "Inverchapel Calls for Unity; Conant Gains British Award," *Crimson*, March 19, 1948. Lowell Hall was originally named "New Lecture Hall," where Inverchapel spoke.

39. The fullest recent account of Greenglass's life and motivations at the time is Sam Roberts, *The Brother* (New York: Random House, 2001).

40. Ibid., 371, 380; on Cohn's ex-parte contact with Kaufman, see Von Hoffman, *Citizen Cohn*, 100–101.

41. Sartre, "Les Animaux malades de la rage," *Libération*, June 22, 1953. Translated in *Writings of Jean-Paul Sartre* (Evanston: Northwestern University Press, 1985), 207–11.

42. David Caute, *The Great Fear* (New York: Simon and Schuster, 1979), 185, 186.

43. *Soviet Atomic Espionage*, 11.
44. Ibid., 12.
45. Ibid. The quotation appears in *On Understanding Science* (7–8) and *Science and Common Sense* (9–10).
46. Quoted in Hershberg, *James B. Conant*, 438.
47. Sartre, "Les Animaux malades de la rage," 209.

Chapter 16. The Thomas Kuhn Experience

1. John S. Coulsen to Kuhn, Aug. 21, 1962; Kuhn to Coulson, Sept. 17, 1962, TSK-MIT, box 25, folder 12.

2. See, e.g. *Mad Men Unbuttoned*, by Natasha Vargas-Cooper (New York: Harper Collins, 2010). See pp. 10–11 for the Marlboro Man.

3. *Renaissance News* 10 (Winter 1957): 217–20; *Journal of Higher Education* 28, no. 9 (Dec. 1957): 514.

4. Friedlander to Kuhn, Feb. 15, 1961, TSK-MIT, box 25, folder 49.

5. Nagel to Kuhn, June 4, 1961, TSK-MIT, box 25, folder 53; Merton to Kuhn, Dec. 13, 1962, TSK-MIT, box 4, folder 12; Charles Gillispie in *Science*, Dec. 14, 1962, 1251–53, 1251; Boring to Kuhn, Nov. 9, 1962, TSK-MIT, box 4, folder 7; Boring in *Contemporary Psychology* (Spring 1963), 180–82. On Boring's support of Kuhn's theories, see for example "Cognitive Dissonance: Its Use in Science," *Science* 145, no. 3633 (Aug. 14, 1964), 680–85.

6. Kuhn to Gombrich, Oct. 24, 1963, TKS-MIT, box 4, folder 9; Kuhn to Carroll Bowen, Dec. 17, 1962, TSK-MIT, box 25, folder 55. Kuhn agreed with his copy editor at the University of Chicago Press that "the book will need to be read from beginning to end" if the reader is to "catch my argument" (Kuhn to Mrs. Nancy C. Romoser, May 14, 1962, UCPR, box 278, folder 4).

7. Kuhn to Carroll Bowen, Dec. 17, 1962, TSK-MIT, box 25, folder 55.

8. "United We Stand," playscript, TSK-MIT, box 1, folder 2. The play was later edited and published by Hessian Hills School teacher George Willison in his book *Let's Make a Play* (New York: Harper, 1940). Willison's edition however does not include Kuhn's or other students' names as playwrights or actors. I thank Juan Vicente Mayoral de Lucas for sharing this discovery. Kuhn's editorial essay is titled "The U.S. As a World Power" (TSK-MIT, box 1, folder 2).

9. Kuhn, "Phi Beta Kappa Address," TSK-MIT, box 1, folder 3.

10. James Miller, *Democracy Is in the Streets, From Port Huron to the Siege in Chicago* (Cambridge: Harvard University Press, 1994), 28–29; Hook, *Out of Step*, 44; Phelps, *Young Sidney Hook*, 163.

11. Miller, *Democracy*, 141.

12. Bob Ross, quoted in Miller, *Democracy*, 144; on SDS's break with LID, 235.

13. Miller, *Democracy*, 146, 51. Hayden quoted on 331.

14. Kuhn, *Structure*, 92, 93.

15. Seth Rosenfeld, *Subversives*; on Oppenheimer, see ch. 1; on Reagan and Kerr, ch. 15; on Savio, see Pt. II.

16. Harrison E. Salisbury, "Rioting in Soviet Reported over Anti-Stalin Campaign," *New York Times*, March 17, 1956; on American communists' reactions to Khrushchev's speech, see Vivian Gornick, *The Romance of American Communism* (New York: Basic Books, 1977); on Kennedy, see Adam Clymer, "When Presidential Words Led to Swift Action," *New York Times*, June 9, 2013.

17. Tom Hayden, draft of Port Huron Statement, "Introduction: Agenda for a Generation." The forty-eight-page draft can be found online at http://www.sds-1960s.org/.

18. Unless noted, the following quotations are from the final Port Huron Statement, in Miller, *Democracy*, 329–74.

19. Kuhn, *Structure*, 93.

20. Ibid.

21. Ibid., 94.

22. On "impossibilists" in the Communist Party of the U.S.A., see for example Theodore Draper, *The Roots of American Communism* (Chicago: Ivan Dee, 1989 [originally published 1957]), 23, 45; *Prairie Fire: The Politics of Revolutionary Anti-Imperialism* (Communications Co., 1974), 9, 7, 24.

23. *Prairie Fire*, 3, 45.

24. Ibid., 3, 1.

25. Miller, *Democracy*, 248–51, 21.

26. Ibid., 3.

27. Ibid., 21; Rick Perlstein, *Nixonland* (Scribners, 2009), 327. Footage of Daley's tirade can be seen on youtube.com.

28. Paul Avrich, *The Haymarket Tragedy* (Princeton: Princeton University Press, 1986), 221, passim.

29. "Police in Chicago Clash with Whites After 3 Marches," *New York Times*, Aug. 15, 1966; "Martin Luther King Jr. in Chicago," *Chicago Tribune*, Aug. 5, 1966.

30. Footage of the exchange is at youtube.com.

31. Nathaniel Sheppard Jr., "2 in Weather Underground Are Bargaining to Surrender," *New York Times*, Nov. 24, 1980.

32. Douglas Robinson, "More Dynamite is Found in Rubble of Townhouse," *New York Times*, March 13, 1970.

33. The Beatles, "Revolution," *The Beatles* (White album), Apple Records, 1968; The Beatles, "I Am the Walrus," *Magical Mystery Tour*, Capitol Records, 1967; Vivian Gornick, *The Romance of American Communism*, 258, 257; Kuhn, *Structure*,122–23.

34. Stephen R. Covey, *The Seven Habits of Highly Effective People*, 29.

35. See for example books and videos by Joel Barker adapting Kuhn's insights for business, the British TV series "The Day the Universe Changed" featuring author

James Burke on cultural and political history, and countless books and articles titled "paradigm shift."

36. SDS leader Paul Potter used a similar strategy when speaking at SDS's 1965 march in Washington, D.C., to protest the Vietnam war: "I talked about the system," he reflected, "not because I was afraid of the term capitalism but because I wanted ambiguity, because I sensed that there was something new afoot in the world that we were part of that made the rejection of the old terminology part of the new hope for radical change in America." Miller, *Democracy*, 233.

37. Barbara Decker Ritchey to Kuhn, Oct. 11, 1968, TSK-MIT, box 4, folder 14; Bernard Colby to Kuhn, May 27, 1971; Werner Cahnman to Kuhn, TSK-MIT, Oct. 4, 1965, Box 4, folder 8.

38. R. A. McConnell to Kuhn, Aug. 4, 1963, TSK-MIT, box 4, folder 12; James Curtis to Kuhn, Nov. 7, 1970, box 4, folder 8.

39. Edward Dewey to Kuhn, May 1, 1963, TSK-MIT, box 4, folder 8; Roy Woodmansee to Kuhn, July 13, 1965, TSK-MIT, box 4, folder 16; Marseille Spetz, MD to Kuhn, n.d., TSK-MIT, box 4, folder 15.

40. Kuhn, *RSS*, 308.

41. David Layton to Kuhn, July 23, 1974, TSK-MIT, box 4, folder 11.

Chapter 17. A Revolution and a New Ideology

1. William Beecher, "Raids in Cambodia by U.S. Unprotested," *New York Times*, May 9, 1969.

2. "Leading U.S. and Soviet Citizens to Hold Talks Here," *New York Times*, March 16, 1970.

3. Peter Grose, "Russians Drop Princeton Talks As Anti-Nixon Protests Erupt," *New York Times*, May 2, 1970.

4. See, for example, "Attendance Is Off at Colleges Here," *New York Times*, May 12, 1970.

5. Jerry Raymond, "Goheen, Falk, Orfield Express Community's Outrage," *Daily Princetonian*, May 5, 1970.

6. The leaflet is quoted in Sarah Bridger, "Scientists and the Ethics of Cold War Weapons Research," PhD dissertation, Columbia University, 2011, 339. Student and faculty resolution transcribed from a broadcast by WPRB, Princeton's student radio station, which covered the event live.

7. Hershberg, *James B. Conant*, 751.

8. Sarah Bridger, "Scientists and the Ethics of Cold War Weapons Research," 339.

9. Kuhn to John Green, April 14, 1970, Kuhn Committee papers, Mudd Library, Princeton University, box 3, folder 21.

10. Sarah Bridger, "Scientists and the Ethics of Cold War Weapons Research," 341, 348–49.

11. Ibid., 342, 344.
12. Ibid., 349.
13. Princeton University President's Report, October 1971, *Princeton Alumni Weekly*, Nov. 9, 1971, 17.
14. Untitled document ("Dear Professor Frank"), TSK-MIT, box 25, folder 53. Kuhn mentions in *Structure*, for example, "the insulation of the scientific community from society" (163).
15. See the epilogue of Hershberg, *James B. Conant*, for an account of these difficult years, as well as Jennet Conant's *Man of the Hour*, ch. 22.
16. Hershberg, *James B. Conant*, 752.
17. Conant, *My Several Lives*, 578–79, 561. On McCarthy's fondness for Hitler's *Mein Kampf* and Hitler's political acumen, see Oshinsky, 29.
18. Conant, *My Several Lives*, 561–79, 561, 578.
19. Hershberg, *James B. Conant*, 748.
20. Ibid., 750, 749.
21. Conant, *On Understanding Science*, 4; Hershberg, *James B. Conant*, 752.
22. Kuhn to Conant, Aug. 5, 1961, TSK-MIT, box 25, folder 53.
23. Kuhn, *Structure*, 121.
24. Kuhn to Susan Abrams, Dec. 14, 1982, MIT-TSK, folder "UC Press Correspondence."
25. In the *Times Literary Supplement* list of the *Most Important 100 Books Since World War II*, published in 1995, Geertz's appears as number 68 and Kuhn's at 58 (Oct. 6, 1995, issue of *The Times Literary Supplement*). Kuhn would likely have met Geertz at the Center for Advanced Study in the Behavioral Sciences, where they both studied in 1958 and 1959; they were also likely to have known each other when Kuhn was at Princeton and Geertz was nearby at the Institute for Advanced Study. Clifford Geertz, "The Legacy of Thomas Kuhn: The Right Text at the Right Time," in *Available Light* (Princeton: Princeton University Press, 2000) 160–66, quote on 160.
26. The remark appears in the paperback edition of 1986.
27. Kuhn, "Second Thoughts on Paradigms," in *The Structure of Scientific Theories*, ed. F. Suppe (Urbana: University of Illinois Press, 1974), 459–82; "Reflections on My Critics," in *Criticism and the Growth of Knowledge*, ed. Lakatos and Musgrave, 231–78; "Postscript-1969," *Structure*, 173–208.
28. Kuhn, "Postscript-1969," 181, 182.
29. Conant to Kuhn, June 5, 1961; Kuhn to Conant, June 29, 1961, both in TSK-MIT, box 25, folder 53.
30. Kuhn, "Postscript-1969," 182–86, 185.
31. "Discoveries as Revolutionary," TSK-MIT, box 4, folder 2, 37.
32. Kuhn, "Reflections on My Critics," 231–32.
33. Ibid., 237–47.
34. Ibid., 247, 252.
35. Ibid., 252; Conant to Kuhn, June 5, 1961, TSK-MIT, box 25, folder 53.

36. Kuhn, "Logic of Discovery or Psychology of Research?" 21.
37. Kuhn, "Reflections on My Critics," 241, 248.
38. Ibid., 248, 241; "Discussion: T. S. Kuhn," in *Scientific Change*, ed. Crombie, 391. Kuhn's notes discuss "group ideology," as described in chapter 11.

Epilogue

1. Kuhn, *RSS*, 299; "Preface," *The Essential Tension*, xviii–xix. Illustrations of interest in Kuhn's "linguistic turn" include Stefano Gattei, *Thomas Kuhn's "Linguistic Turn" and the Legacy of Logical Empiricism* (Farnham, UK: Ashgate, 2008); Amani Albeda, "A Gadamerian Critique of Kuhn's Linguistic Turn: Incommensurability Revisited," *International Studies in the Philosophy of Science* 20, no. 3 (Oct. 2006): 323–45; Bird, Thomas Kuhn, ch. 5; K. Brad Wray, "Kuhn's Evolutionary Social Epistemology," 24–29.
2. Kuhn, "Reflections on My Critics," 240–41, 272.
3. Kuhn, foreword to Paul Hoyningen-Huene, *Reconstructing Scientific Revolutions*, xiii.
4. Kuhn, "Afterwords," in *RSS*, 241, 242; *Structure*, 19.
5. Kuhn, "The Relations between the History and the Philosophy of Science," 18, 17.
6. Kenneth Pietrzak to Kuhn, April 6, 1973; Kuhn to Kenneth Pietrzak, April 17, 1973, both in TSK-MIT, box 10, "general correspondence." In fact, the connection was real and substantial; but the influence had gone the other way, from politics to the historiography of science. In the postscript Kuhn mentions, he noted that his insights were in some ways "borrowed from other fields" in which histories of "literature, of music, of the arts, of political development, and many other human activities" are often parsed "as a succession of tradition-bound periods punctuated by non-cumulative breaks." If this is the "bit of information" to which Kuhn referred, it seems to point to Rosenstock-Huessy's episodic and revolutionary view of human history on which Kuhn had taken careful notes.
7. *Structure*, 207; Fleck, *Genesis and Development*, ix.
8. See Kuhn, *RSS*, 1, 2, note 1.
9. Kuhn, "Introduction: Textbook Science and Creative Science," 8.
10. Kuhn, letter to Steve Fuller, June 30, 1993. TSK-MIT box 20, folder 23.
11. This question (including the emphasis on "legitimate") and the subsequent comment quoted below, came from anonymous reviewers.
12. "It has not been in my power to read more than a fraction of it," the philosopher Alexander Bird wrote of the philosophical literature on Kuhn (*Thomas Kuhn*, x). One exception is Peter Novick's survey of *Structure*'s impact in his *That Noble Dream* (ch. 15, but also in subsequent chapters). One recent metastudy belongs to Andrew Abbott, "*Structure* As Cited; *Structure* As Read," in *Kuhn's* Structure of Scientific Revolutions *at Fifty* (Chicago: University of Chicago Press, 2016), 167–94.

13. *Structure*, 173.

14. According to Peter Novick, this is emphatically (and ironically) true for the American historical profession and its central plank of objectivity—"the rock on which the venture was constituted, its continuing raison d'être" (*That Noble Dream*, 1). At the end of his book, Novick summarizes, "As I have attempted to show, the evolution of historians' attitudes on the objectivity question has always been closely tied to changing social, political, cultural, and professional contexts" (628).

15. Kuhn to Professor David Owen, Jan. 6, 1951, TSK-MIT, box 3 folder 10.

16. Other works exploring these convergences include Sigmund Diamond's *Compromised Campus*; Ellen Schrecker's *No Ivory Tower*; Noam Chomsky et al., *The Cold War and the University* (New York: Free Press, 1997); Jane Sanders, *The Cold War on Campus* (Seattle: University of Washington Press, 1979); Paul Lazarsfeld and Wagner Thielens, *The Academic Mind* (New York: Free Press, 1958).

17. Kuhn to Morris, Oct. 13, 1960, TSK-MIT, box 25, folder 53; *RSS*, 292.

18. See for example, John McCumber, *Time in the Ditch*; Don Howard, "Two Left Turns Make a Right: On the Curious Political Career of North American Philosophy of Science at Midcentury," in *Logical Empiricism in North America*, ed. Gary L. Hardcastle and Alan W. Richardson (Minneapolis: University of Minnesota Press, 2003), 25–96; John Capps, "Pragmatism and the McCarthy Era," *Transactions of the Charles S. Peirce Society* 39, no. 1 (Winter 2003): 61–76. On the rise of game theory and rational choice theory in professional philosophy, see John McCumber, *The Philosophy Scare*; and Paul Erickson et al., *How Reason Almost Lost Its Mind: The Strange Career of Cold War Rationality* (Chicago: University of Chicago Press, 2013).

19. In his survey of political and scientific revolutions, I. Bernard Cohen noted the surge of scholarly interest in scientific revolutions in the wake of *Structure*'s publication as well as the abstraction of the concept from that of political revolutions: "I doubt whether scholars have always had in mind the analogy with a particular political or social revolution when referring to a revolution in science" (*Revolution in Science*, 23, 24).

20. Quoted in Paul Hoyningen-Huene, "Two Letters," 371.

21. *Structure*, 4.

22. See, for example, Louis Menand's *The Metaphysical Club* (New York: Farrar, Straus and Giroux, 2001) which construes the origins of American pragmatism as a response to the civic and moral upheavals of the American Civil War. For a survey of how American historians of the nineteenth and twentieth centuries responded to slavery, wars, the Holocaust, and the exigencies of the cold war, see Peter Novick's *That Noble Dream*. For a survey of how works in American literature and philosophy responded to the crises posed by the rise of Nazism, World War II, and the Holocaust, see Mark Greif, *The Age of the Crisis of Man* (Princeton: Princeton University Press, 2015).

Bibliography

Archival Collections Consulted

Charles Morris Papers, Center for American Thought, Indiana University Purdue University Indianapolis, Indianapolis, IN.
Harvard University Archives, Harvard University, Cambridge, MA.
National Archives and Records Administration, College Park, MD.
Niels Bohr Library and Archives, American Institute of Physics, College Park, MD.
Papers of Harlow Shapley, 1906–1966. Harvard University Archives.
Papers of James B. Conant, 1862–1987 (personal papers). Harvard University Archives.
Papers of Philipp Frank, 1943–1995. Harvard University Archives.
Records of the President of Harvard University, James Bryant Conant, 1933–1955. Harvard University Archives.
Rockefeller Foundation Papers, Rockefeller Archive Center, Tarrytown, NY.
Rudolf Carnap Papers, Archives of Scientific Philosophy, University of Pittsburgh, Department of Special Collections, Hillman Library, Pittsburgh, PA.
San Francisco History Center, San Francisco Public Library, San Francisco, CA.
Sidney Hook Papers, Hoover Institution, Stanford University, Stanford, CA.
Special Committee on Sponsored Research Records (Kuhn Committee); Princeton University Archives; Department of Rare Books and Special Collections; Princeton University Library, Princeton, NJ.
Swarthmore College Peace Collection, Swarthmore, PA.
Thomas S. Kuhn Papers, Massachusetts Institute of Technology, Institute Archives and Special Collections, Manuscript Collection MC240, Cambridge, MA.
Unity of Science Movement Records, Special Collections Research Center University of Chicago Library. Chicago, IL.
University of Chicago Press Records, Special Collections Research Center University of Chicago Library, Chicago, IL.
The Wisconsin Historical Society, Madison WI.

Sources Cited

Abbott, Andrew. "*Structure* As Cited; *Structure* As Read." In *Kuhn's* Structure *at Fifty*, edited by Richards and Daston, 167–94. Chicago: University of Chicago Press, 2016.

Adbusters, "An Open Letter to the Ecological Economics Movement," July 16, 2009.

Albeda, Amani. "A Gadamerian Critique of Kuhn's Linguistic Turn: Incommensurability Revisited." *International Studies in the Philosophy of Science* 20, no. 3 (Oct. 2006): 323–45.

Andresen, Jensine. "Crisis and Kuhn," *Isis* 90, supplement (1999): S43–S67.

Arendt, Hannah. *The Origins of Totalitarianism*, New York: Harcourt, 1985.

Aristotle. *The Physics*, trans. Philip H. Wicksteed and Francis M. Cornford. New York: G. P. Putnam, 1929.

Aronova, Elena. "The Congress for Cultural Freedom, 'Minerva,' and the Quest for Instituting 'Science Studies' in the Age of Cold War." *Minerva* 50, no. 3 (2012): 307–37.

Avrich, Paul. *The Haymarket Tragedy*. Princeton: Princeton University Press, 1986.

Axtell, Guy. "In the Tracks of the Historicist Movement: Re-assessing the Carnap-Kuhn Connection." *Studies in History and Philosophy of Science* 24, no. 1 (1993): 119–146.

Barber, Bernard. "Resistance by Scientists to Scientific Discovery." *Science*, Sept. 1, 1961, 596–602.

Beecher, William. "Raids in Cambodia by U.S. Unprotested." *New York Times*, May 9, 1969.

Bell, Daniel. *The End of Ideology: On the Exhaustion of Political Ideas in the Fifties*, Glencoe, IL: Free Press, 1960.

Bernal, John Desmond. *The Social Function of Science*, New York: Macmillan, 1939.

Biddle, Justin. "Putting Pragmatism to Work in the Cold War: Science, Technology and Politics in the Writings of James B. Conant." *Studies in History and Philosophy of Science* 42 (2011): 552–61.

Biderman, Albert D. "The Image of 'Brainwashing.'" *Public Opinion Quarterly* 26, no. 2 (1962): 547–63.

———, and Herbert Zimmer. *The Manipulation of Human Behavior*. New York: John Wiley and Sons, 1961.

Bird, Alexander. *Thomas Kuhn*. Princeton: Princeton University Press, 2000.

Birmingham, Stephen. *"Our Crowd": The Great Jewish Families of New York*. New York: Dell, 1967.

Black, Max. *Critical Thinking: An Introduction to Logic and Scientific Method*, New York: Prentice-Hall, 1946.

Bloom, Samuel. *The Word as Scalpel*. Oxford: Oxford University Press, 2002.

Blumenthal, Ralph. "Vietnam Backers Urged to 'Shout.'" *New York Times*, Nov. 29, 1965.

Boring, Edwin. "Cognitive Dissonance: Its Use in Science." *Science*, Aug. 14, 1964, 680–85.
Brinton, Crane. *French Revolutionary Legislation on Illegitimacy 1789–1804*. Cambridge: Harvard University Press, 1936.
———. *The Anatomy of Revolution*. New York: Random House, 1956.
Brooklyn Daily Eagle. "Brooklyn Boys Create, Manage Manufacturing Concerns that Flourish in Big League Style." Nov. 20, 1932, B11.
Brooklyn Eagle. "Ex-Kin of Remington Seized on Foreign Agent Charge." March 30, 1951.
Brosseau, Grace L. H. "Annual Message of the President General." *Proceedings of the Thirty-Eighth Continental Congress, National Society of the DAR*, 1929.
Buckley, William F. Jr. *God and Man at Yale*, Washington, DC: Regnery, 1951.
Burnham, James. *The Coming Defeat of Communism*, New York: John Day, 1950.
Bush, Vannevar. *Science: The Endless Frontier*, Washington, DC: National Science Foundation, 1945.
Butterfield, Herbert. *Origins of Modern Science*, New York: Macmillan, 1949.
Campbell, Katherine Moos. *An Experiment in Education: The Hessian Hills School, 1925–1952*. PhD Dissertation, Boston University, 1984.
Capps, John. "Pragmatism and the McCarthy Era." *Transactions of the Charles S. Peirce Society* 39, no. 1 (Winter 2003): 61–76.
Carruthers, Susan. "Redeeming the Captives: Hollywood and the Brainwashing of America's Prisoners of War in Korea." *Film History* 10, no. 3, "The Cold War and the Movies" (1998): 275–94.
Caute, David. *The Fellow Travellers: Intellectual Friends of Communism*. New Haven: Yale University Press, 1973.
———. *The Great Fear*. New York: Simon and Schuster, 1979.
Chambers, Whittaker. *Witness*. Washington, DC: Regnery, 1952.
Chase, Stuart. *The Tyranny of Words*. New York: Harcourt Brace, 1938.
Chen, Theodore. *Thought Reform of the Chinese Intellectuals*. Hong Kong: Oxford University Press, 1960.
Chicago Tribune. "Martin Luther King Jr. in Chicago." Aug. 5, 1966.
Chomsky, Noam, et al. *The Cold War and the University*. New York: Free Press, 1997.
Citizenship Educational Service. "Footprints of the Trojan Horse: Some Methods Used by Foreign Agents within the United States." Pamphlet, New York, 1942.
Clymer, Adam. "When Presidential Words Led to Swift Action." *New York Times*, June 9, 2013.
Cohen, I. B. *Revolution in Science*. Cambridge: Harvard University Press, 1985.
Cohen-Cole, Jamie. "Instituting the Science of Mind: Intellectual Economies and Disciplinary Exchange at Harvard's Center for Cognitive Studies." *British Journal for the History of Science* 40, no. 4 (2007): 567–97.
———. "The Creative American: Cold War Salons, Social Science, and the Cure for Modern Society." *ISIS* 100 (2009): 219–62.

———. *The Open Mind: Cold War Politics and the Sciences of Human Nature.* Chicago: University of Chicago Press, 2014.
Collodel, Matteo. "A Note on Kuhn and Feyerabend at Berkeley (1956–1964)." Unpublished.
Conant, Grace Richards. "The Cold War of the Mind: Regimentation in East Germany." *Modern Age* 5, no. 2 (Spring 1961): 117–24.
Conant, James B. "Free Inquiry of Dogma." *Atlantic Monthly*, April 1935, 436–42.
———. "No Retreat for the Liberal Arts." *New York Times*, Feb. 21, 1943.
———. "Wanted: American Radicals." *Atlantic Monthly*, May 1943, 41–45.
———. "Science and the National Welfare." *The Journal of Higher Education* 15, no. 8 (1944): 399–406.
———. Letter to the Editor ("National Research Argued"). *New York Times*, Aug. 13, 1945.
———. "Civil Courage." In *Representative American Speeches, 1945–1946*, edited by A. Craig Baird, 223–28. New York: H. W. Wilson, 1946.
———. *On Understanding Science.* New Haven: Yale University Press, 1947.
———. *Education in a Divided World: The Function of the Public Schools in our Unique Society.* New York: Greenwood Press, 1948.
———. "Science and Politics in the Twentieth Century." *Foreign Affairs* 28, no. 2 (Jan. 1950): 189–202.
———. "Greetings to the National Conference of the Institute for the Unity of Science. Boston, Massachusetts: April, 1950." *Proceedings of the American Academy of Arts and Sciences* 80, no. 1 (1951): 9–13.
———. *Science and Common Sense.* New Haven: Yale University Press, 1951.
———. *Modern Science and Modern Man.* New York: Columbia University Press, 1952.
———. *The Citadel of Learning.* New Haven: Yale University Press, 1956.
———. "Foreword." In Thomas S. Kuhn, *The Copernican Revolution*, xiii–xviii. 1957.
———. *Germany and Freedom: A Personal Appraisal.* Cambridge: Harvard University Press, 1958.
———. *Two Modes of Thought.* New York: Trident, 1964.
———. *My Several Lives: Memoirs of a Social Inventor.* New York: Harper and Row, 1970.
———, ed. *Harvard Case Studies in Experimental Science.* 2 vols. Cambridge: Harvard University Press, 1957.
Conant, Jennet. *Man of the Hour: James Conant, Warrior Scientist.* New York: Simon and Schuster, 2017.
Conklin, William R. "Soviet Is Attacked at Counter Rally." *New York Times*, March 27, 1949.
Counts, George S. *The Soviet Challenge to America.* New York: John Day, 1931.
———. "Dare the School Build a New Social Order." New York: John Day, 1932.
——— et al. "A Call to the Teachers of the Nation." New York: John Day, 1933.

Covey, Stephen R. *The Seven Habits of Highly Effective People.* New York: Free Press, 1989.
Cremin, Lawrence. *Transformation of the School,* New York: Vintage, 1961.
The Crimson, articles without byline.
———. "Conant Letter Urges Faculty to Sign Oath, but Criticizes Bill," Oct. 8, 1935.
———. "Student Council Urges Repeal of Teachers Oath Bill in Resolution." March 10, 1936.
———. "Text of President's Baccalaureate Address." June 21, 1937.
———. "Pacifists Plan to Display Feelings in Rallies Fixed for Armistice Weekend." Nov. 9, 1939.
———. "American Interests Jeopardized if U.S. Intervenes in Europe's War, McKay Warns." Nov. 24, 1939.
———. "Harvard's United Front." Dec. 14, 1939.
———. "Youth Talks Back." Feb. 28, 1940.
———. "Split Keeps HSU From Joining Walkout Called by National Organization on Over 100 Campus." Oct. 11, 1940.
———. "Lead, Kindly White." Oct. 25, 1940.
———. "The President's Message." Jan. 7, 1941.
———. "Thus Far and No Farther." Feb. 12, 1941.
———. "Colonel Jay Reviews ROTC Regiment of 600." Dec. 13, 1941.
———. "E for Effort." April 30, 1942.
———. "Pacifist Group Holds Fast Amid Quickening War Fever,." Aug. 19, 1942.
———. "Forecast for '47." Oct. 7, 1942.
———. "Free Society Conant Topic." Jan. 7, 1943.
———. "Excerpts from Conant Valedictory Address." Jan. 11, 1943.
———. "Student Opinion Wins First Victory in Tutorial See-Saw." April 27, 1946.
———. "Poll of Students Uphold General Education Plan." Jan. 22, 1947.
———. "Conant Advises Cut in University's Enrollment." Jan. 22, 1948.
———. "Barnes Fails to See Cause for Alarm." Feb. 6, 1948.
———. "Inverchapel Calls for Unity; Conant Gains British Award." March 19, 1948.
———. "Conant Urges Marshall Plan, UMT to Halt Soviet Advance." March 25, 1948.
———. "Rally No Red Front, Shapley States." March 24, 1949.
———. "Text of Conant's Speech." June 23, 1949.
———. "University Mobilized Rapidly in '42, Was Naval Training Camp by '43." Feb. 7, 1951.
———. "East, West View of Life Merging, Lecturer Asserts." Feb. 7, 1952.
———. "Conant to Teach New Course in Philosophy." April 9, 1952.
———. "Adenauer, Conant Visit Here Today." April 17, 1953.
———. "Conant Claims German Nazism Dead." Jan. 25, 1956.

Crombie, A. C., ed. *Scientific Change: Historical Studies in the Intellectual, Social, and Technical Conditions for Scientific Discovery and Technical Invention, from Antiquity to the Present.* New York: Basic Books, 1963.

Crozier, Brian. "Why the Cold War?" In *The Collapse of Communism*, edited by Lee Edwards, 141–58. Stanford: Hoover Institute, 2000.

Cushman, Elisabeth. "Wisdom on War Flows from Students' Mouths." *The Daily Argus* (Mt. Vernon, NY), Nov. 18, 1935.

Daily Princetonian. "Thomas Mann to Lecture on the Problem of Freedom." Feb. 13, 1940, 4.

Dallin, David. "The Making of the Russian Mind." *New York Times*, July 2, 1950.

Davies, Lawrence E. "Conant Sees Peace Under Atom Pact." *New York Times*, Sept. 9, 1947.

Davis, Elmer. "Four-Year-Old Educational Experiment." *New York Times*, Aug. 28, 1921.

Decock, Lieven. *Trading Ontology for Ideology.* The Netherlands: Kluwer, 2002.

Dewey, John. *The School and Society.* Chicago: University of Chicago Press, 1900.

———. *Experience and Education.* New York: Simon and Schuster, 1938.

——— et al. *Not Guilty: Report of the Commission of Inquiry into the Charges Made Against Leon Trotsky in the Moscow Trials.* New York: Harper and Brothers, 1938.

Diamond, Sigmund. *Compromised Campus: The Collaboration of Universities with the Intelligence Community, 1945–1955.* New York: Oxford University Press, 1992.

Dilling, Elizabeth. *The Red Network.* Kenilworth, IL: Elizabeth Dilling, 1934.

Draper, Theodore. *The Roots of American Communism.* Chicago: Ivan Dee, 1989 (orig. 1957).

Earman, John. "Carnap, Kuhn, and the Philosophy of Scientific Methodology." In *World Changes*, edited by Paul Horwich, 9–36. Cambridge: MIT Press, 1993.

Egan, Kieran. *Getting It Wrong from the Beginning: Our Progressivist Inheritance from Herbert Spencer, John Dewey, and Jean Piaget.* New Haven: Yale University Press, 2004.

Engerman, David C. *Know Your Enemy: The Rise and Fall of America's Soviet Experts.* New York: Oxford University Press, 2009.

Erickson, Christine K. "'I have Not had One Fact Disproven': Elizabeth Dilling's Crusade against Communism in the 1930s." *Journal of American Studies* 36, no. 3 (Dec. 2002): 473–89.

Erickson, Paul, et al. *How Reason Almost Lost Its Mind: The Strange Career of Cold War Rationality.* Chicago: University of Chicago Press, 2013.

Erk, Frank C. "H. Bentley Glass." *Proceedings of the American Philosophical Society* 153 (2009): 327–39.

Fairfield, William S. "FBI's Activities Spread Fear at Yale." *Crimson*, June 4, 1949.

Falk, Julius, and Gordon Haskell. "Civil Liberties and the Philosopher of the Cold War." *The New International* 19 (1953): 184–227.

Faust, Drew Gilpin. " 'My Dear Sir:' A Sealed Letter from the University Archives Reaches Drew Faust on the Occasion of Her Inauguration." *Harvard University Library Letters* 2, no. 5 (Fall 2007/Winter 2008): 5.

Fenton, John H. "Pusey Is 'Unaware' of Reds on Faculty." *New York Times*, Nov. 10, 1953.

Feyerabend, Paul. "Consolations for the Specialist." In *Criticism and the Growth of Knowledge*, edited by Lakatos and Musgrave, 197–230. Cambridge: Cambridge University Press, 1970.

———. *Killing Time*. Chicago: University of Chicago Press, 1995.

Fine, Benjamin. "Conant Sees Peril to U.S. Education." *New York Times*, April 8, 1952.

Finney, John W. "U.S. Missile Experts Shaken By Sputnik." *New York Times*, Oct. 13, 1957.

Fishbein, Leslie. *Rebels in Bohemia: The Radicals of* The Masses, *1911–1917*. Chapel Hill: University of North Carolina Press, 1982.

Fisher, John. "NY Rally Part of Red Move to Snarl Artists and Scientists." *Nashua* (NH) *Telegraph*, April 7, 1949, 16.

Fleck, Ludwik. *The Genesis and Development of a Scientific Fact*. Chicago: University of Chicago Press, 1979.

Forrester, John. "On Kuhn's Case: Psychoanalysis and the Paradigm." *Critical Inquiry* 33 (Summer 2007): 782–819.

Frank, Philipp. "Introductory Remarks." *Contributions to the Analysis and Synthesis of Knowledge, Proceedings of the American Academy of Arts and Sciences* 80, no. 1 (1951): 5–8.

———. "The Variety of Reasons for the Acceptance of Scientific Theories." In *The Validation of Scientific Theories*, edited by Frank, 3–18. Boston: Beacon Press, 1957.

Friedman, Michael. "Remarks on the History of Science and the History of Philosophy." In *World Changes: Thomas Kuhn and the Philosophy of Science*, edited by P. Horwich, 37–54. Cambridge: MIT Press, 1992.

———. "Kant, Kuhn, and the Rationality of Science." *Philosophy of Science* 69, no. 2 (2002): 171–90.

Fuller, Steve. *Thomas Kuhn: A Philosophical History for Our Time*. Chicago: University of Chicago Press, 2000.

———. *Popper vs. Kuhn: The Struggle for the Soul of Science*. New York: Columbia, 2004.

Galison, Peter. "Practice All the Way Down." In *Kuhn's* Structure *at Fifty*, edited by Richards and Daston, 42–69. Chicago: University of Chicago Press, 2016.

Gattei, Stefano. *Thomas Kuhn's "Linguistic Turn" and the Legacy of Logical Empiricism*. Farnham, UK: Ashgate 2008.

Geertz, Clifford. "The Legacy of Thomas Kuhn: The Right Text at the Right Time." In *Available Light*, 160–66. Princeton: Princeton University Press, 2000.

Gentry, Curt. *J. Edgar Hoover: The Man and His Secrets*. New York: Norton, 2001.
Glass, H. Bentley. "Academic Freedom and Tenure in the Quest for National Security." *Bulletin of the Atomic Scientists* 12, no. 6 (June 1956): 221–23, 226.
———. "Liberal Education in a Scientific Age." *Bulletin of the Atomic Scientists* 14, no. 11 (Nov. 1958): 346–53.
———. "The Establishment of Modern Genetical Theory as an Example of the Interaction of Different Models, Techniques, and Inferences." In *Scientific Change*, edited by A. C. Crombie, 521–41. New York: Basic Books, 1963.
———. "Discussion: H. Bentley Glass." In *Scientific Change*, edited by A. C. Crombie, 381–82. New York: Basic Books, 1963.
———. *Progress or Catastrophe: The Nature of Biological Science and Its Impact on Human Society*. New York: Praeger, 1985.
Gleason, Abbott. *Totalitarianism*. New York: Oxford University Press, 1995.
Goheen, Robert. "Princeton University President's Report, Oct. 1971." *Princeton Alumni Weekly*, Nov. 9, 1971, 17.
Gornick, Vivian. *The Romance of American Communism*. New York: Basic Books, 1977.
Gouldner, Alvin. *The Coming Crisis of Western Sociology*. New York: Avon, 1971.
Gordon, Dean. "Excerpts From Testimony of Leading Witnesses in Security Hearings on Oppenheimer." *New York Times*, June 17, 1954.
Greif, Mark. *The Age of the Crisis of Man*. Princeton: Princeton University Press, 2015.
Grose, Peter. "Russians Drop Princeton Talks As Anti-Nixon Protests Erupt." *New York Times*, May 2, 1970.
Groves, Leslie R. *Now It Can Be Told: The Story of the Manhattan Project*. New York: Da Capo Press, 1983.
Grutzner, Charles. "Pickets to Harass Cultural Meeting; Delegates Arrive." *New York Times*, March 24, 1949.
Gutek, Gerald. *George S. Counts and American Civilization*. Macon, GA: Mercer University Press, 1982.
Halberstam, David. "Russian Center Studies Soviet Social System." *Crimson*, Oct. 9, 1953.
Hall, A. R. *The Scientific Revolutions, 1500–1800*. London: Longmans, Green, 1954.
———. "Commentary by A. Rupert Hall." In *Scientific Change*, edited by A. C. Crombie, 370–75. New York: Basic Books, 1963.
Hamlin, Christopher. "The Pedagogical Roots of the History of Science: Revisiting the Vision of James Bryant Conant." *Isis* 107, no. 2 (2016): 282–308.
Harvard Alumni Bulletin. "Prescription for Cambridge." Sept. 22, 1945, 29.
Harvard University, Committee on the Objectives of a General Education in a Free Society. *The Objectives of a General Education in a Free Society*. Cambridge: Harvard University Press, 1945.
Hayakawa, S. I. *Language in Action*. New York: Harcourt Brace, 1941.
Hayek, Friedrich. *The Road to Serfdom*. London: Routledge, 1944.
Herald Statesman (Yonkers, NY). "Educator Reports on Second China Visit." March 29, 1976, 16.

Hershberg, James. *James B. Conant: Harvard to Hiroshima and the Making of the Nuclear Age*. New York: Knopf, 1993.
Hesse, Mary. "Review of *The Structure of Scientific Revolutions*." *Isis* 54 (June 1963): 286–87.
Heyl, Barbara S. "The Harvard 'Pareto Circle.' " In *Talcott Parsons: Critical Assessments Vol. 1*, edited by P. Hamilton, 29–49. London: Routledge, 1992.
Hofstadter, Richard. *Anti-Intellectualism in American Life*. New York: Vintage, 1962.
———. "The Paranoid Style in American Politics." *Harper's Magazine*, Nov. 1964, 77–86. (Reprinted in *The Paranoid Style in American Politics and other essays*. New York: Vintage, 2008.)
Hollinger, David. *Science, Jews, and Secular Culture*. Princeton: Princeton University Press, 1996.
Homans, George. *Sentiments and Activities*, Glencoe, IL: Free Press, 1962.
———. *Coming to My Senses: The Autobiography of a Sociologist*. New Brunswick, NJ: Transaction Books, 1984.
Hook, Sidney. "The Philosophy of Dialectical Materialism. I." *Journal of Philosophy* 25, no. 5 (March 1, 1928): 113–14.
———. "The Philosophy of Dialectical Materialism. II." *Journal of Philosophy* 25, no. 6 (March 15, 1928): 141–55.
———. "Conant on Politics and Education." *New York Times*, Oct. 24, 1948.
———. "Should Communists Be Permitted to Teach?" *New York Times*, Feb. 27, 1949.
———. "International Communism." *Dartmouth Alumni Magazine*, March, 1949, 13–20.
———. "Mindless Empiricism." *Journal of Philosophy* 49, no. 4 (Feb. 14, 1952): 89–100.
———. *Heresy, Yes—Conspiracy, No*, New York: John Day, 1953.
———. Letter to the Editor. *New York Times*, March 15, 1953.
———. *Out of Step*, New York: Carroll and Graf, 1987.
Hoover, J. Edgar. "The Communist Menace." In "Bills to Curb or Outlaw the Communist Party of the United States," part 2, United States House Committee on Un-American Activities, 80th Congress, 1st sess., March 26, 1947, 33–50. Washington, DC: U.S. Government Printing Office.
———. *Masters of Deceit*. New York: Henry Holt, 1958.
———. *On Communism*. New York: Random House, 1969.
Horney, Karen. *Our Inner Conflicts: Toward a Constructive Theory of Neurosis*. New York: Norton, 1945.
Howard, Don. "Two Left Turns Make a Right: On the Curious Political Career of North American Philosophy of Science at Midcentury." In *Logical Empiricism in North America*, edited by Gary L. Hardcastle and Alan W. Richardson, 25–96. Minneapolis: University of Minnesota Press, 2003.
Hoyningen-Huene, Paul. *Reconstructing Scientific Revolutions*. Chicago: University of Chicago Press, 1993.

———. "Two Letters of Paul Feyerabend to Thomas S. Kuhn on a Draft of *The Structure of Scientific Revolutions.*" *Studies in History and Philosophy of Science* 26, no. 3 (1995): 353–87.

———. "More Letters by Paul Feyerabend to Thomas S. Kuhn on *Proto-Structure.*" *Studies in History and Philosophy of Science* 37 (2006): 610–32.

Hufbauer, Karl. "From Student of Physics to Historian of Science: T. S. Kuhn's Education and Early Career." *Physical Perspectives* 14 (2012): 421–70.

Hunter, Edward. *Brainwashing in Red China: The Calculated Destruction of Men's Minds.* New York: Vanguard, 1951.

———. *Brainwashing: The Story of Men Who Defied It.* New York: Farrar, Strauss and Cudahy, 1956.

———. Letter to the Editor, *New York Times*, March 29, 1959.

"I Want to Be Like Stalin" From the Russian Text on Pedagogy by B. P. Yesipov and N. K. Goncharov. Translated by George S. Counts and Nucia P. Lodge. New York: John Day, 1947.

Ilin, M. [pseud.]. *New Russia's Primer: The Story of the Five-Year Plan.* Translated by George S. Counts and Nucia P. Lodge. New York: Houghton Mifflin, 1931.

Indianapolis Recorder. "Du Bois Acquitted in Foreign Agent Trial." Nov. 24, 1951, 1.

Inkeles, Alex. *Public Opinion in Soviet Russia: A Study in Mass Persuasion.* Cambridge: Harvard University Press, 1950.

Irzik, Gurol, and Teo Grünberg. "Carnap and Kuhn: Arch Enemies or Close Allies?" *British Journal for the Philosophy of Science* 46, no. 3 (Sept. 1995): 285–307.

Isaac, Joel. "Kuhn's Education: Wittgenstein, Pedagogy, and the Road to *Structure.*" *Modern Intellectual History* 9, no. 1 (2012): 89–107.

———. *Working Knowledge: Making the Human Sciences from Parsons to Kuhn.* Cambridge: Harvard University Press, 2012.

Isherwood, Paul. "A Failed Elite: The Committee on the Present Danger and the Great Debate of 1951." MA thesis, Ohio University, March 2009.

Jacobs, Struan. "Michael Polanyi and Thomas Kuhn: Priority and Credit." *Tradition and Discovery: The Polanyi Society Periodical* 33, no. 2 (2006/2007): 25–36.

———. "Thomas Kuhn's Memory." *Intellectual History Review* 19 (2009): 83–101.

James, William. *Principles of Psychology v. 1.* New York: Henry Holt, 1890.

———. *The Varieties of Religious Experience: A Study in Human Nature.* New York: Modern Library, 1902.

———. *Pragmatism.* Indianapolis: Hackett, 1981 (orig. Longmans, Green, 1907).

Jewell, Edward Alden. "Art." *New York Times*, May 12, 1931.

Mac R. Johnson, "Hook Invades Shapley's Room to Ask Apology." *New York Herald Tribune*, March 26, 1949.

Jorden, William J. "Soviet Fires Earth Satellite into Space." *New York Times*, Oct. 5, 1957.

Jumonville, Neil. *Critical Crossings: The New York Intellectuals in Postwar America.* Berkeley: University of California Press, 1990.

———. "Polemics, Open Discussion, and Tolerance." In *Sidney Hook Reconsidered*, edited by M. Cotter, 225–44. Amherst, NY: Prometheus Books, 2004.
Kaempffert, Waldemar. "Toward Bridging the Gaps Between the Sciences." *New York Times*, Aug. 7, 1938.
———. "Science in Review." *New York Times*, Oct. 1, 1944.
———. "Science in Review: Further Arguments in Favor of Research Organized on a National Scale." *New York Times*, Aug. 19, 1945.
———. "Science in Review: Organization and Planning Held Necessary to Close Gaps in Our Knowledge." *New York Times*, Sept. 9, 1945.
———. "'For a Hierarchy of Scientists,' review of Bush, *Endless Horizons* (Public Affairs Press, 1946)." *New York Times*, March 17, 1946.
Kaiser, David. "A Mannheim for All Seasons: Bloor, Merton, and the Roots of the Sociology of Scientific Knowledge." *Science in Context* 11, no. 1 (1998): 51–87.
———. "Thomas Kuhn and the Psychology of Scientific Revolutions." In *Kuhn's Structure at Fifty*, edited by Richards and Daston, 71–95. Chicago: University of Chicago Press, 2016.
Kallen, Horace. "The Meanings of 'Unity' Among the Sciences." *Educational Administration and Supervision* 26, no. 2 (1940): 81–97.
———. "The Significance of the Unity of Science Movement: Reply." *Philosophy and Phenomenological Research* 6, no. 4 (June 1946): 515–26.
Kelly, Daniel. *James Burnham and the Struggle for the World*. Wilmington, DE: ISI Books, 2002.
Kennan, George. Telegram to George Marshall ("Long Telegram"), February 22, 1946. Harry S. Truman Administration File, Elsey Papers, www.trumanlibrary.org.
———. [Mr. X] "The Sources of Soviet Conduct." *Foreign Affairs* 25 (July 1947): 566–78, 580–82.
———. *Russia, the Atom, and the West*. London: Oxford University Press, 1958.
———. *At a Century's Ending: Reflections 1982–1995*. New York: Norton, 1996.
Kinkead, Eugene. *In Every War but One*. New York: Norton, 1959.
Kirker, Harold, and Burleigh Taylor Wilkins. "Beard, Becker and the Trotsky Inquiry." *American Quarterly* 13, no. 4 (Winter 1961): 516–25.
Koestler, Arthur. "Arthur Koestler." In *The God that Failed*, edited by Richard Crossman, 15–75. New York: Columbia University Press, 2001 (orig. London: Hamilton, 1950).
Korzybski, Alfred. *Science and Sanity*. Lakeville, CT: International Non-Aristotelian Library, 1933.
Kridel, Craig. "Towards an Understanding of Progressive Education and 'School': Lee Dick's 1939 Documentary Film on the Hessian Hills School." Unpublished.
Kubie, Lawrence S. "Some Unsolved Problems of the Scientific Career." *American Scientist* 41, no. 4 (1953): 596–613; 42, no. 1 (1954): 104–12.

Kuhn, Thomas S. "Subjective View. Thomas S. Kuhn, on Behalf of the Recent Students, Reflects Undergraduate Attitude." *Harvard Alumni Bulletin*, Sept. 22, 1945, 29–30.

———. *The Copernican Revolution*. Cambridge: Harvard University Press, 1957.

———. "Discussion: T. S. Kuhn." In *Scientific Change*, edited by A. C. Crombie, 386–95. New York: Basic Books, 1963.

———. "The Function of Dogma in Scientific Research." In *Scientific Change*, edited by A. C. Crombie, 347–69. New York: Basic Books, 1963.

———. "Postscript-1969." In *The Structure of Scientific Revolutions*, 173–208. Chicago: University of Chicago Press, fourth edition, 2012 (orig. 1962).

———. "Logic of Discovery or Psychology of Research?" In *Criticism and the Growth of Knowledge*, edited by Lakatos and Musgrave, 1–23. Cambridge: Cambridge University Press, 1970.

———. "Reflections on My Critics." In *Criticism and the Growth of Knowledge*, edited by Lakatos and Musgrave, 231–78. Cambridge: Cambridge University Press, 1970.

———. "Second Thoughts on Paradigms." In *The Structure of Scientific Theories*, edited by F. Suppe, 459–82. Urbana: University of Illinois Press, 1974.

———. *The Essential Tension*. Chicago: University of Chicago Press, 1977.

———. "The Essential Tension: Tradition and Innovation in Scientific Research." In *The Essential Tension*, 225–39. Chicago: University of Chicago Press, 1977.

———. "A Function for Thought Experiments." In *The Essential Tension*, 240–65. Chicago: University of Chicago Press, 1977.

———. "The Relations between the History and the Philosophy of Science." In *The Essential Tension*, 3–20. Chicago: University of Chicago Press.

———. "Foreword." In Ludwik Fleck, *Genesis and Development of a Scientific Fact*. Chicago: University of Chicago Press, 1979.

———. "Foreword." In Paul Hoyningen-Huene, *Reconstructing Scientific Revolutions*. Chicago: University of Chicago Press, 1993.

———. "Afterwords." In *World Changes: Thomas Kuhn and the Philosophy of Science*, edited by P. Horwich, 311–41. Cambridge: MIT Press, 1993; reprinted in *The Road Since Structure*, 224–52.

———. *The Road Since Structure*. Edited by James Conant and John Haugland. Chicago: University of Chicago Press, 2000.

———. *The Structure of Scientific Revolutions*. Chicago: University of Chicago Press, fourth edition, 2012 (orig. 1962).

Lakatos, I. "Falsification and the Methodology of Scientific Research Programmes." In *Criticism and the Growth of Knowledge*, edited by I. Lakatos and A. Musgrave, 91–195. Cambridge: Cambridge University Press, 1970.

———, and A. Musgrave, eds. *Criticism and the Growth of Knowledge*, Cambridge: Cambridge University Press, 1970.

Lazarsfeld, Paul, and Wagner Thielens. *The Academic Mind*. New York: Free Press, 1958.

Laurence, William L. "'One-World' Ideal Is Seen in Science." *New York Times*, Sept. 24, 1946.
———. "Scientist Decries Curb on Condon, Physicist a Victim of Rumor and Anxiety." *New York Times*, Dec. 29, 1954.
Leviero, Anthony. "Eisenhower Backed on Book Ban Talk." *New York Times*, June 17, 1953.
———. "New Code Orders P.O.W.'s to Resist in 'Brainwashing.'" *New York Times*, Aug. 18, 1955.
———. "Thinking to Order," review of Hunter, *Brainwashing: The Story of Men Who Defied It*. *New York Times*, May 20, 1956.
Life magazine. "Dupes and Fellow Travellers Dress up Communist Fronts." April 4, 1949, 39–43.
Lifton, Robert Jay. *Thought Reform and the Psychology of Totalism: A Study of "Brainwashing" in China*. Chapel Hill: University of North Carolina Press, 1989 (orig. New York: Norton, 1961).
Lillico, Stuart. "'Black-Out in Red China,' review of Hunter, *Brainwashing in Red China*." *New York Times*, Oct. 21, 1951.
Little, Douglas. "Red Scare, 1936: Anti-Bolshevism and the Origins of British Non-Interventionism in the Spanish Civil War." *Journal of Contemporary History*, 23, no. 2 (1988): 291–311.
Long, Jancis. "The Unforgiven: Imre Lakatos' Life in Hungary." In *Appraising Lakatos*, edited by G. Kampis, L. Kvasz, and M. Stöltzner, 263–302. New York: Springer, 2002.
LOOK magazine. "How to Identify an American Communist," March 4, 1947.
Lowe, Victor. "A Resurgence of 'Vicious Intellectualism.'" *Journal of Philosophy* 48, no. 14 (July 5, 1951): 435–47.
McCullough, David. *Truman*. New York: Simon and Schuster, 1992.
McCumber, John. *Time in the Ditch*. Evanston: Northwestern University Press, 2001.
———. *The Philosophy Scare*. Chicago: University of Chicago Press, 2016.
McLaughlin, Martin. "Prewar Student Activism: A Profile." In *American Students Organize: Founding the National Student Association after World War II*, 15–18. Santa Barbara, CA: Praeger, 2006.
Maier, William L. Letter to the Editor. *New York Times*, March 15, 1953.
Main, Thomas J. "Fearless Sidney Hook." *Policy Review*, Aug. 1, 2003.
Mandelbaum, Maurice. "A Note on Thomas Kuhn's "*The Structure of Scientific Revolutions*." *The Monist* 60, no. 4 (1977): 445–52.
Mann, Thomas. "The Problem of Freedom." *Vital Speeches of the Day* 5, no. 18 (July 1, 1939): 547–50.
Mannheim, Karl. *Ideology and Utopia: An Introduction to the Sociology of Knowledge*. Translated by Louis Wirth and Edward Shils. New York: Harcourt, Brace, 1954 (orig. 1936).
Marbury, William L. "The Hiss-Chambers Libel Suit." *Maryland Law Review* 41, no. 1 (1981): 75–102.

Marcum, James A. *Thomas Kuhn's Revolution: An Historical Philosophy of Science*. New York: Continuum, 2005.
Marks, John. *The Search for the "Manchurian Candidate": The CIA and Mind Control*. New York: New York Times Books, 1979.
Marshall, S. L. A. " 'P.O.W.'s in North Korea: The Way They Were and Why,' review of *In Every War But One*, by Eugene Kinkead." *New York Times*, Feb. 22, 1959.
Masterman, Margaret. "The Nature of a Paradigm." In *Criticism and the Growth of Knowledge*, edited by Lakatos and Musgrave, 59–89. Cambridge: Cambridge University Press, 1970.
May, Gary. *Un-American Activities: The Trials of William Remington*. New York: Oxford University Press, 1994.
Mayoral, Juan V. "Five Decades of *Structure*: A Retrospective View." *Theoria* 27, no. 3 (2013): 261–80.
Meerloo, Joost. "Pavlov's Dog and Communist Brainwashers." *New York Times*, May 9, 1954.
Meiklejohn, Alexander. "Should Communists Be Allowed to Teach?" *New York Times*, March 27, 1949.
Menand, Louis. *The Metaphysical Club*. New York: Farrar, Straus and Giroux, 2001.
Miami Daily News. "Soviet Spy Lists Reds in Key U.S. Jobs." Associated Press, Aug. 1, 1948.
Miller, James. *Democracy Is in the Streets, From Port Huron to the Siege in Chicago*. Cambridge: Harvard University Press, 1994.
Milosz, Czeslaw. "The Happiness Pill." *Partisan Review* 18, no. 5 (Sept.-Oct. 1951): 540–56.
Milwaukee Journal. "Cohn and Schine Get Headlines in London's Papers." AP news, March 13, 1954, 2.
Moleski, Martin. "Polanyi vs. Kuhn: Worldviews Apart." *Tradition and Discovery: The Polanyi Society Periodical 33*, no. 2 (2006/2007): 8–24.
Moore Jr., Barrington. *Terror and Progress USSR*. Cambridge: Harvard University Press, 1954.
Moos, Elizabeth. "A Woman's Place Is in the Factory." *New China*, Spring 1975.
Morris, Charles W. *Signs, Language, Behavior*. New York: George Braziller. 1946.
Morse, Philip M. "Edward Uhler Condon, 1902–1974." Washington, DC: National Academy of Sciences, 1976.
Navasky, Victor. *Naming Names*. New York: Viking Press, 1980.
Needell, Allan. " 'Truth Is Our Weapon': Project TROY, Political Warfare, and Government-Academic Relations in the National Security State." *Diplomatic History* 17 (1993): 399–420.
Neurath, Otto, Otto Neurath, *Modern Man in the Making*. New York: Alfred A. Knopf, 1939.
———. "Foundations of the Social Sciences," *International Encyclopedia of Unified Science* 2, no. 1, Chicago: University of Chicago Press, 1944.

———. "After Six Years." *Synthese* 5 (1946): 77–82.
———. "Personal Life and Class Struggle." In *Empiricism and Sociology*, edited by R. S. Cohen and Marie Neurath, 249–98. Boston: Reidel, 1973.
———. "Encyclopedia as 'Model.'" In Neurath, *Philosophical Papers 1913–1946*, edited by R. S. Cohen and Marie Neurath, 145–58. Boston: Reidel, 1983.
———. "The New Encyclopedia of Scientific Empiricism." In Neurath, *Philosophical Papers 1913–1946*, edited by R. S. Cohen and Marie Neurath, 189–199. Boston: Reidel, 1983.
New York Sun. "Junior Craft Clubs Open Their Exhibit at Rumsey Studio." Jan. 26, 1933, 20.
The New York Times, articles without byline.
———. "Teaching Three R's Held Time Waster." March 11, 1928.
———. "Progressive Schools Cut 'Three R' Drills." May 26, 1929.
———. "Tradition Ignored in School Design." May 3, 1931.
———. "Hessian Hills School Opens $1,000,000 Drive: Lincoln Steffens, Floyd Dell and Stuart Chase Laud Progressive Education Aims at Dinner." May 5, 1931.
———. "Smith, Wagner, and Thomas Write Views on Disarmament for Schoolboy Editor, 14." Feb. 3, 1932.
———. "Benefit for Hessian Hills School." May 25, 1933.
———. "Child-Aid Leaders to Convene May 2." April 22, 1935.
———. "Dr. Counts Assails 'Liberty's Enemies.'" Feb. 26, 1936.
———. "Mrs. Borg Heads Board." Jan. 13, 1939.
———. "Conant Urges Aid to Allies at Once." May 30, 1940.
———. "Dr. Conant Urges Universal Military Service." June 13, 1940.
———. "5th Column Methods Bared in Pamphlet." Sept. 20, 1940.
———. "President Roosevelt's Message to Congress on the State of the Union." Jan. 7, 1941.
———. "Conant Sets Value of Colleges in War." Jan. 15, 1942.
———. "Bundist Round-Up Nets 72 More Here." July 9, 1942.
———. "College Days Few, Harvard is Told." Oct. 7. 1942.
———. "Mass Bund Trials Now in Prospect." Oct. 21, 1942.
———. "Willkie Lifts His Voice for Liberal Arts Study." Jan 17, 1943.
———. "Kuhn is Ordered Deported to Reich." May 19, 1945.
———. "Research for Defence." July 21, 1945.
———. "Truman on Research Bill." Oct. 30, 1945.
———. "Truman Aid Asked for Magnuson Bill." Nov. 27, 1945.
———. "Russians Blame Byrnes." March 29, 1946.
———. "This Week's Events." March 31, 1946.
———. "Days of Seclusion Ended by Gromyko." April 5, 1946.
———. "Atomic Education Urged by Einstein." May 25, 1946.
———. "Sponsors of the World Peace Conference." March 24, 1949.
———. "200 Sponsors Join Culture Unit Foes." March 25, 1949.

———. "Hook Confronts Shapley in Latter's Hotel Room." March 26, 1949.
———. "Hook Cites Letters in Shapley Dispute." March 27, 1949.
———. "Panel Discussions of the Cultural Conference Delegates Cover a Wide Range of Subjects." March 27, 1949.
———. "Chain Reaction." Feb. 12, 1950.
———. "'World Peace' Plea Is Circulated Here." July 14, 1950.
———. "Moscow Decrees 'Peace' Sabotage." July 22, 1950.
———. "Text of Group's Statement on Present Peril." Dec. 13, 1950.
———. "Foreign Aid Charge Denied by Mrs. Moos." April 3, 1951.
———. "Mrs. Simon Kuhn, 83, A Cincinnati Leader." April 23, 1952.
———. "The Present Danger" (editorial). Dec. 22, 1952.
———. "Conant Discusses Role of Scholars." Jan. 26, 1953.
———. "President Denies Clemency to Rosenbergs in Spy Case." Feb. 12, 1953.
———. "British Press Pokes Fun at McCarthy Investigators." April 26, 1953.
———. "McCarthy Blamed by Kaghan in Bonn." May 13, 1953.
———. "The Nation: Two in Lewisburg." Nov. 28, 1954.
———. "Excerpts from Reports on Congressional Investigation of Tax-Exempt Foundations." Dec. 20, 1954.
———. "Spotting Communists" (editorial). June 14, 1955.
———. "2 Challenge View on Brainwashing." Sept. 22, 1956.
———. "Legal Perfectionist: Edward S. Greenbaum." Jan 27, 1957.
———. "Topics: A Hobby Headquarters." April 17, 1959.
———. "Police in Chicago Clash with Whites After 3 Marches." Aug. 15, 1966.
———. "Leading U.S. and Soviet Citizens to Hold Talks Here." March 16, 1970.
———. "Attendance Is Off at Colleges Here." May 12, 1970.
———. "Edward S. Greenbaum, Lawyer, Is Dead at 80." June 13, 1970.
Novick, Peter. *That Noble Dream*. Cambridge University Press, 1988.
Nye, Mary Jo. *Michael Polanyi and His Generation*. Chicago: University of Chicago Press, 2011.
O'Doherty, E. F. "Brainwashing." *Studies: An Irish Quarterly Review* 52, no. 205 (Spring 1963): 1–15.
Olson, Lynn. *Those Angry Days*. New York: Random House, 2013.
Orwell, George. *1984*. New York: Knopf, 1987.
Oshinsky, David. *A Conspiracy So Immense: The World of Joe McCarthy*. New York: MacMillan, 1983.
Ossining Citizen Register. "Croton School Sponsors Rivera Lecture on Art." May 13, 1933.
Packard, Vance. *The Hidden Persuaders*. New York: McKay, 1957.
Parker, Francis W. "Obstacles in Our Way." *Journal of Education* 15 (May 18, 1882): 315.
Peale, Norman Vincent. *The Power of Positive Thinking*. New York: Prentice-Hall, 1952.
Pearson, Drew. "Murder Highlights Prison Vice," "Washington Merry-Go-Round" (syndicated column). Dec. 4, 1954.

Pearson, Karl. *The Grammar of Science*. Mineola, NY: Dover, 2004.
Pels, Dick. *Autonomy and Reflexivity in the Social Theory of Knowledge*. Liverpool: Liverpool University Press, 2003.
Perlstein, Rick. *Nixonland*. New York: Scribners, 2009.
Phelps, Christopher. *Young Sidney Hook*. Ithaca, Cornell University Press, 2005.
———. "Why Wouldn't Sidney Hook Permit the Republication of His Best Book?' *Historical Materialism* 11, no. 4 (2003): 305–15.
Philbrick, Herbert A. *I Led 3 Lives*. New York: Grosset and Dunlap, 1952.
Poincaré, Henri. *Science and Method*. New York: Science Press, 1905.
Popper, Karl. *The Open Society and Its Enemies*, 2 vols. Princeton: Princeton University Press, 1971 (orig: London: Routledge, 1945).
———. *Poverty of Historicism*. Boston: Beacon, 1957.
———. *The Logic of Scientific Discovery*. New York: Basic Books, 1959.
———. "Normal Science and Its Dangers." In *Criticism and the Growth of Knowledge*, edited by Lakatos and Musgrave, 51–58. Cambridge: Cambridge University Press, 1970.
———. "Unended Quest: An Intellectual Biography." In *The Philosophy of Karl Popper*. La Salle, IL: Open Court, 1976,
Porter, Russell B. "Reds Here Shift in Stand on War." *New York Times*, July 27, 1941.
Post, Robert. "A Very Special Relationship: SHOT and the Smithsonian's Museum of History and Technology." *Technology and Culture* 42 (2001): 401–35.
Prairie Fire: The Politics of Revolutionary Anti-Imperialism. Communications Co., 1974.
Quine, W. V. O. "Truth By Convention." In *Philosophical Essays for A. N. Whitehead*, edited by O. H. Lee. New York: Longmans, 1936.
———. "On What There Is," *Review of Metaphysics* 2, no. 5 (September 1948): 21–38.
———. "Ontology and Ideology." *Philosophical Studies* 2, no. 1 (Jan. 1951): 11–15.
———. *The Time of My Life*. Cambridge: MIT Press, 1985.
Ranzal, Edward. "McCarthy Charges 'Mess' at Harvard." *New York Times*, Nov. 6. 1953.
Raymond, Jerry. "Goheen, Falk, Orfield Express Community's Outrage." *Daily Princetonian*, May 5, 1970.
Reisch, George A. "Did Kuhn Kill Logical Empiricism?" *Philosophy of Science* 58, no. 2 (June 1991).
———. "Planning Science: Otto Neurath and the International Encyclopedia of Unified Science." *British Journal for the History of Science* 27 (1994): 153–75.
———. "Epistemologist, Economist . . . and Censor? On Otto Neurath's Infamous *Index Verborum Prohibitorum*." *Perspectives on Science* 5, no. 3 (1997): 452–80.
———. *How the Cold War Transformed Philosophy of Science: To the Icy Slopes of Logic*. New York: Cambridge University Press, 2005.
———. "Telegrams and Paradigms: On Cold War Geopolitics and *The Structure of Scientific Revolutions*." In *Science Studies during the Cold War and Beyond: Paradigms Defected*, edited by Elena Aronova and Simone Turchetti, 23–53. New York: Palgrave, 2016.

———. "Pragmatic Engagements: Philipp Frank and James Bryant Conant on Science, Education, and Democracy." *Studies in East European Thought* 69 (2017): 227–44.

———. "On the Couch with Freud and Kuhn." *Metascience* 27, no. 1 (March 2018): 37–46.

Richardson, Robert D. *William James: In the Maelstrom of American Modernism.* New York: Houghton Mifflin, 2007.

Roberts, Sam. *The Brother.* New York: Random House, 2001.

Robinson, Douglas. "More Dynamite Is Found in Rubble of Townhouse." *New York Times*, March 13, 1970.

Rorty, Richard. "Afterword." In *Sidney Hook Reconsidered*, edited by M. J. Cotter. Amherst, NY: Prometheus Books, 2004.

Rosberg, Gerald M. "War Protest at Harvard Is Not New; Pacifists Got Support in '16 and '41." *Crimson*, June 16, 1966.

Rosenfeld, Seth. *Subversives: The FBI's War on Student Radicals and Reagan's Rise to Power.* New York: Farrar Strauss and Giroux, 2012.

Rosenstock-Huessy, Eugen. *Die Europäischen Revolutionen: Volkscharaktere und Staatenbildung.* Jena: E. Diederichs, 1931.

———. *Out of Revolution: Autobiography of Western Man.* Providence, RI: Berg, 1993 (orig. New York: William Morrow, 1938).

Rossi, John. "Farewell to Fellow Traveling: The Waldorf Peace Conference of March 1949." *Continuity* 10 (1985): 1–31.

Rudolph, John. *Scientists in the Classroom.* New York: Palgrave McMillan, 2002.

Salisbury, Harrison. "Rioting in Soviet Reported Over Anti-Stalin Campaign." *New York Times*, March 17, 1956.

———. "The Strange Correspondence of Morris Ernst and J. Edgar Hoover." *The Nation*, Dec. 1, 1984.

Sanders, Jane. *The Cold War on Campus,* Seattle: University of Washington Press, 1979.

Sartre, Jean-Paul. "Les Animaux malades de la rage." *Libération*, June 22, 1953, translated in *Writings of Jean-Paul Sartre*, 207–11. Evanston: Northwestern University Press, 1985.

Schein, Edgar H. "Brainwashing and Totalitarianization in Modern Society." *World Politics* 11, no. 3 (April 1959): 430–41.

———. "From Brainwashing to Organizational Therapy." *Organization Studies* 27, no. 2 (2006): 287–301.

———. "Brainwashing." Cambridge, MA: Center for international Studies, MIT, Dec. 1960.

———, Inge Schneier, and Curtis H. Barker. *Coercive Persuasion: A Psycho-Social Analysis of the "Brainwashing" of American Civilian Prisoners by the Chinese Communists.* New York: Norton, 1961.

Schorner, Michael. "The Reception of Kuhn's *Structure of Scientific Revolutions* in West and East Germany." Unpublished ms.

Schrecker, Ellen. *No Ivory Tower: McCarthyism and the Universities*. New York: Oxford University Press, 1986.
———. *The Age of McCarthyism: A Brief History with Documents*. New York: Bedford/St. Martins, 2002.
Schwarz, Fred. *Communism: Diagnosis and Treatment*, Los Angeles: World Vision, n.d.
Seigel, Kalman. "Ex-Wife Identifies Remington as Red." *New York Times*, Dec. 27, 1950.
———. "College Freedoms Being Stifled by Student's Fear of Red Label." *New York Times*, May 10, 1951.
Shapere, Dudley. " 'The Structure of Scientific Revolutions' (review)." *The Philosophical Review* 73, no. 3 (July 1964): 383–94.
Shapiro, Edward, ed. *Letters of Sidney Hook: Democracy, Communism, and the Cold War*. Armonk, NY: A. E. Sharpe, 1995.
Shapley, Harlow. "A Word from Harlow Shapley." In *Speaking of Peace*. New York: National Council of the Arts, Sciences, and Professions, 1949.
Sheppard, Nathaniel Jr. "2 in Weather Underground Are Bargaining to Surrender." *New York Times*, Nov. 24, 1980.
Steel, Lewis M. "College Life during World War II Based on Country's Military Needs." *Crimson*, Dec. 7, 1956.
Stopes-Roe, H.V. "*The Structure of Scientific Revolutions* by Thomas S. Kuhn." *The British Journal for the Philosophy of Science* 15, no. 58 (Aug. 1964): 158–61.
Story, Walter R. Letter to the Editor. *New York Times*, March 15, 1953.
Swerdlow, Amy. *Women Strike for Peace: Traditional Motherhood and Radical Politics in the 1960s*. Chicago: University of Chicago Press, 1993.
Time magazine. "Science: Stargazer." Aug. 29, 1949.
Toulmin, Stephen. "Discussion: S. E. Toulmin." In *Scientific Change*, edited by Crombie, 382–84. New York: Basic Books, 1963.
———. "Does the Distinction between Normal and Revolutionary Science Hold Water?" In *Criticism and the Growth of Knowledge*, edited by Lakatos and Musgrave, 39–48, Cambridge: Cambridge University Press, 1970.
Trenn, Thaddeus J. "Preface." In Fleck, *The Genesis and Development of a Scientific Fact*. Chicago: University of Chicago Press, 1979.
Trussell, C. P. "Woman Links Spies to U.S. War Offices and White House." *New York Times*, July 31, 1948.
———. "Some Books Literally Burned after Inquiry, Dulles Reports." *New York Times*, June 16, 1953.
Uebel, Thomas. "The Enlightenment Ambitions of Epistemic Utopianism." In *Origins of Logical Empiricism*, edited by R. N. Giere and A. W. Richardson. Minneapolis: University of Minnesota Press, 1996.
———. *Empiricism at the Crossroads*, Chicago: Open Court, 2007.
United States Congress, Joint Committee on Atomic Energy. *Soviet Atomic Espionage*. Washington, DC: Government Printing Office, April 1951.

United States House of Representatives, Committee on Un-American Activities. *Testimony of Dr. Edward U. Condon: Hearing before the Committee on Un-American Activities, House of Representatives, Eighty-second Congress, second session.* Sept. 5, 1952. Washington, DC: Government Printing office, 1952.

———, Select Committee to Investigate Tax-Exempt Foundations and Comparable Organizations. Final report. Washington, DC: U.S. Government Printing Office, 1953.

———. "Communist Methods of Infiltration (Education)." Hearings before the Committee on Un-American Activities, House of Representatives, Feb. 25, 26, and 27, 1953, Washington, DC: Government Printing Office.

———. "Communist Psychological Warfare (Brainwashing). Consultation with Edward Hunter." Washington, DC: Government Printing Office, 1958.

———. "Hearings before the Committee on Un-American Activities, House of Representatives," Dec. 11–13, 1962. Washington, DC: U.S. Government Printing Office, 1963.

United States Senate. *Executive Sessions of the Senate Permanent Subcommittee on Investigations of the Committee on Government Operations.* Vol. 2. Washington, DC: Government Printing Office, 1953.

Vargas-Cooper, Natasha. *Mad Men Unbuttoned.* New York: Harper Collins, 2010.

Von Hoffman, Nicholas. *Citizen Cohn.* New York: Doubleday, 1988.

Waelder, Robert. "Demoralization and Re-education." *World Politics* 14, no. 2 (Jan. 1962): 375–85.

Wall, Wendy L. *Inventing 'The American Way': The Politics of Consensus from the New Deal to the Civil Rights Movement.* New York: Oxford University Press, 2008.

Wang, Jessica. *American Scientists in an Age of Anxiety.* Chapel Hill: University of North Carolina Press, 1999.

Wartofsky, Marx. *Models: Representation and the Scientific Understanding*, Dordrecht: Reidel, 1979.

Watkins, J. W. N. "Against Normal Science." In *Criticism and the Growth of Knowledge*, edited by Lakatos and Musgrave, 25–37. Cambridge: Cambridge University Press, 1970.

Weaver, Warren. "Free Science Sought." Letter to Editor. *New York Times*, Sept. 2, 1945.

Weeks, Edward. *The Lowells and their Institute.* Boston: Little Brown, 1966.

Weidlich, Thomas. *Appointment Denied: The Inquisition of Bertrand Russell.* Amherst, NY: Prometheus Books, 2000.

Willison, George. *Let's Make a Play.* New York: Harper, 1940.

Wirth, Louis. "Preface." In Karl Mannheim, *Ideology and Utopia: An Introduction to the Sociology of Knowledge*, translated by Louis Wirth and Edward Shils. New York: Harcourt, Brace, 1954 (orig. 1936).

Wolfe, Audra. "The Organization Man and the Archive: A Look at the Bentley Glass Papers." *Journal of the History of Biology* 44 (2011): 147–51.

Woodger, Joseph. *The Technique of Theory Construction*. Chicago: University of Chicago Press, 1939.
Wray, K. Brad. "Kuhn and the Discovery of Paradigms." *Philosophy of the Social Sciences* 41, no. 3 (2011): 380–97.
———. *Kuhn's Evolutionary Social Epistemology*. Cambridge: Cambridge University Press, 2011.
———. "The Influence of James B. Conant on Kuhn's *Structure of Scientific Revolutions*." HOPOS 6, no. 1 (2016): 1–23.
Zimbalist, Stephanie. "Streetscapes/Readers' Questions." *New York Times*, Sept. 7, 2003.

Index

Acheson, Dean, xvii, 310
Adenauer, Conrad, xvii
Adler, Alfred, 299
Allen, Woody, 164–65
Alliluyeva, Svetlana
   Twenty Letters to a Friend, 137
American Civil Liberties Union, 136
American Committee to Defend Leon
   Trotsky, 93
The American Education Fellowship,
   23
Americans for Intellectual Freedom, 84
American Student Union, 3, 4, 32,
   320–21
American Workers Party, 93
American Youth Congress, 19
anticommunism
   literature, 113–14
   targeting philosophers, 116
anticommunist paranoia, xv–xvii, xxx,
   102
anti-"truth" alliance, 211
Archimedes, 287, 289
Arendt, Hannah
   The Origins of Totalitarianism, 120
Aristotle, 49, 62, 158, 160, 188, 193,
   219, 223, 282, 287, 299, 319,
   361, 364
   Physics (*De Physica*), 65–66, 153–54
"atom spies," 312–13
Ayer, Alfred, 163
   Language, Truth, and Logic, 160

Bailyn, Bernard
   *The Ideological Origins of the
   American Revolution*, 222
Barber, Bernard, 131, 206, 250
Barnes, Clarence, 76–77
"Battle of the Pickets" newsreel, 103
Bauer, Raymond, 122
The Beatles
   "I Am the Walrus," 332
Bell, Daniel, xiii–xiv
   The End of Ideology, xiii
Bentley, Elizabeth, 19, 72
Berkeley's People's Park, 330
Bernal, John Desmond
   The Social Function of Science, 29
Biddle, George, 9
Biderman, Albert, 122
   The Manipulation of Human Behavior,
   122
"big lie" technique, 343
Biological Sciences Curriculum Study
   (BSBC), 256
Bird, Alexander, xxxi
Bismarck, Otto von, 145
Black, Algernon, 83–84
Black Panthers, 328
Bloomfield, Leonard
   Linguistic Aspects of Science, 207
Boas, Franz, 93
Bogdanov, Alexander, 116
Bonnet, Charles, 266
Boring, Edwin, 319

Boudin, Kathy, 332
Bowen, Carroll, 249–52, 283
Boyle, Robert, 50, 191
brainwashing, 116–24
Brando, Marlon, 83
Bridgman, Perry, 45–46, 140
Brinton, Crane, 221–23
    The Anatomy of Revolution, 221, 308
Bruner, Jerome, 128
Buck, Paul, 35, 41, 44, 146
Buckley, William F.
    God and Man at Yale, xvi
Burnett, Leo, 317–18
Burnham, James, xliii, 58–59, 61–62, 90, 93, 102, 116, 179, 328
    The Coming Defeat of Communism, 58
    The Struggle for the World, 58
Bush, Vannevar, 30, 181–82, 191
Byrnes, James, 59

Cadogan, Alexander, 59
Caldwell, Otis, 7
Camus, Albert
    The Rebel, 322
captive soldiers, code of conduct, 118
Carlson, A. J., 84
Carnap, Rudolf, xxxi, 84, 86, 94, 99, 102, 180, 206–208, 211, 249–50, 252, 311
    Hook's warning to, 94–95, 104
Caute, David
    The Great Fear, 312
Cavendish, Henry, 130
Chamberlain, Neville, 69
Chambers, Whittaker, 72–73
Chaplin, Charlie, 83
Chase, Stuart, 193–94
    The Tyranny of Words, 9, 193
Chaucer, Geoffrey, 39
Chemical Warfare Service, 27

Chen, Theodore, 122, 125
    Thought Reform of the Chinese Intellectuals, 121
Chiang Kai-shek, 73
The China Monthly Review, 123
Churchill, Winston, 96
The Citizens Committee for Peace with Freedom in Vietnam, 345
Citizenship Education Service, 137
Clagett, Marshall, 302
Clio, 242
Clucas, Lowell, xix, xxiii
Cohen, I. Bernard, 207, 302
Cohen-Cole, Jamie, xxxiv–xxxv
Cohn, Roy, xvii–xx, xxii, xxix, xl, 21, 108, 294, 309–11, 315, 344
cold war, xxvii–xxviii, 255
    and American academy, xxx
    and anticommunism, xxx
    in Britain, 293–94
    and ideology, 198
    political conformity in, 146
    and science, xiii
college campus tensions, in 1951, 173
Committee on the Present Danger, 75, 97, 133–34, 136, 195, 201, 297, 315, 345
Committee Supporting the Bush Report, 183
Committee to Defend America by Aiding the Allies, 32, 34
The Communist (magazine), 90, 177
Communist Party, 18, 112
Communist Party line, ever changing, 91–92
"Communist philosophy" as threat, 114–16
Conant, James Bryant, xiv–xix, xxvi–xxvii, xxix, 25, 27–28, 82–83, 92, 102, 104, 108, 120, 126–27, 137, 140, 157, 165, 175, 185–86, 191, 194, 196–98, 200, 202, 211–13,

216, 222, 229, 232, 235, 248,
  250, 257, 262, 266, 271, 295–98,
  300–301, 304, 310, 315, 338,
  340, 352–53, 355–59, 361, 363,
  365
on academic freedom, 76–79
alliance with Hook, 95–99
and anticommunism, xvii, 72,
  147–48
and atomic bomb, 30, 69–70,
  73–74
The Citadel of Learning, 275, 277,
  288, 291
as cold warrior, 71, 74–76, 88, 97,
  146, 345
on competition, 28, 31
on conceptual schemes, xxv–xxvi,
  49–50, 127, 162, 261
"conversion" of, 70–71, 74, 79
disagreement with Kaempffert, 182,
  264, 269
Education and Liberty, 98
Education in a Divided World, 53,
  56–57, 59, 72–73, 77, 89, 95,
  134, 143, 147
on fragmentation in America, 41
"Free Inquiry or Dogma," 71
General Education Committee, 167
general education project, xxiv, xxvii,
  xxx, 52, 61–62, 141–42, 177,
  195, 200, 203, 256, 258, 260; on
  democracy, 143; search for unity,
  141–42
in Germany, xviii, 28, 224, 275,
  343
Germany and Freedom (lectures), 224
Godkin Lectures, 224
helping war effort, 34–35, 39–40
as High Commissioner of Germany,
  xvi
on intellectual freedom, 275–77,
  288

on intervention, 69
and Kuhn: critique of, xxxv, xxxix,
  28, 80, 135, 272–74, 278–81,
  285–88, 349; differences between,
  xxxv, xxxviii, 28; influence on,
  xxiii–xxvi, xxxiv; views moving
  apart, 275, 278
letter to the future, 133–34
on liberal arts, 40
liberalism of, 29, 149
on liberal tradition, 41–42
on loyalty oaths, 71–72
on mass psychosis, 145
and McCarthy, xx–xxiii, xxxvi–
  xxxvii, 79, 343–45
military language of, 51, 147–48
as military scientist, 27, 30–31
My Several Lives, 27, 343
on national ideals, 53–54
Natural Sciences 4 course, 184
on nuclear technology, 343
Objectives report, 142, 147
On Understanding Science, xxiv–xxv,
  43, 46, 48, 51, 66–67, 130, 147,
  178, 191, 207, 280, 282, 302,
  313–14
and open society, 314
on paradigms, 273, 287, 290–91,
  301, 349
on pluralism, 276–77
policy opposing Communist faculty,
  146
as president of Harvard, 27–28
on public education reform, 53–54,
  97–98, 255
as public leader, 134
on the red menace, 72
response to Truman, 183
in run-up to war, 31–33
on science: and art, 289; case
  histories, 49; as cumulative,
  67–68, 159, 282; education,

444 | Index

Conant, James Bryant *(continued)*
  42–43; guiding, 178; and practical problems, 279–80; and progress, 261, 276–77, 287, 289; as social process, 49, 52
  Science and Common Sense, 47, 313
  on scientific dogma, 130–31
  on scientific revolutions, 278–79, 281
  on Soviet ideology, 57–58, 60–61
  on totalitarianism, 143–46, 291
  on unity, 143–44, 147–48
  and unity of science movement, 200–201
  on Vietnam, 345–46
  on worldviews, dangers of, 289–90
Conant, Patty, 27–28, 165, 298
conceptual schemes. *See* Conant, James Bryant: on conceptual schemes
Condon, Edward, 227
Condon, Richard
  The Manchurian Candidate, 117, 129
Congress for Cultural Freedom, 315
Copernicus, Nicolaus, 361
Coughlin, Charles, 15
counterculture, 332
Counts, George X., xl, xlii, 6, 11, 14–16, 21, 25, 71, 102, 328
  as anticommunist, 84
  "Dare Progressive Education Be Progressive?" 8, 10
  defection from the left, 15–16
  "Education through Indoctrination," 8
  "Freedom, Culture, Social Planning and Leadership," 8
  The Soviet Challenge to America, 8–9
Covey, Stephen, 333
  The Seven Habits of Highly Effective People, xxvi, 333
Crombie, Alistair, 252, 264
Cronkite, Walter, 331
"creative revolution" (Burnett), 318

Crosley, George, 73
Croton-on-Hudson News, 16
Cultural and Scientific Conference for World Peace (Waldorf Conference), 82–84
Currie, Lauchlin, 19
Cutter, Richard Ammi, 75

Dahlberg, Winifred, 23
Daily Worker, 18
Daley, Richard J., 330–31
Dalton, John, 191
Dante Alighieri, 67
Darwin, Charles, 202, 287, 289
Daughters of the American Revolution (DAR), 14–15
Davis, Herbert, 84–85
Davis, Robert Gorham, xliii, 37–39, 55, 109, 153, 156–57, 173, 203, 227, 357, 364
  testimony to HUAC, 105–107
"Days of Rage," 331
Decker-Ritchey, Barbara, 334
de Gaulle, Charles, 141
Dell, Floyd, 9
Democratic Party, convention of 1968, 330–31
Denkkollektiv (thought collective), 220, 358
Descartes, René, 153
Dewey, Edward, 335
Dewey, John, 6–8, 10, 24–25, 47, 89, 93, 142, 211, 321, 328
  Education and Democracy, 7
  School and Society, 7
Dewey commission, 93–94
Dick, Leo, 17
Dilling, Albert, 12
Dilling, Elizabeth, xliii, 16, 25, 111, 174
  Red Channels, 136
  The Red Network, 12–15

Douglas, Paul, 345–46
Douglas, William O., 287
Du Bois, W. E. B., 21–22
Dubos, René, 287
Dubridge, Lee, 73
Dulles, John Foster, xix–xx
Duranty, Walter, 94
Dylan, Bob, 317

Earhart, Amelia, 11
Earman, John, 296
Eastman, Max, 9, 11, 15, 102
Eddington, Arthur, 299
Edison, Thomas, 262
Einstein, Albert, 12, 31, 50–51, 65, 68, 174, 278–79, 299
  "The Fight Against War," 14
  on nuclear weapons, 14
Eisenhower, Dwight, xvi–xvii, xx, 32, 298, 311
Emerson, Ralph Waldo, 144
  "On Self Reliance," 70
Encyclopedia of the Social Sciences, 208
Engels, Friedrich, 177, 247, 275
Enriques, Federique, 207
Ernst, Morris L., 136
Euclid, 289

Fadayev, A. A., 82
falsification, 300
Feyerabend, Paul, xxxix, xliii, 250, 271, 301, 306, 347–48, 350–53, 359, 361, 363, 366
  critique of Kuhn, 301–306
Finney, John, 255
Fleck, Ludwik, xxxi, 131, 220–21, 225, 358–59
  The Genesis and Development of Scientific Fact, 176, 220–21, 358
  "harmony of illusions," 220, 228, 235
  on "thought collective," 220

"Footprints of the Trojan Horse" (pamphlet), 137
Foreign Agents Registration Act, 22
Franco, Francisco, 98
Frank, Jerome, 9
Frank, Philipp, xliv, 29–30, 45–47, 174, 200, 207–209, 213, 218, 342, 353
  "Research Project in the Sociology of Science," 202
  on science: and ideology, 201; and social forces, 201–204
  Science, Facts, and Values, 201
  under FBI investigation, 206
Frankenstein, Victor, 348
Freedom House conference, 100, 102–103
Free Speech Movement, 324
Freud, Sigmund, 120, 164–66, 245, 299
Friedlander, Mort, 318–19
Fromm, Erich, 287
Fuchs, Klaus, 74, 310, 312–14
Fuller, Steve
  Thomas Kuhn: A Philosophical History for Our Time, xxxv
Furry, Wendell, xxiv, xxx, xliv, 107–109, 115, 173, 208, 227, 357

Galileo, 47, 50, 66–67, 158, 289, 361, 364
Geertz, Clifford
  The Interpretation of Culture, 348
General Education Board, 6
General Education Committee, 45
German-American Bund, 94–95
Gestalt psychology, 165
Gide, Andre, 247
Gillispie, Charles, 319
Glass, H. Bentley, xxxix, xliv, 266–71, 279, 284, 286, 295, 297, 321, 353, 359, 361, 363
  on intellectual freedom, 266
  scientific education reform, 267
  on scientific progress, 267–68

The God that Failed: A Confession (essays), 247
Goheen, Robert, 338–40, 342
Golos, Jacob, 19
Gombrich, E. H., 286, 319
Goodman, Paul, 326
Gornick, Vivian, 332
Greenbaum, Edward S., xl, xliv, 136–39
Greenglass, David, 310–12
Greenglass, Ruth, 311
Griswold, Erwin, 99
Gromyko, Andrei, 59–60
Groves, Leslie, 30, 73, 293, 314

Hall, A. Rupert, 266
Hammett, Dashiell, 83
Hanson, Norwood Russell, xxxi, xxxvi, 250–52, 283, 296, 302, 304
  Patterns of Discovery, 250
Harvard University
  noninterventionism debates at, 32–33, 35
  and war effort, 34–35
Harvard Crimson, xvi, 32, 35, 38
  on run-up to war, 32–33
Harvard Society of Fellows, xxxiii
Hatfield, Margaret, 9, 193
Hayakawa, S. I.
  Language in Action, 193
Hayden, Tom, xliv, 320–24, 329, 331–32
  "The Port Huron Statement," 321–28
Hayek, Friedrich, 178
  The Road to Serfdom, 178
Haymarket Square, 331
Hearst, William Randolph, 15
Hellman, Lillian, 83, 104
Henderson, Lawrence J., xxxiii–xxxiv, 279
Hendrix, Jimi, 332–33

Hersey, John
  "Hiroshima," 14
Hershberg, James, 28, 35, 70, 78, 343
Hessian Hills School, 3, 9–11, 15–18, 20, 23, 25, 38, 112, 174, 193, 320, 357, 364
  closing of, 24
Hiss, Alger, xvi–xvii, 72–73, 97, 134, 345, 357
Hitch, Charles, 341
Hitler, Adolf, 11–12, 15–17, 25, 30, 32–33, 40, 69, 89–90, 105, 113, 119–20, 137, 139, 145, 179–80, 220, 224, 293, 310, 343–46
Hofstadter, Richard, 118
Hollinger, David, xxxiii
"Hollywood Ten," 81
Homans, George, xxxiv
Hook Sidney, xxix, xxx, xxxvi, xxxviii, xl, 6, 8, 14, 21, 102, 105, 109, 114, 116, 148, 173–74, 179–80, 208, 248, 266–67, 309–10, 321, 328, 356, 358, 362, 365
  alliance with Conant, 95–99
  as anticommunist, 81–90, 94–97
  challengers to 100–101
  on Communist Party members, 101
  confronting Shapley, 81–87, 103
  conversion of, 92–95
  as critic of American Left, 94
  defection from the left, 15
  Education for Modern Man, 98
  on "fellow travelers," 99–100
  Freedom House conference, 100
  From Hegel to Marx, 15
  hatred of Stalin, 91, 94
  Heresy, Yes—Conspiracy, No, 90, 97, 99
  "International Communism—There Is Nothing Mysterious about Its Core of Belief or about What It Is Trying to Do," 96

as Marxist, 93–94
in New York Times Magazine, 89, 91
Towards the Understanding of Karl Marx, 15, 95
warning to Carnap, 94–95, 104
Hoover, Herbert, 18
Hoover, J. Edgar, xiii, xxxvi, 20, 22, 113–14, 136, 146, 174–75, 206, 208, 227, 309, 311, 321, 324, 329, 339
Masters of Deceit, 114
Horney, Karen, 164
House Un-American Activities Committee (HUAC), 22, 39, 72, 103, 121, 324, 331
inquisition of Harvard faculty, 104–108
Hovde, Bryn, 102
Howe, Irving, 94
"How to Identify an American Communist" (*Look* article), 113
Hoyningen-Huene, Paul, xxxii
Hume, David, 47, 294
Hunter, Edward, xliv, 116–17, 121–25, 127–28
Brainwashing: The Story of the Men Who Defied It, 117
Brainwashing in Red China, 117
Huxley, Aldous
Brave New World Revisited, 124

ideology. *See also* Kuhn: and ideology; Kuhn: *The Structure of Scientific Revolutions*, on
ideology
as controversial word, 208
in science, 199
and war, 198
Imandt, Robert, 9
"infection" metaphor, in postwar America, 309–10

Inkeles, Alex
Public Opinion in Soviet Russia, 60
Institute for the Unity of Science, 200, 208
intellectual history, and politics, 361
Intercollegiate Socialist Society, 321
International Encyclopedia of Unified Science, xxxi, 29, 178–79, 199, 200, 206–208, 211, 250
Iron Curtain, 58–59
Isaac, Joel, xxxiv
Working Knowledge, xxxiii
I Want to Be Like Stalin, 16

James, William, 48, 79, 154–55, 157, 164, 179–80, 192, 211
influence on Kuhn, 154–55
"The Many and the One," 176
Pragmatism, 142, 154
"Pragmatism and Truth," 176
on truth, 211
The Varieties of Religious Experience, xxxii, 70, 154, 246
on "vicious intellectualism," 101
Jaspers, Karl, 287
Jenner, William, 107
Jewish Board of Guardians, 136
Jews in America, cold war fears of, 174
"The Jimi Hendrix Experience," 333
Johnson, Lyndon, 338
Jones, Ernest, 164
Journal of Marketing, 317

Kaempffert, Waldemar, 178, 181–86, 191, 199, 212
on national foundation for scientific research, 181–83
on unified science, 184–85
Kaghan, Theodore, xix, xxiii
Kallen, Horace, 30, 180, 183–84, 194, 200, 208
critique of unifying science, 179–85

448 | Index

Kant, Immanuel, 67
Kaufman, Arnold, 322
Kaufman, Irving, 311
Kearney, Bernard W., 107
Keats, John, 67
Kennan, George F., xliv, 60–62, 88, 137, 194, 198, 310
    "long telegram," xliv, 56–58
    on Soviet operatives, 57
Kennedy, Bobby, 330
Kennedy, John F., xxxvi, 325, 338
Kent State University shooting, 331, 340
Kepler, Johannes, 361
Kerr, Clark, 324
Khrushchev, Nikita, 324
Kilgore, Harley, 181
Killian, James, 96
King, Martin Luther Jr., 328, 330–31
Kinkead, Eugene
    In Every War but One, 118, 120
Kluckhorn, Clyde, 60
Koestler, Arthur, 247–48, 358
    The Sleepwalkers, 247
Korea, and American soldier collaboration, 117–18
Korzybski, Alfred, 194
    Science and Sanity, 193
Koyré, Alexander, xxxvi, 302
Kranzberg, Melvin, 232
Kristallnacht, 69
Kubie, Lawrence, xlv, 165–68, 245, 336, 359
Kuhn, Abraham, 6
Kuhn, Minette Stroock, 5–6, 11, 164, 319
Kuhn, Roger, 6, 11, 45
Kuhn, Samuel, 6
Kuhn, Samuel L., 5–6, 55, 136
Kuhn, Setty Swartz, 6, 12, 14
Kuhn, Simon, 6
Kuhn, Thomas, 11, 149, 340–41
    antiwar politics of, xl–xli, 3–5
    on Aristotle, 65–68

Aristotle experience of, xxxii, xxxviii, 68, 70, 79, 109, 131, 153–54, 156–58, 160, 162, 169, 177, 195, 203, 216–17, 220, 222–23, 231, 242, 246–47, 265, 282, 348, 355, 358, 363
    on beliefs, shaping of, 38–39
    Black Body Theory and the Quantum Discontinuity, 348
    and Carnap, 211
    and cold war politics, xiii–xiv, xxvi, 227
    on commitment, functions of, 217
    and Conant, 256, 271–78;
        differences between, xxxv, xxxviii, 134–35; influence of, xxiii–xxiv, xxvi, 51, 55, 188; similarities between, 46; views moving apart, 274–75, 278
    on Conant's critique, 273–78, 281–86
    on Conant's war effort, 36
    on conceptual schemes, 261
    on connotations, infinity of, 158–59, 166–67
    on consensus, 135, 140
    conversion of, 37
    The Copernican Revolution, xxiv, 196, 215, 243, 247, 260–62, 274, 277–78, 285, 318, 335, 348;
        Conant's foreword, 277
    crisis of, 37, 39
    "The Crisis in Democracy," 135, 137, 140, 357
    and data invalidation, 157
    on democracy, 138–39
    on disciplinary matrix, 348
    on disunity, 135
    on dogma, 38
    "The Essential Tension: Tradition and Innovation in Scientific Research," 288
    on fringe meanings, 158–59, 192–94, 207, 211, 238

"The Function of Dogma in Scientific Research," 262, 264–65, 269, 271, 279, 308, 349, 351, 359
and general education program, 44–48, 54, 141, 157, 203, 256–58
geopolitical influences on, 61–62
on gestalts, 221, 356
at *Harvard Crimson*, 35–36
Hessian Hills influence on, 24
and ideology, xiii–xiv, 197, 209, 213, 217–19, 225–27, 229, 352
influence of, xxvi
influences on, xxiii–xxiv, xxvi, 24, 51, 55, 154–55, 160–63, 169, 221–23, 232
as interdisciplinary, xxxiii–xxxiv, 25, 45
"International Morale and a United States Declaration of War," 105
and *International Encyclopedia of Unified Science*, 206–208
as interventionist, 36–37
letter to Frank, 203–206
liberalism of, 37
linguistic turn, 355
and logical empiricism, xxxi
on Lowell Institute ads, 171–76, 185–86, 189
Lowell Lectures, 171–72, 176, 186–96, 204, 207–10, 212, 216, 221, 238, 240, 243, 329, 357; and McCarthyism, 192, 196
on meanings, 195
"The Metaphysical Possibilities of Physics," 156
military work of, 43–44, 141
monograph, 109, 208–10, 213, 215, 218, 222, 225–26, 233, 242–45, 248, 250, 264
on natural language, 157–59
as pacifist, 36–38, 137

on paradigms, xxv–xxvi, xxxv, 126–27, 154, 166, 168–69, 187, 212, 223, 239–41, 243–46, 248, 264–66, 269, 280–81, 285–86, 291, 295–96, 303, 315, 348–49, 352–53
on perception, 282–83
Phi Beta Kappa address, 320
on philosophy of history, 357
Piaget's influence on, 160–63, 169
and political revolutions, 326–27
and politics, 358, 364–65
on Popper, 306
and popular culture, 333
on Princeton committee on sponsored research, 341
on propaganda, 139
and psychoanalytics, 164–67
"The Quest for Physical Theory" (lecture), 171, 177
as radical, 24
reader letters to, 334–35
"Reflections on My Critics," 348, 350–52, 355
on revolutionaries, 138
on revolutionary divides, 61–62
The Road Since Structure, 359
on science: and art, 228–29, 286; classical, 187; community nature of, 351; and consensus, 140, 227–30, 233, 238–40; and conversion experiences, 245–46; crisis stage in, 187, 191; and cumulativeness, 230–33; 'cuts' and 'boxes' in, 160, 166–67, 192, 209; and dogma, 25, 127–28, 131, 261–70; external influences on, 342–43, 346–47; and fringe meanings, 192–95, 209; and group-unconscious, 166–67, 245; history of, 139, 220; as ideology, xxvii, xxix, 209–10; immaturity/ maturity phases of, 239–40; individual

Kuhn, Thomas *(continued)*
   autonomy in, 352–53; and the layperson, 258–60, 262; normal, xxxv, 127–29, 141, 187, 227–29, 244–46, 263, 285, 289, 352; and philosophy, 127; and politics, xxvii; predispositions in, 167–69; and progress, 188–89, 233, 282, 286; revolutionary, 141; sociology of, 202–205, 211; textbook, 188, 217, 225, 241, 257, 362; and totalitarianism, xxxiii; and truth, 286, 288; and the unknown, 261, 264–65
"The Sciences in the Harvard General Education Program," 256
on scientific method, 186
on scientific process, and "psychological real worlds," 161–62
on scientific revolutions, 79, 128, 163, 168, 209–10, 213, 216–17, 219, 221–22, 225, 228–29, 265; invisibility of, 129–30, 286
"Second Thoughts on Paradigms," 348
on semantic boxes, 158–59
and semantics movement, 193–94
shift in language, 274
"Sputnik and the American Public Mind," 262
The Structure of Scientific Revolutions, xiii–xiv, xxiv–xxvii, xxix, 24–25, 39, 51–52, 68, 79–80, 125–26, 141, 149, 156, 158–59, 163, 168–69, 186, 188, 193, 196–97, 210–11, 215–16, 225–26, 232–33, 243, 264, 269, 271, 282, 289, 342, 356; and cold war, 212; and collectivism, 359; on "the confirmation debate," 229–30; Conant's critique of, xxxv, xxxix, 80, 135, 272–74, 278–81, 284–88: Kuhn's response to, 273–78, 281–86; on consensus in science, 236–42; as conservative book, 336; "conversion" language in, xxxii, xxxvii–xxxviii, 126, 363; as critique of liberalism, 262; critiques and reviews of, xxxv, xxxix, 223, 251, 295–96, 298–99, 301–309, 316, 318–19, 347, 350–51, 355, 360, 366; on cumulativeness, in science, 235–36; dedication of, xxiv, 284–85, 347; drafting of, 215–18, 225, 227–32, 235–44, 294, 334; final chapter, 286; Fleck's influence on, 221; on ideology, 25, 56, 62, 79, 130, 212, 349; influence of, xxx, xxxvi, 360–61; influences on, 221–23; invisibility of politics in, xxix–xxx; as "monster," 348; on normal science, 187; on paradigms, xxv–xxvi, xxxv, 126–27, 154, 166, 168–69, 187, 212, 223, 239–40, 285–86, 295–96; on political revolutions, 323; and politics, xxxiii–xxxvii, 62, 356–59, 361–65; and pop culture, 333, 336; postscript to, 348, 352; and postwar America, 309; publication of, 252; 'purple passages' in, xxxii–xxxiii, xxxvii; reader responses to, 334–35; and relativism, 295; revolution in, 25, 323, 328; on science/art contrast, 235; on scientific progress, 235; on science textbooks, 188; on scientific revolutions, 166, 223, 235–43, 323, 358; on specialization, 212; title of, xxix
on theory as ideology, 217–19, 221, 223, 226–27, 246, 261

theory of language, 192
"The War and My Crisis," 37, 92, 105, 109, 135, 153, 212
and William James, 154–55
on words, power of, 194
Kushner, Tony
Angels in America, 311

Lakatos, Imre, 348, 350, 352, 355
critique of Kuhn, 305–306
Lamont, Corliss, 94
Langer, Suzanne
Philosophy in a New Key, 160
Lawrence, William, 177
Layton, David, 336
League for Industrial Democracy, 321
Leary, Timothy, 333
Lenin, Vladimir, 15, 54, 60, 116, 247
Materialism and Empiriocriticism, 275
Lennon, John, 332
Leo Burnett Agency, 317
Leviero, Anthony, 118
Lewis, Helen B., 114
Lewis Naphtali, 115
Lewisite, 27
liberalism, 89
Lifton, Robert Jay
on brainwashing tactics, 123–24
Thought Reform and the Psychology of Totalism, 121
Lindbergh, Charles, 31
Locke, John, 47
Loeb, Solomon, 6
logical empiricism, and science, 207
Lowe, Victor, 101
Lowell, Abbott Lawrence, 176
Lowell, John, 176
Lowell, Ralph, 171–72, 364
Lowell Institute, 171–72, 176, 185
loyalty oaths, in universities, 77
Łukasiewicz, Jan, 207
Lysenko, Trofim, 178, 263

MacDonald, Dwight, 23
"The Dream World of Soviet Totalitarianism," 24
Mach, Ernst, 161
MacLeish, Archibald, 79, 91, 144
"Speech to the Scholars," 69
Malisoff, William, 9, 174, 177
Malthus, Thomas, 289
The Manchurian Candidate (Condon), 117, 129
Manhattan Project, xv, 69, 181, 227
Mann, Thomas, 83
on democracy, 138–39
"The Problem of Freedom," 138
Mannheim, Karl, xxxi, 176, 197–98, 202, 221
Ideology and Utopia, 197
Mao Zedong, 73, 220–21, 329, 346
Marbury, William, 72–73, 134
Marlboro Man campaign, 318
Marshall, Samuel, 120–21
Marshall Plan, 58, 315
Martin, George, 332
Marx, Karl, xxxiii, 54, 56, 93, 198, 247, 275, 330
Marx-Engels Institute, 93
The Masses (magazine), 9, 102
Maupertuis, Pierre Louis, 266
May, Alan, 312–13
McCarthy, Joseph, xiii, xv–xix, xxiv, xxvi–xxvii, xxix, xxxvi, 12, 22, 25, 74, 76, 79, 102, 104, 109, 112, 173–75, 192, 194, 208, 224, 294, 310–11, 315, 321, 324, 343–45, 365
"Communist philosophy" interrogations, 114–16
confronting Conant, xx–xxiii
on Furry, 108
on Harvard, xvi, 108
shift of views against, 344
on universities, xvi

McCarthyism, xxxvii, 21, 174, 192, 344
   and book burning, xix–xx, xxii
McCumber, John, 116
McGovern, George, 330
McKay, Donald C., 32
McLuhan, Marshall, 335
McMahon, Brien, 313–14
Meerloo, Joost, 119, 122, 125
   The Rape of the Mind, 124
Meiklejohn, Alexander, 100–101
Meier, Annaliese, 302
Merton, Robert, xxxi, 202, 205, 213, 250, 283, 319
   Science, Technology and Society in Seventeenth-Century England, 160
Michelangelo, 67
Miller, Arthur, 83
Miller, James, 322, 329–30
Mills, C. Wright, 322–23
Milosz, Czeslaw, 290
Moos, Elizabeth, xl, xlv, 9, 11, 15, 17–18, 20, 25, 38, 71, 84, 112, 164, 173, 193, 222, 320, 328, 334, 338–39, 357
   acquittal of, 22
   arrest of, 22
   indictment of, 21
   pacifism of, 21, 84, 92
   "Steps toward the American Dream," 10
   testimony to HUAC, 22–23
Morris, Charles, xxxi, xlv, 29–30, 109, 180, 199, 206–208, 210–11, 213, 215–18, 225–26, 233, 239, 242, 248–50, 252, 297, 301, 321, 353
   Signs, Language, and Behavior, 193
Murrow, Edward R., 76, 344

Nagel, Ernest, 202, 205, 213, 250, 271–72, 283, 301, 318
Nash, Leonard, 257

National Council of American-Soviet Friendship, 21
National Defense Education Act, 256, 262
National Lawyers Guild, 136
Nazi-Soviet nonaggression pact (Hitler-Stalin pact), 12, 15–16, 68, 92, 105, 112, 324
Neurath, Otto, 29–30, 178–79, 185–86, 192, 199–200, 206–208, 211–12
   demise of, 180–81
   encyclopedia of science, 199
   on fragmentation in science, 199
   index verborum prohibitorum, 211
   Kullen's attack on, 200
   Modern Man in the Making, 199
   on unified science, 178–80
   unity of science movement, 178–79, 184
Newman, Louis, 87
New Masses, 18
Newton, Isaac, 7, 47, 50, 65–68, 153, 157, 160, 219, 238, 278, 282, 289, 294, 361
Niebuhr, Reinhold, 93
Nietzsche, Friedrich, 330
Nittle, Alfred M., 22
Nitze, Paul, 58
Nixon, Richard, 72–73, 331, 337–38, 340–41, 346–47
   invading Cambodia, 337–38
North, Joe, 18–19
Not Guilty: Report of the Commission of Inquiry into the Charges Made against Leon Trotsky in in the Moscow Trials, 93
Novick, Peter, xxxiii
Noyes, Pierre, 250
NSC-68, 76

Office of Scientific Research and Development (OSRD), 181–83

Oppenheimer, J. Robert, xv, 107, 227, 324
Ormandy, Eugene, 83
Ornstein, Martha, 160
Orwell, George, xxxix, 120, 225, 286
  1984, xxxi, xxxvii, 129–30, 213, 220
Osborn, Frederick, 59–60, 62
Owen, David, 167–69, 173, 186, 204, 209, 245
Oxford Pledge (Oxford Oath), 3, 38

Palmer, A. Mitchell, 111
paradigms. *See* Conant, James Bryant: on paradigms; Kuhn, Thomas: on paradigms
"paranoid style in American politics" (Hofstadter), 118
Pareto, Vilfredo, xxxiii–xxxiv
Pareto Circle, xxxiii
Parker, Francis W., 6–7
Participatory Democracy, 321–22
Pascal, Blaise, 40, 289
Patterson, Robert, 136
Paul (Biblical), 144
Pavlov, Ivan, 121
Peace Information Center, 21–22
Peale, Norman Vincent, 120
Pearson, Karl
  The Grammar of Science, 49, 186
"Peekskill Riot," 23
perception, and anomalies, 128–29
Philbrick, Herbert
  I Led Three Lives: Citizen, "Communist," Counterspy, 22, 113
philosophers of science, anticommunist attention on, 174
Philosophy and Phenomenological Research journal, 180
"philosophy" as subversive, 116
Philosophy of Science journal, 174, 177
Physical Sciences Study Committee (PSSC), 256

Piaget, Jean, 25, 160–63, 192
Pietrzak, Kenneth, 358, 364
Plato, 300, 361
playing card experiment, 128–29
Poincaré, Henri, 156
Polanyi, Michael, xxxvi, 178
  "The Magic of Marxism," 290
Pontecorvo, Bruno, 312
Popper, Karl, xlv, 178, 251–52, 271, 294, 299, 305–306, 309, 314–15, 347–48, 350, 352, 358, 361
  critique of Kuhn, 306–307
  "Logic of Discovery or Psychology of Research?" 306
  on normal science, 307
  The Open Society and Its Enemies, 300, 307
  The Poverty of Historicism, 300
Popper, Martin, 81, 87
popular culture, paradigm shifts in, 333
Port Huron Statement, 321–28
Postman, Leo, 128
pragmatic philosophers, 142
Priestly, Joseph, 130
"Princeton Plan," 340
progressive education, 6–8
Progressive Education Association (PEA), 9, 12–15, 23
Progressive Education journal, 10
Progressive Party, 23
Proust, Marcel, 153
Pusey, Nathan, 108, 340

Quine, W. V. O., 46, 55, 208, 287
  "Ontology and Ideology," 208

Radio Research Lab, 140
Radó, Sándor
  Psychodynamics as a Basic Science, 164
Ransome, Arthur, 111

Rather, Dan, 331
Reagan, Ronald, 76, 324
"Red Spy Queen," 19
Reece Commission, 208
Reed, John
   Ten Days that Shook the World, 111
reformism versus revolution, 328
Reichenbach, Hans, 220
Reinhardt, Ad, 83
Reisch, George
   How the Cold War Transformed Philosophy of Science, xiii
Rembrandt, 67, 307
Remington, Ann, 18–20
Remington, William, xxiv, xlv, 18–20, 24, 72, 194, 222, 357
   death of, 22
   prison sentence, 21
Renoir, Jean
   The Grand Illusion (film), 32
Ribicoff, Abraham, 330
Richards, Grace "Patty" Thayer, 27–28
Rivera, Diego, 11, 16, 93
Robeson, Paul, 23, 83
Rockefeller, Foundation, 184, 206, 208
Roentgen, Wilhelm, 126
Roosevelt, Franklin Delano, 29, 30–31, 34, 48, 53, 181–82, 191
   on "slackers," 33, 35
   State of the Union speech, 33
Rorty, Richard
   Philosophy and the Mirror of Nature, 361
Rosenberg, Ethel, xvii, xx, xlv, 174, 209, 310–11, 315–16
Rosenberg, Julius, xvii, xx, xlv, 174, 194, 209, 310–11, 315–16
Rosenberg, Michael, 312
Rosenberg, Robert, 312
Rosenstock-Huessy, Eugen, xxxix, xlv, 222–25, 228, 235
   Die Europäischen Revolutionen, 222

Out of Revolution: Autobiography of Western Man, 222
Ross, Bob, 322
Roth, Philip
   Portnoy's Complaint, 164
Rougier, Louis, 207
Russell, Bertrand, 45, 116, 140
Russian Research Center, 60–61, 197, 221
Russian Revolution, 111

Sartre, Jean-Paul, 311
Savio, Mario, 324
Saypol, Irving, 311
Scaiky, Leon, 17–18
Schein, Edgar, 122, 220
   on brainwashing tactics, 124–26
   Coercive Persuasion, 121, 125
Schine, G. David, xvii–xx, xxii, xxix, 294, 309–10
School: A Film about Progressive Education (Lee), 17–18
Schrecker, Ellen, xxx
science. *See also* Kuhn, Thomas: on science
   and betterment of world as goal, 177
   and cold war, xiii
   controversies about guiding it, 178–85
   fragmentation in, 199
   historical understandings of, xxxvi
   and ideology, xiii
   and logical empiricism, 207
   and paradigms, xiii
Science and Society journal, 177
Science Service, 83
Science, the Endless Frontier (report), 181
Shakespeare, William, 7, 39
Shapere, Dudley, 295–96, 304, 347
Shapley, Harlow, xvi, 21, 81–87, 94, 97, 99, 102–103, 173, 206, 209, 357
   confronted by Hook, 81–87, 103

Index | 455

as internationalist, 88
news coverage of, 104
Shelley, Mary, 348
Sherr, Rubby, 107
Shils, Edward, 197
Shipler, Guy Emery, 84–85
Shostakovich, Dmitri, 82, 103
Sinclair, Upton, 321
Smith, Alfred, 10
Sobell, Morton, 174
Social and Economic Museum, 199
Socrates, 116
Soviet and Western ideologies, as irreconcilable, 57–61
"Soviet Atomic Espionage" (Congressional report), 293
Soviet intellectuals, visiting Princeton University, 338–39
Soviet invasion, fears of, 75–76
Soviet spies, in Britain, 293
Special Committee on Sponsored Research at Princeton, 341–42
Spinoza, Baruch, 67
Sputnik satellite, 255–56, 262–63, 267
Sproul, Robert Gordon, 77
Stalin, Joseph, xiii, xx, 8, 11–12, 15–16, 25, 52, 56, 75, 77, 83, 89, 92–94, 98, 113, 137, 177, 220, 275–76, 310, 324
Steffens, Lincoln, 9
Steichen, Edward, 10
Stenbuck, Mr., 173, 177, 185, 189, 191, 195–96
Stevin, Simon, 289
"Stockholm Appeal," 21
Stopes-Roe, Harry, 298–99
Stroock, Minette, 5–6
"struggle for men's minds," xxvii–xxix, xxxviii, 120
Student League for Industrial Society (SLID), 321
student movements, in 1960s, 318, 320–32

Students for Democratic Society (SDS), 318, 320–22, 324, 328–31, 340
Sykes, Charles Henry, 112

Tavenner, Frank S., 105–107
Thomas, Norman, 10
"Too Slow for Me" (cartoon), 111–12
Torricelli, Evangelista, 47, 50
totalitarianism, 120
postwar fears of, 180
Toulmin, Stephen, xxxi, 266, 350
critique of Kuhn, 308
Trotsky, Leon, 15, 93–95, 116
Truman, Harry, 23, 48, 52, 58, 73–74, 82, 183, 191
Tunney, Gene, 11

UNESCO, 83
United States Information Service libraries, investigation of, xvii–xx
unity of science movement, 29–30, 178
critics of, 178–85
U.S. China Peoples Friendship Association, 21

van der Lubbe, Marinus, 119
Van Vleck, John, 140, 153
Vienna Circle, 142, 199, 300
Vorhees, Tracy, 75

Waelder, Robert, 122
"Waldorf Conference," 21, 82–84, 95, 97, 103–104
Wallace, Alfred Russel, 202
Wallace, Henry, 22–23, 83
Warren, Alvin, 18
Watkins, John, 315, 348, 350, 355
critique of Kuhn, 309
Watt, James, 130
Weaver, Warren, 184–86, 194, 212

Weathermen, 318, 328–32
    Prairie Fire: The Politics of Revolutionary Anti-Imperialism, 328
Welch, Joseph, 344
White, Harry Dexter, 19
White, William Allen, 32–34
Whitney, Eli, 262
Whorf, Benjamin Lee, 157
    Language, Thought, and Reality, 193
Wilkie, Wendell, 39–40
Williams, Tennessee, 165
Wilson, Edmund, 93
Wirth, Louis, 197–99, 207
Wolfe, Glenn, xviii–xix, xxiii
Women's International League for Peace and Freedom, 22
Women Strike for Peace, 22, 338–39
Woodger, Joseph, 192–94
    Techniques of Theory Construction, 207
Woodmansee, Roy, 335
World Peace Council, 21
Wray, K. Brad, xxxi
Wright, Benjamin, 45–46
Wright, Frank Lloyd, 83
Wright, Richard, 247

Yale Daily News, xvi
Young Communist League, 92, 137

Zimmer, Herbert
    The Manipulation of Human Behavior, 122
Zuckerman, Max, 14

www.ingramcontent.com/pod-product-compliance
Lightning Source LLC
Chambersburg PA
CBHW051843300426
44117CB00006B/245